GENETIC MAPS

Locus Maps of Complex Genomes

FIFTH EDITION

Stephen J. O'Brien, Editor

Laboratory of Viral Carcinogenesis, National Cancer Institute

BOOK 4

Nonhuman Vertebrates

COLD SPRING HARBOR LABORATORY PRESS 1990

GENETIC MAPS

First edition, March 1980
Second edition, June 1982
Third edition, June 1984
Fourth edition, July 1987
Fifth edition, February 1990

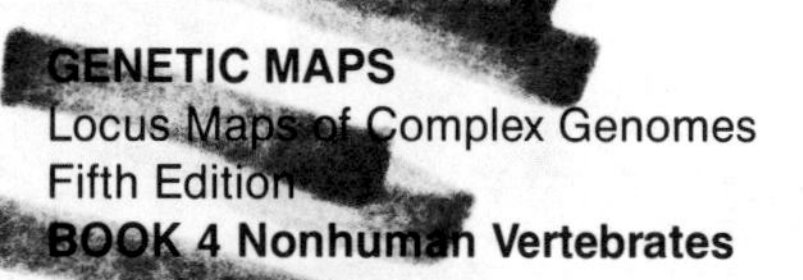

GENETIC MAPS
Locus Maps of Complex Genomes
Fifth Edition
BOOK 4 Nonhuman Vertebrates

Printed in the United States of America

Library of Congress Catalog Card Number 84-644938
ISBN 0-87969-345-2
ISSN 0738-5269

Cover design by Leon Bolognese

All Cold Spring Harbor Laboratory Press publications may be ordered directly from Cold Spring Harbor Laboratory, Box 100, Cold Spring Harbor, New York 11724. Phone 1-800-843-4388. In New York (516) 367-8423.

PREFACE

"The map is a sensitive indicator of the changing thought of man, and few of his works seem to be such an excellent mirror of culture and civilization."

Norman J.W. Thrower
Maps and Men, 1972

The early geographic explorers are revered by their countrymen and descendants throughout the world for pioneering the discovery and charting of foreign lands. The precision and attention to detail accorded to their charts and maps are almost always taken for granted today. Astronauts, and then those who saw their photographs from space, have marveled at the resemblance of these real images of the Florida peninsula, the boot of Italy, and the British Isles to the maps we had grown up with. John Noble Wilford, in his fascinating monograph entitled *The Mapmakers,* reminds us that it is only within the last quarter of the twentieth century that it could be said that the earth has been mapped. It is humbling to consider how important these maps have been to our history, our culture, and our way of life today; in this context, the early cartographers were indeed intrepid.

The new explorers of the eukaryote and prokaryote genome will certainly one day be considered with equivalent veneration, since the topography of their charts and maps may prove as daunting as the earth's surface. The animal and plant genomes are products of several hundred million years of biological evolution, and there are clearly far more secrets to be deciphered in the genomes that can be anticipated in future generations. However, genetic detectives have begun to unravel some of the mysteries of genetic organization, and as we enter the final decade of the century, the scientific community (as well as the enlightened public) has begun to grasp the enormous value of generating and expanding genetic maps of man, of agriculturally significant plants and animals, and of model systems that allow us a glimpse of how genes are organized, replicated, and regulated.

The first genetic map was formulated by A.H. Sturtevant in 1913 and consisted of five genes arranged in a linear fashion along the X chromosome of the fruit fly, *Drosophila melanogaster.* In the ensuing decades, gene mapping of numerous species has proceeded deliberately and cumulatively in organisms as diverse as flies, corn, wheat, mink, apes, man, and bacteria. All of these maps, whether based on recombination, restriction, physical or DNA sequence, are predicated on Sturtevant's logical notion that gene order on a chromosome could be displayed as a linear array of genetic markers. The results of these efforts in more than 100 genetically studied organisms are the basis of *GENETIC MAPS: Locus Maps of Complex Genomes.*

During the preparation of the previous edition, it occurred to me that the rate of growth of the gene mapping effort was so rapid that we should prepare to publish future editions in multiple volumes. The fifth edition is a realization of that idea. *Genetic Maps* now consists of six smaller books, each based on an arbitrary subdivision of biological organisms. These are BOOK 1 Viruses; BOOK 2 Bacteria, Algae, and Protozoa; BOOK 3 Lower Eukaryotes; BOOK 4 Nonhuman Vertebrates; BOOK 5 Human Maps; BOOK 6 Plants. Each of these is available in paperback at a modest cost from Cold Spring Harbor Laboratory Press. The entire compendium can be purchased as a hardback volume of 1098 map pages suitable for libraries and research institutions.

Our original intent was to publish in one volume complete, referenced genetic maps of every organism with a substantive group of assigned loci. We intentionally excluded DNA sequences, since these are easily available in several computer databanks (Genbank, EMBL, and others). Text was to be kept at a minimum, and the maps would be both comprehensive and concise. The collection was to be updated every 2–3 years, and each new volume would contain a complete revision, rendering the previous volume obsolete.

The original publication of *Genetic Maps* was an experiment, which I believe now can be judged a success. The execution of such a venture depended heavily on the support and cooperation of thousands of geneticists throughout the worldwide scientific community. This cooperation was graciously extended, and the result was an enormously valuable and accessible collection. On behalf of my colleagues in the many fields that we call genetic biology, I gratefully acknowledge the numerous geneticists, scientists, and readers who have contributed to or corrected these maps. To ensure the continuation of these heroic efforts in the future, readers are encouraged to send to me any suggestions for improvement, particularly suggestions for new maps to be included, as well as constructive criticisms of the present maps. When new organisms are recommended, I would also appreciate names and addresses of prospective authors. In addition, readers are encouraged to supply corrections, reprints, and new mapping data to the appropriate authors who may be included in future editions.

The compilation of each genetic map, like the drafting of the first geographic maps, is an extremely tedious, yet important, assignment. All of the authors deserve special thanks for the large efforts they have expended in contributing their maps. Even in this computer age, we often feel like we are proofreading the telephone book, but I, for one, believe that the final product makes the effort worthwhile. Finally, I acknowledge specifically my editorial assistants, Patricia Johnson and Virginia Frye, who have cheerfully and expeditiously carried the bulk of the editorial activities for the present volume, and Annette Kirk, Lee Martin, and John Inglis of Cold Spring Harbor Laboratory Press for support and advice on the preparation of this edition.

Stephen J. O'Brien, *Editor*

CONTENTS

COMPLETE CONTENTS OF THE FIFTH EDITION

PROTOZOA

Paramecium tetraurelia	K.J. Aufderheide and D. Nyberg
Tetrahymena thermophila	P.J. Bruns

BOOK 3 Lower Eukaryotes

FUNGI

Dictyostelium discoideum	P.C. Newell
Neurospora crassa	
nuclear genes	D.D. Perkins
mitochondrial genes	R.A. Collins
restriction polymorphism	R.L. Metzenberg and J. Grotelueschen
Saccharomyces cerevisiae	
nuclear genes	*R.K. Mortimer, D. Schild, C.R. Contopoulou, and J.A. Kans*
mitochondrial DNA	*L.A. Grivell*
Podospora anserina	D. Marcou, M. Picard-Bennoun, and J.-M. Simonet
Sordaria macrospora	G. Leblon, V. Haedens, A.D. Huynh, L. Le Chevanton, P.J.F. Moreau, and D. Zickler
Coprinus cinereus	J. North
Magnaporthe grisea (blast fungus)	D. Z. Skinner, H. Leung, and S.A. Leong
Phycomyces blakesleeanus	M. Orejas, J.M. Diaz-Minguez, M.I. Alvarez, and A.P. Eslava
Schizophyllum commune	C.A. Raper
Ustilago maydis	A.D. Budde and S.A. Leong
Ustilago violacea	A.W. Day and E.D. Garber
Aspergillus nidulans	
nuclear genes	A.J. Clutterbuck
mitochondrial genome	T.A. Brown

NEMATODE

Caenorhabditis elegans	M.L. Edgley and D.L. Riddle

INSECTS

Drosophila melanogaster	
biochemical loci	G.E. Collier
in situ hybridization data	E. Kubli and S. Schwendener
cloned genes	J. Merriam, S. Adams, G. Lee, and D. Krieger
Aedes (Stegomyia) aegypti	L.E. Munstermann
Aedes (Protomacleaya) triseriatus	L.E. Munstermann
Drosophila pseudoobscura	W.W. Anderson
Anopheles albimanus	S.K. Narang and J.A. Seawright
Anopheles quadrimaculatus	S.K. Narang, J.A. Seawright, and S.E. Mitchell
Nasonia vitripennis	G.B. Saul, 2nd

BOOK 4 Nonhuman Vertebrates

RODENTS

Mus musculus (mouse)	
nuclear genes	M.T. Davisson, T.H. Roderick, D.P. Doolittle, A.L. Hillyard, and J.N. Guidi
DNA clones and probes, RFLPs	J.T. Eppig
retroviral and cancer-related genes	C.A. Kozak
Rattus norvegicus (rat)	G. Levan, K. Klinga, C. Szpirer, and J. Szpirer
Cricetulus griseus (Chinese hamster)	
nuclear genes	R.L. Stallings, G.M. Adair, and M.J. Siciliano
CHO cells	G.M. Adair, R.L. Stallings, and M.J. Siciliano
Peromyscus maniculatus (deermouse)	W.D. Dawson
Mesocricetus auratus (Syrian hamster)	R. Robinson
Meriones unquiculatus (Mongolian gerbil)	R. Robinson

OTHER MAMMALS

Felis catus (cat)	S.J. O'Brien
Canis familiaris (dog)	P. Meera Khan, C Brahe, and L.M.M. Wijnen
Equus caballus (horse)	L.R. Weitkamp and K. Sandberg
Sus scrofa domestica L. (pig)	G. Echard
Oryctolagus cuniculus (rabbit)	R.R. Fox
Ovis aries (sheep)	G. Echard
Bos taurus (cow)	J.E.Womack
Mustela vison (American mink)	O.L. Serov and S.D. Pack
Marsupials and Monotremes	J.A. Marshall Graves

PRIMATES

Primate Genetic Maps	N. Creau-Goldberg, C. Cochet, C. Turleau, and J. de Grouchy
Pan troglodytes (chimpanzee)	
Gorilla gorilla (gorilla)	
Pongo pygmaeus (orangutan)	
Hylobates (Nomascus) concolor (gibbon)	
Macaca mulatta (rhesus monkey)	
Papio papio, hamadryas, cynocephalus (baboon)	
Cercopithecus aethiops (African green monkey)	
Cebus capucinus (capuchin monkey)	
Microcebus murinus (mouse lemur)	
Saquinus oedipus (cotton-topped marmoset)	P.A. Lalley
Aotus trivirgatus (owl monkey)	N.S.-F. Ma

FISH

Salmonid fishes *Salvelinus, Salmo, Oncorhynchus*	B. May and K.R. Johnson
Non-Salmonid fishes *Xiphophorus, Poeciliopsis, Fundulus, Lepomis*	D.C. Morizot

AMPHIBIAN

Rana pipiens (leopard frog)	D.A. Wright and C.M. Richards

BIRD

Gallus gallus (chicken)	R.G. Somes, Jr., and J.J. Bitgood

BOOK 5 Human Maps

HUMAN MAPS (Homo sapiens)

BOOK 6 Plants

PLANTS

BOOK 4

NONHUMAN VERTEBRATES

LOCUS MAP OF THE MOUSE (*Mus musculus/domesticus*) **2n=40**

August 1989

Muriel T. Davisson
Thomas H. Roderick
Donald P. Doolittle
Alan L. Hillyard
John N. Guidi
The Jackson Laboratory
Bar Harbor, ME 04069

The locus map of the mouse is a cumulative summary of the efforts of many investigators from all over the world. It contains information accumulated since the first linkage group was described in the mouse in 1915. It contains linkage data from wild-derived subspecies and interspecific crosses with *Mus spretus* as well as from laboratory mice. The data that supports the locus map of the mouse is stored in GBASE, a computerized database at The Jackson Laboratory. GBASE is available for on-line access via Telenet. GBASE contains published genetic information for the mouse, including chromosomal mapping data, information on unmapped loci, allelic information and mouse/human homologies, genetic maps of the mouse genome compiled from these data, characterization of mouse strains for polymorphic loci (including biochemical and DNA RFLP polymorphisms) and references. Through GBASE one can request a laser-printed map of the mouse genome, a map of the mouse genome with all known human homologies overlaid, custom maps tailored to an individual researcher's needs for a particular view of the mouse genome and a list of known mouse loci with references. Allelic characterization of strains and strain distribution patterns of loci can be obtained in various ways along with several analysis options. The names of the loci are listed in Table 1 and full descriptions of most of them can be found in Green (1).

CONVENTIONS USED ON THE MAP

Solid vertical bars represent the chromosomes and are drawn to their proportional lengths based on an estimated total haploid length of 1600cM. Chromosome numbers are shown above the centromeres which are represented by knobs. Nucleolus organizers are symbolized by **NO**, except on Chr 12 where a ribosomal RNA locus has been mapped using a DNA polymorphism and is symbolized **Rnr12**. Locus symbols are given at the right of the chromosome bars. Numbers to the left of the chromosomes are recombination percentages or cM and distances are given as distance in cM from the centromere. The resolution of the current map is 1cM. The distances between centromeres and proximal markers in most chromosomes have been determined using Robertsonian chromosomes and may be underestimated. Loci listed at the bottoms of chromosomes have been assigned to those chromosomes by parasexual methods or are just known to be linked to those chromosomes. The map is compiled from female and male linkage data and recombinant inbred strain data. We calculate a weighted average using the reciprocal of the variance.

Some further conventions are used to show the relative certainty of position of a locus and the relationships of loci. Positions of some loci are well known from three-point crosses and extensive data. These **anchor** loci are indicated by lines extending through the chromosome bars and beyond all others to the left and the symbol is printed in boldface. The shorter lines through the chromosome indicate loci whose order is known with less assurance. When a locus is mapped with respect to only one other locus, the symbol of the locating locus is added in parentheses. When a locus is known to be near another but recombination values are not known, the new locus is placed next to the linked locus but no line is drawn to the chromosome. Characters in square brackets indicate the cytological (band) location of parasexually mapped loci. When more than one locus maps to the same position, the loci are listed on the same line and, if one is an **anchor** locus, its symbol is given first. An upward caret (^) means the locus or loci following it belong in the line above; two carets (^^) means two lines above, etc. The construct **locus A (@locus B)** indicates that locus A has been mapped relative to locus B by physical mapping methods.

ADDITIONAL MAPPING INFORMATION NOT PRESENTED ON THE MAP

The following list contains ordering and position information that is known, but not illustrated on the map because of the 1cM limit of resolution. The symbol => indicates order and a position less than 1cM from the other locus in the pair; the arrow points to the more distal locus.

Chromosome 1

Grmp => *Atpb* => *T3z*
Ly-17 => *Bcm-1*

Chromosome 2

B2m => *H-3*
Il-1 => *Prn*

Chromosome 3

Ly-38 => *Cacy*
Atpa-1 => *Ampd-1, Tshb, Ngfb*

Chromosome 5

Rw => *W (?= Kit)*

Chromosome 6

Hd => [*Igk-V11, Igk-V24, Igk-V9-26*] =>
[*Igk-V1, Igk-V9*] =>
[*Igk-V4, Igk-V8, Igk-V10, Igk-V12,13, Igk-V19*] =>
[*Igk-V28, Rn7s-6*] => *Igk-V23* =>
[*Igk-V21, Igk-J, Igk-C*] => [*Ly-2, Ly-3*] => *wa-1*

Chromosome 9

Emv-3 => *se*

Chromosome 17

Tcp-1 => *Fu*
Tla => *Ce-2* => *Pgk-2* => *Tpx-1* => *Mep-1*

X Chromosome

Cybb => *Syn-1* => *Timp* => *DXWas68* => *DXSmh172*
Araf => *Xlr-1*
Hprt => *Xlr-2*
DXPas8 => *DXPas14* => *DXPas13*
Gdx => *G6pd*
Rsvp => *Cf-8*

Published linkage data, to be combined, must include the following: the type of cross, the phase (coupling or repulsion) of the loci involved when it is appropriate, the sex of the F1 parent in backcrosses, the number of recombinants, and the total number examined. Please indicate whether new data include formerly published information or information in *Mouse News Letter* so that we will not use data twice. We appreciate preprints or reprints of articles or abstracts that have mapping data. Your assistance helps greatly in keeping the map current and complete. Please quote original sources when referring to linkage data. Maintenance of the map is supported by a contract from the Howard Hughes Medical Institute, grant BSR 8418828 from the National Science Foundation, Cancer Core Center grant CA 34196 from the National Cancer Institute, and a grant from the Pew Memorial Trust. We thank Peter D'Eustachio, Margaret C. Green, Peter A. Lalley, Josephine Peters, Frank H. Ruddle and Benjamin A. Taylor for their help with the map.

1. Green MC. Catalog of mutant genes and polymorphic loci. In: Lyon, MF and Searle AG (eds), *Genetic Variants and Strains of the Laboratory Mouse*, 2nd ed. Oxford University Press. Oxford. 1989.

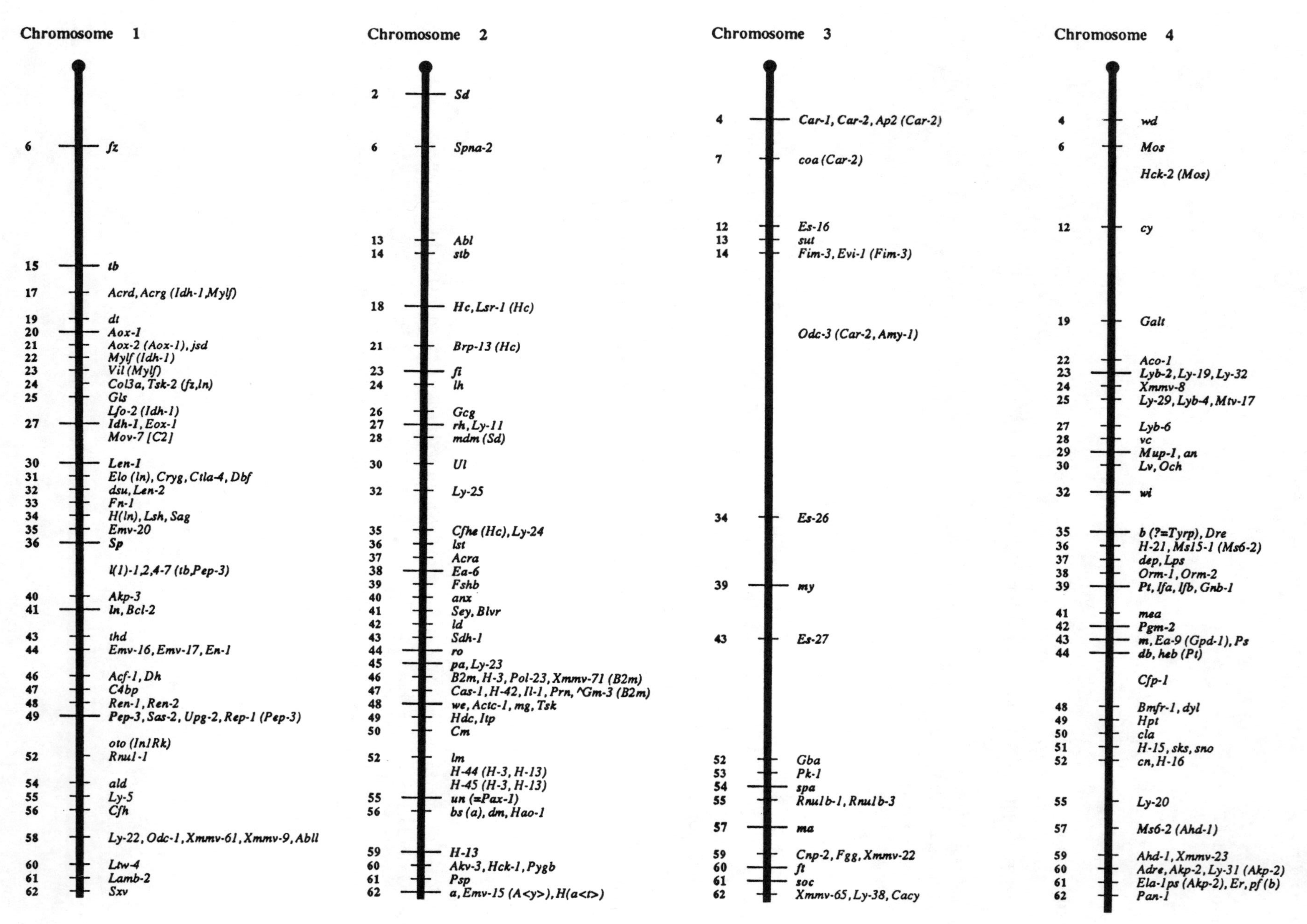
Chromosome 1
6 fz
15 tb
17 Acrd, Acrg (Idh-1,Mylf)
19 dt
20 Aox-1
21 Aox-2 (Aox-1), jsd
22 Mylf (Idh-1)
23 Vil (Mylf)
24 Col3a, Tsk-2 (fz,ln)
25 Gls
Lfo-2 (Idh-1)
27 Idh-1, Eox-1
Mov-7 [C2]
30 Len-1
31 Elo (ln), Cryg, Ctla-4, Dbf
32 dsu, Len-2
33 Fn-1
34 H(ln), Lsh, Sag
35 Emv-20
36 Sp
l(1)-1,2,4-7 (tb,Pep-3)
40 Akp-3
41 ln, Bcl-2
43 thd
44 Emv-16, Emv-17, En-1
46 Acf-1, Dh
47 C4bp
48 Ren-1, Ren-2
49 Pep-3, Sas-2, Upg-2, Rep-1 (Pep-3)
oto (In1Rk)
52 Rnul-1
54 ald
55 Ly-5
56 Cfh
58 Ly-22, Odc-1, Xmmv-61, Xmmv-9, Abll
60 Ltw-4
61 Lamb-2
62 Sxv
Chromosome 2
2 Sd
6 Spna-2
13 Abl
14 stb
18 Hc, Lsr-1 (Hc)
21 Brp-13 (Hc)
23 fi
24 lh
26 Gcg
27 rh, Ly-11
28 mdm (Sd)
30 Ul
32 Ly-25
35 Cfhe (Hc), Ly-24
36 lst
37 Acra
38 Ea-6
39 Fshb
40 anx
41 Sey, Blvr
42 ld
43 Sdh-1
44 ro
45 pa, Ly-23
46 B2m, H-3, Pol-23, Xmmv-71 (B2m)
47 Cas-1, H-42, Il-1, Prn, ^Gm-3 (B2m)
48 we, Actc-1, mg, Tsk
49 Hdc, Itp
50 Cm
52 lm
H-44 (H-3, H-13)
H-45 (H-3, H-13)
55 un (=Pax-1)
56 bs (a), dm, Hao-1
59 H-13
60 Akv-3, Hck-1, Pygb
61 Psp
62 a, Emv-15 (A<y>), H(a<t>)
Chromosome 3
4 Car-1, Car-2, Ap2 (Car-2)
7 coa (Car-2)
12 Es-16
13 sut
14 Fim-3, Evi-1 (Fim-3)
Odc-3 (Car-2, Amy-1)
34 Es-26
39 my
43 Es-27
52 Gba
53 Pk-1
54 spa
55 Rnulb-1, Rnulb-3
57 ma
59 Cnp-2, Fgg, Xmmv-22
60 ft
61 soc
62 Xmmv-65, Ly-38, Cacy
Chromosome 4
4 wd
6 Mos
Hck-2 (Mos)
12 cy
19 Galt
22 Aco-1
23 Lyb-2, Ly-19, Ly-32
24 Xmmv-8
25 Ly-29, Lyb-4, Mtv-17
27 Lyb-6
28 vc
29 Mup-1, an
30 Lv, Och
32 wi
35 b (?=Tyrp), Dre
36 H-21, Ms15-1 (Ms6-2)
37 dep, Lps
38 Orm-1, Orm-2
39 Pt, lfa, lfb, Gnb-1
41 mea
42 Pgm-2
43 m, Ea-9 (Gpd-1), Ps
44 db, heb (Pt)
Cfp-1
48 Bmfr-1, dyl
49 Hpt
50 cla
51 H-15, sks, sno
52 cn, H-16
55 Ly-20
57 Ms6-2 (Ahd-1)
59 Ahd-1, Xmmv-23
60 Adre, Akp-2, Ly-31 (Akp-2)
61 Ela-1ps (Akp-2), Er, pf (b)
62 Pan-1

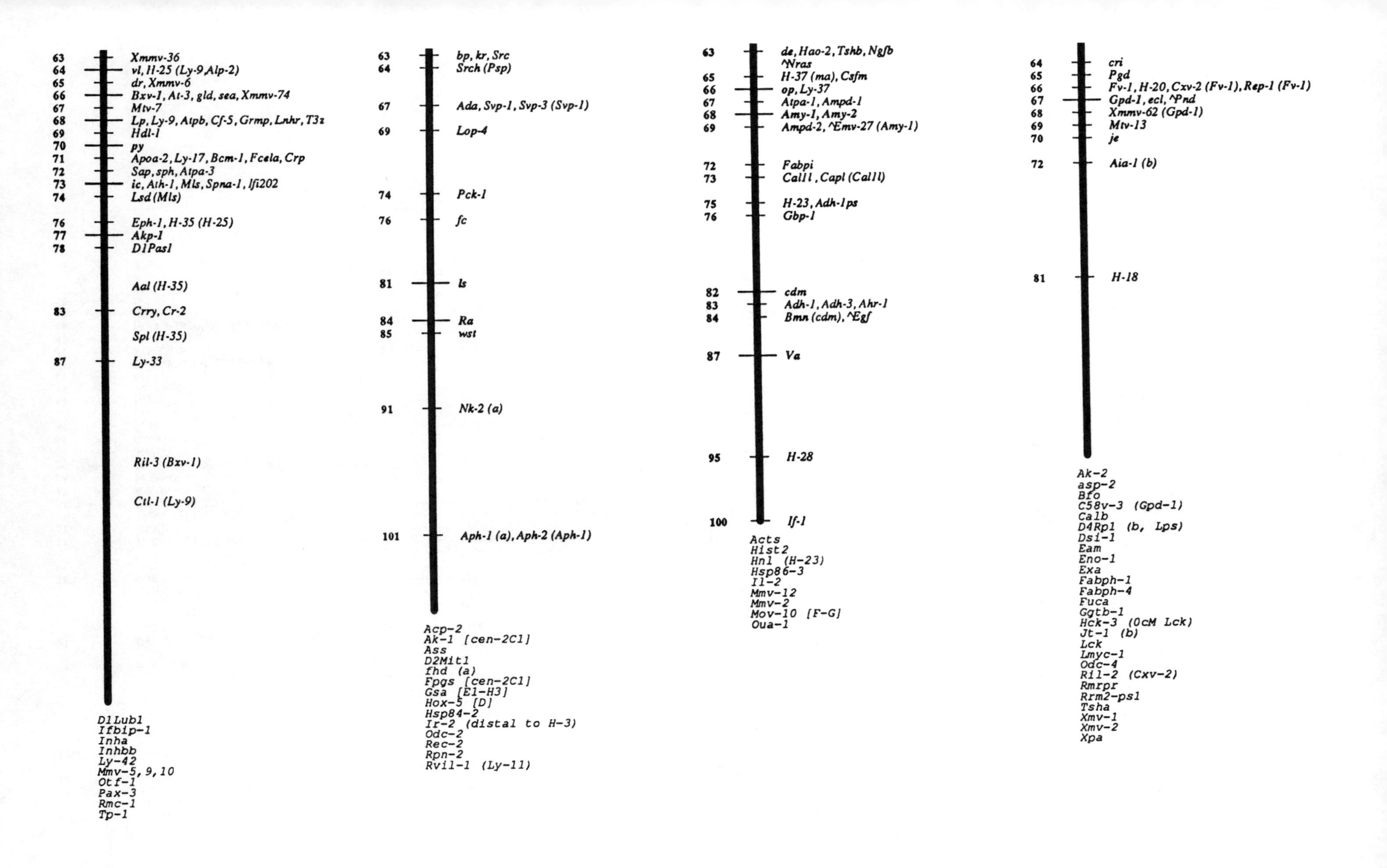
63 Xmmv-36
64 vl, H-25 (Ly-9,Alp-2)
65 dr, Xmmv-6
66 Bxv-1, At-3, gld, sea, Xmmv-74
67 Mtv-7
68 Lp, Ly-9, Atpb, Cf-5, Grmp, Lnhr, T3z
69 Hdl-1
70 py
71 Apoa-2, Ly-17, Bcm-1, Fcela, Crp
72 Sap, sph, Atpa-3
73 ic, Ath-1, Mls, Spna-1, Ifi202
74 Lsd (Mls)
76 Eph-1, H-35 (H-25)
77 Akp-1
78 D1Pas1
Aal (H-35)
83 Crry, Cr-2
Spl (H-35)
87 Ly-33
Ril-3 (Bxv-1)
Ctl-1 (Ly-9)
D1Lub1
Ifbip-1
Inha
Inhbb
Ly-42
Mmv-5, 9, 10
Otf-1
Pax-3
Rmc-1
Tp-1
63 bp, kr, Src
64 Srch (Psp)
67 Ada, Svp-1, Svp-3 (Svp-1)
69 Lop-4
74 Pck-1
76 fc
81 ls
84 Ra
85 wst
91 Nk-2 (a)
101 Aph-1 (a), Aph-2 (Aph-1)
Acp-2
Ak-1 [cen-2C1]
Ass
D2Mit1
fhd (a)
Fpgs [cen-2C1]
Gsa [E1-H3]
Hox-5 [D]
Hsp84-2
Ir-2 (distal to H-3)
Odc-2
Rec-2
Rpn-2
Rvil-1 (Ly-11)
63 de, Hao-2, Tshb, Ngfb
^Nras
65 H-37 (ma), Csfm
66 op, Ly-37
67 Atpa-1, Ampd-1
68 Amy-1, Amy-2
69 Ampd-2, ^Emv-27 (Amy-1)
72 Fabpi
73 Call1, Capl (Call1)
75 H-23, Adh-1ps
76 Gbp-1
82 cdm
83 Adh-1, Adh-3, Ahr-1
84 Bmn (cdm), ^Egf
87 Va
95 H-28
100 If-1
Acts
Hist2
Hnl (H-23)
Hsp86-3
Il-2
Mmv-12
Mmv-2
Mov-10 [F-G]
Oua-1
64 cri
65 Pgd
66 Fv-1, H-20, Cxv-2 (Fv-1), Rep-1 (Fv-1)
67 Gpd-1, ecl, ^Pnd
68 Xmmv-62 (Gpd-1)
69 Mtv-13
70 je
72 Aia-1 (b)
81 H-18
Ak-2
asp-2
Bfo
C58v-3 (Gpd-1)
Calb
D4Rp1 (b, Lps)
Dsi-1
Eam
Eno-1
Exa
Fabph-1
Fabph-4
Fuca
Ggtb-1
Hck-3 (0cM Lck)
Jt-1 (b)
Lck
Lmyc-1
Odc-4
Ril-2 (Cxv-2)
Rmrpr
Rrm2-ps1
Tsha
Xmv-1
Xmv-2
Xpa

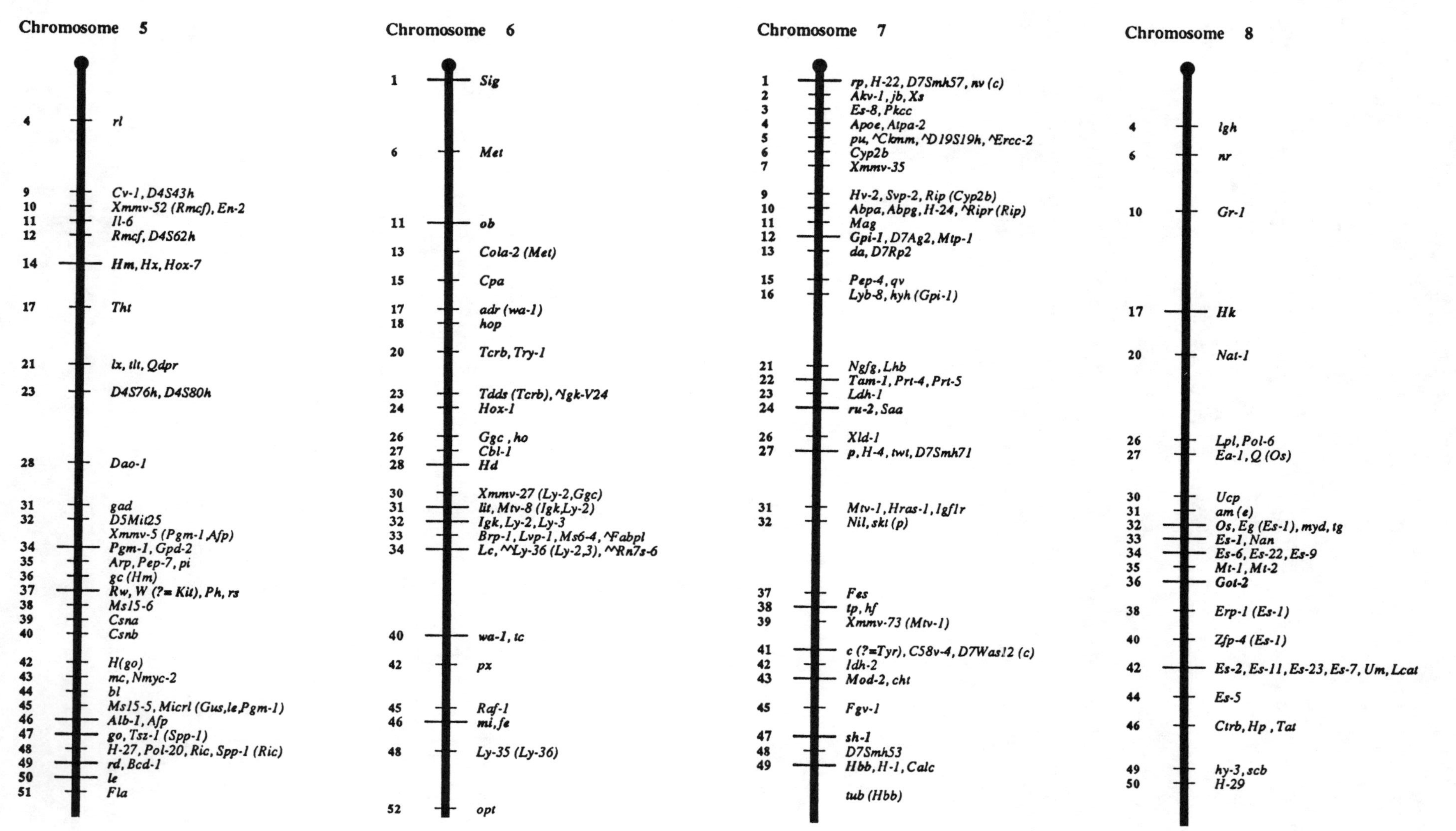
Chromosome 5
4 rl
9 Cv-1, D4S43h
10 Xmmv-52 (Rmcf), En-2
11 Il-6
12 Rmcf, D4S62h
14 Hm, Hx, Hox-7
17 Tht
21 lx, tlt, Qdpr
23 D4S76h, D4S80h
28 Dao-1
31 gad
32 D5Mit25
Xmmv-5 (Pgm-1 Afp)
34 Pgm-1, Gpd-2
35 Arp, Pep-7, pi
36 gc (Hm)
37 Rw, W (?= Kit), Ph, rs
38 Ms15-6
39 Csna
40 Csnb
42 H(go)
43 mc, Nmyc-2
44 bl
45 Ms15-5, Micrl (Gus,le,Pgm-1)
46 Alb-1, Afp
47 go, Tsz-1 (Spp-1)
48 H-27, Pol-20, Ric, Spp-1 (Ric)
49 rd, Bcd-1
50 le
51 Fla
Chromosome 6
1 Sig
6 Met
11 ob
13 Cola-2 (Met)
15 Cpa
17 adr (wa-1)
18 hop
20 Tcrb, Try-1
23 Tdds (Tcrb), ^Igk-V24
24 Hox-1
26 Ggc, ho
27 Cbl-1
28 Hd
30 Xmmv-27 (Ly-2,Ggc)
31 lit, Mtv-8 (Igk,Ly-2)
32 Igk, Ly-2, Ly-3
33 Brp-1, Lvp-1, Ms6-4, ^Fabpl
34 Lc, ^^Ly-36 (Ly-2,3), ^^Rn7s-6
40 wa-1, tc
42 px
45 Raf-1
46 mi, fe
48 Ly-35 (Ly-36)
52 opt
Chromosome 7
1 rp, H-22, D7Smh57, nv (c)
2 Akv-1, jb, Xs
3 Es-8, Pkcc
4 Apoe, Atpa-2
5 pu, ^Ckmm, ^D19S19h, ^Ercc-2
6 Cyp2b
7 Xmmv-35
9 Hv-2, Svp-2, Rip (Cyp2b)
10 Abpa, Abpg, H-24, ^Ripr (Rip)
11 Mag
12 Gpi-1, D7Ag2, Mtp-1
13 da, D7Rp2
15 Pep-4, qv
16 Lyb-8, hyh (Gpi-1)
21 Ngfg, Lhb
22 Tam-1, Prt-4, Prt-5
23 Ldh-1
24 ru-2, Saa
26 Xld-1
27 p, H-4, twt, D7Smh71
31 Mtv-1, Hras-1, Igf1r
32 Nil, ski (p)
37 Fes
38 tp, hf
39 Xmmv-73 (Mtv-1)
41 c (?=Tyr), C58v-4, D7Was12 (c)
42 Idh-2
43 Mod-2, cht
45 Fgv-1
47 sh-1
48 D7Smh53
49 Hbb, H-1, Calc
tub (Hbb)
Chromosome 8
4 Igh
6 nr
10 Gr-1
17 Hk
20 Nat-1
26 Lpl, Pol-6
27 Ea-1, Q (Os)
30 Ucp
31 am (e)
32 Os, Eg (Es-1), myd, tg
33 Es-1, Nan
34 Es-6, Es-22, Es-9
35 Mt-1, Mt-2
36 Got-2
38 Erp-1 (Es-1)
40 Zfp-4 (Es-1)
42 Es-2, Es-11, Es-23, Es-7, Um, Lcat
44 Es-5
46 Ctrb, Hp, Tat
49 hy-3, scb
50 H-29

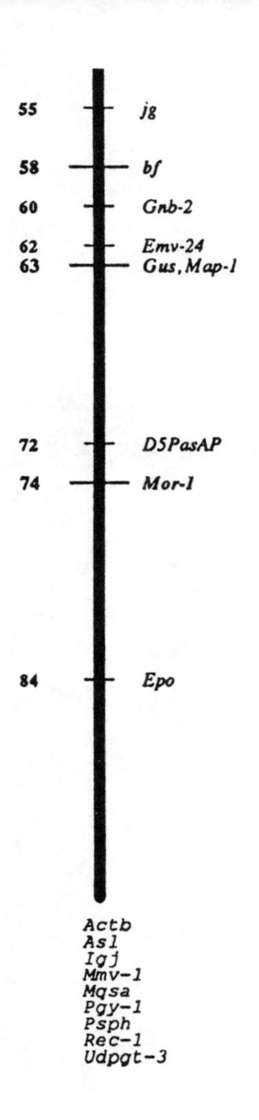
55 jg
58 bf
60 Gnb-2
62 Emv-24
63 Gus, Map-1
72 D5PasAP
74 Mor-1
84 Epo
Actb
Asl
Igj
Mmv-1
Mqsa
Pgy-1
Psph
Rec-1
Udpgt-3

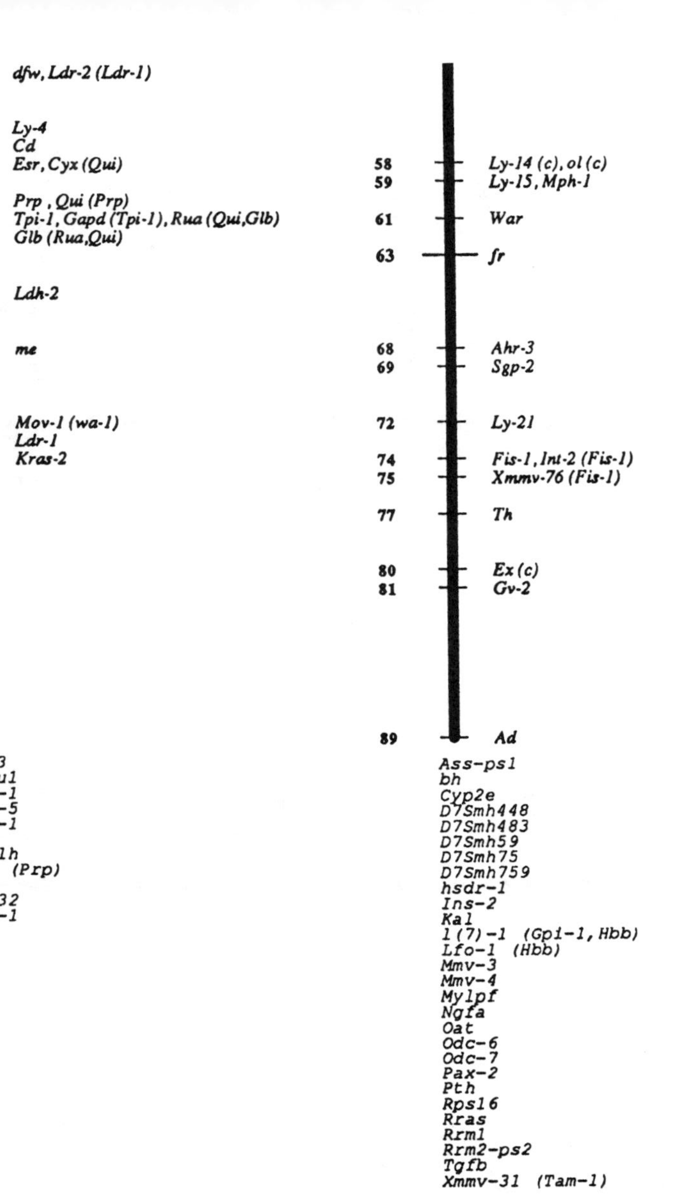
53 dfw, Ldr-2 (Ldr-1)
56 Ly-4
57 Cd
58 Esr, Cyx (Qui)
60 Prp , Qui (Prp)
61 Tpi-1, Gapd (Tpi-1), Rua (Qui,Glb)
62 Glb (Rua,Qui)
65 Ldh-2
68 me
72 Mov-1 (wa-1)
73 Ldr-1
74 Kras-2
Cyp3
D6Mu1
Ins-1
Odc-5
Pcp-1
Por
Pthlh
Qui (Prp)
Rho
Rpl32
Rpn-1
58 Ly-14 (c), ol (c)
59 Ly-15, Mph-1
61 War
63 fr
68 Ahr-3
69 Sgp-2
72 Ly-21
74 Fis-1, Int-2 (Fis-1)
75 Xmmv-76 (Fis-1)
77 Th
80 Ex (c)
81 Gv-2
89 Ad
Ass-ps1
bh
Cyp2e
D7Smh448
D7Smh483
D7Smh59
D7Smh75
D7Smh759
hsdr-1
Ins-2
Kal
l(7)-1 (Gpi-1, Hbb)
Lfo-1 (Hbb)
Mmv-3
Mmv-4
Mylpf
Ngfa
Oat
Odc-6
Odc-7
Pax-2
Pth
Rps16
Rras
Rrm1
Rrm2-ps2
Tgfb
Xmmv-31 (Tam-1)

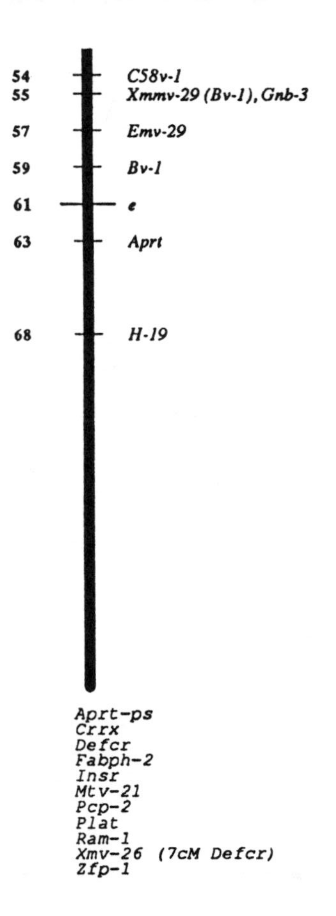
54 C58v-1
55 Xmmv-29 (Bv-1), Gnb-3
57 Emv-29
59 Bv-1
61 e
63 Aprt
68 H-19
Aprt-ps
Crrx
Defcr
Fabph-2
Insr
Mtv-21
Pcp-2
Plat
Ram-1
Xmv-26 (7cM Defcr)
Zfp-1

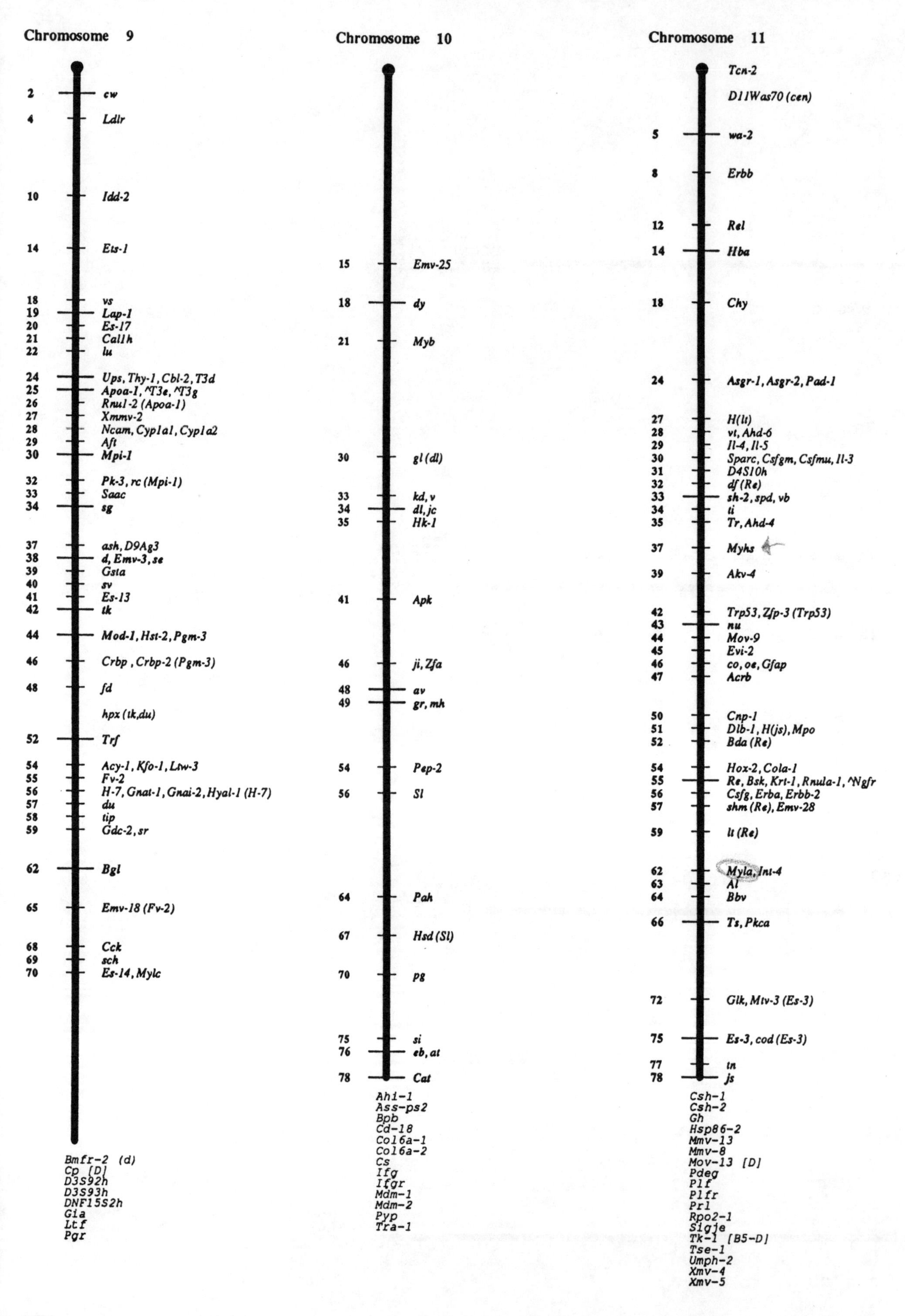

Chromosome 9
2 cw
4 Ldlr
10 Idd-2
14 Ets-1
18 vs
19 Lap-1
20 Es-17
21 Cal1h
22 lu
24 Ups, Thy-1, Cbl-2, T3d
25 Apoa-1, ^T3e, ^T3g
26 Rnu1-2 (Apoa-1)
27 Xmmv-2
28 Ncam, Cyp1a1, Cyp1a2
29 Aft
30 Mpi-1
32 Pk-3, rc (Mpi-1)
33 Saac
34 sg
37 ash, D9Ag3
38 d, Emv-3, se
39 Gsta
40 sv
41 Es-13
42 tk
44 Mod-1, Hst-2, Pgm-3
46 Crbp, Crbp-2 (Pgm-3)
48 fd
hpx (tk,du)
52 Trf
54 Acy-1, Kfo-1, Ltw-3
55 Fv-2
56 H-7, Gnat-1, Gnai-2, Hyal-1 (H-7)
57 du
58 tip
59 Gdc-2, sr
62 Bgl
65 Emv-18 (Fv-2)
68 Cck
69 sch
70 Es-14, Mylc
Bmfr-2 (d)
Cp [D]
D3S92h
D3S93h
DNF15S2h
Gia
Ltf
Pgr
Chromosome 10
15 Emv-25
18 dy
21 Myb
30 gl (dl)
33 kd, v
34 dl, jc
35 Hk-1
41 Apk
46 ji, Zfa
48 av
49 gr, mh
54 Pep-2
56 Sl
64 Pah
67 Hsd (Sl)
70 pg
75 si
76 eb, at
78 Cat
Ahi-1
Ass-ps2
Bpb
Cd-18
Col6a-1
Col6a-2
Cs
Ifg
Ifgr
Mdm-1
Mdm-2
Pyp
Tra-1
Chromosome 11
Tcn-2
D11Was70 (cen)
5 wa-2
8 Erbb
12 Rel
14 Hba
18 Chy
24 Asgr-1, Asgr-2, Pad-1
27 H(lt)
28 vt, Ahd-6
29 Il-4, Il-5
30 Sparc, Csfgm, Csfmu, Il-3
31 D4S10h
32 df (Re)
33 sh-2, spd, vb
34 ti
35 Tr, Ahd-4
37 Myhs
39 Akv-4
42 Trp53, Zfp-3 (Trp53)
43 nu
44 Mov-9
45 Evi-2
46 co, oe, Gfap
47 Acrb
50 Cnp-1
51 Dlb-1, H(js), Mpo
52 Bda (Re)
54 Hox-2, Cola-1
55 Re, Bsk, Krt-1, Rnula-1, ^Ngfr
56 Csfg, Erba, Erbb-2
57 shm (Re), Emv-28
59 lt (Re)
62 Myla, Int-4
63 Al
64 Bbv
66 Ts, Pkca
72 Glk, Mtv-3 (Es-3)
75 Es-3, cod (Es-3)
77 tn
78 js
Csh-1
Csh-2
Gh
Hsp86-2
Mmv-13
Mmv-8
Mov-13 [D]
Pdeg
Plf
Plfr
Prl
Rpo2-1
Sigje
Tk-1 [B5-D]
Tse-1
Umph-2
Xmv-4
Xmv-5

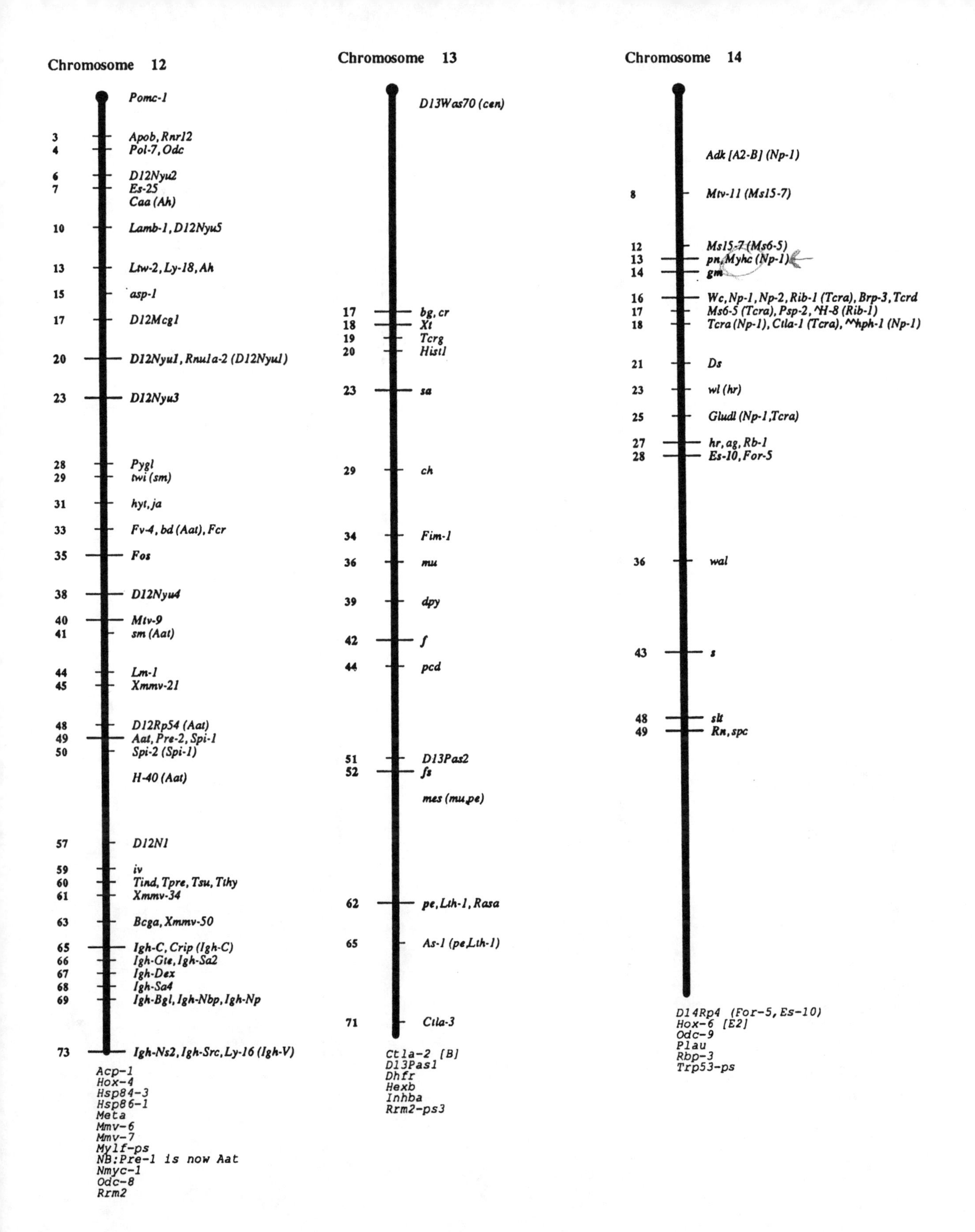
Chromosome 12
Pomc-1
3 Apob, Rnr12
4 Pol-7, Odc
6 D12Nyu2
7 Es-25
Caa (Ah)
10 Lamb-1, D12Nyu5
13 Ltw-2, Ly-18, Ah
15 asp-1
17 D12Mcg1
20 D12Nyu1, Rnu1a-2 (D12Nyu1)
23 D12Nyu3
28 Pygl
29 twi (sm)
31 hyt, ja
33 Fv-4, bd (Aat), Fcr
35 Fos
38 D12Nyu4
40 Mtv-9
41 sm (Aat)
44 Lm-1
45 Xmmv-21
48 D12Rp54 (Aat)
49 Aat, Pre-2, Spi-1
50 Spi-2 (Spi-1)
H-40 (Aat)
57 D12N1
59 iv
60 Tind, Tpre, Tsu, Tthy
61 Xmmv-34
63 Bcga, Xmmv-50
65 Igh-C, Crip (Igh-C)
66 Igh-Gte, Igh-Sa2
67 Igh-Dex
68 Igh-Sa4
69 Igh-Bgl, Igh-Nbp, Igh-Np
73 Igh-Ns2, Igh-Src, Ly-16 (Igh-V)
Acp-1
Hox-4
Hsp84-3
Hsp86-1
Meta
Mmv-6
Mmv-7
Mylf-ps
NB:Pre-1 is now Aat
Nmyc-1
Odc-8
Rrm2
Chromosome 13
D13Was70 (cen)
17 bg, cr
18 Xt
19 Tcrg
20 Hist1
23 sa
29 ch
34 Fim-1
36 mu
39 dpy
42 f
44 pcd
51 D13Pas2
52 fs
mes (mu,pe)
62 pe, Lth-1, Rasa
65 As-1 (pe,Lth-1)
71 Ctla-3
Ctla-2 [B]
D13Pas1
Dhfr
Hexb
Inhba
Rrm2-ps3
Chromosome 14
Adk [A2-B] (Np-1)
8 Mtv-11 (Ms15-7)
12 Ms15-7 (Ms6-5)
13 pn, Myhc (Np-1)
14 gm
16 Wc, Np-1, Np-2, Rib-1 (Tcra), Brp-3, Tcrd
17 Ms6-5 (Tcra), Psp-2, H-8 (Rib-1)
18 Tcra (Np-1), Ctla-1 (Tcra), hph-1 (Np-1)
21 Ds
23 wl (hr)
25 Glud1 (Np-1,Tcra)
27 hr, ag, Rb-1
28 Es-10, For-5
36 wal
43 s
48 slt
49 Rn, spc
D14Rp4 (For-5, Es-10)
Hox-6 [E2]
Odc-9
Plau
Rbp-3
Trp53-ps

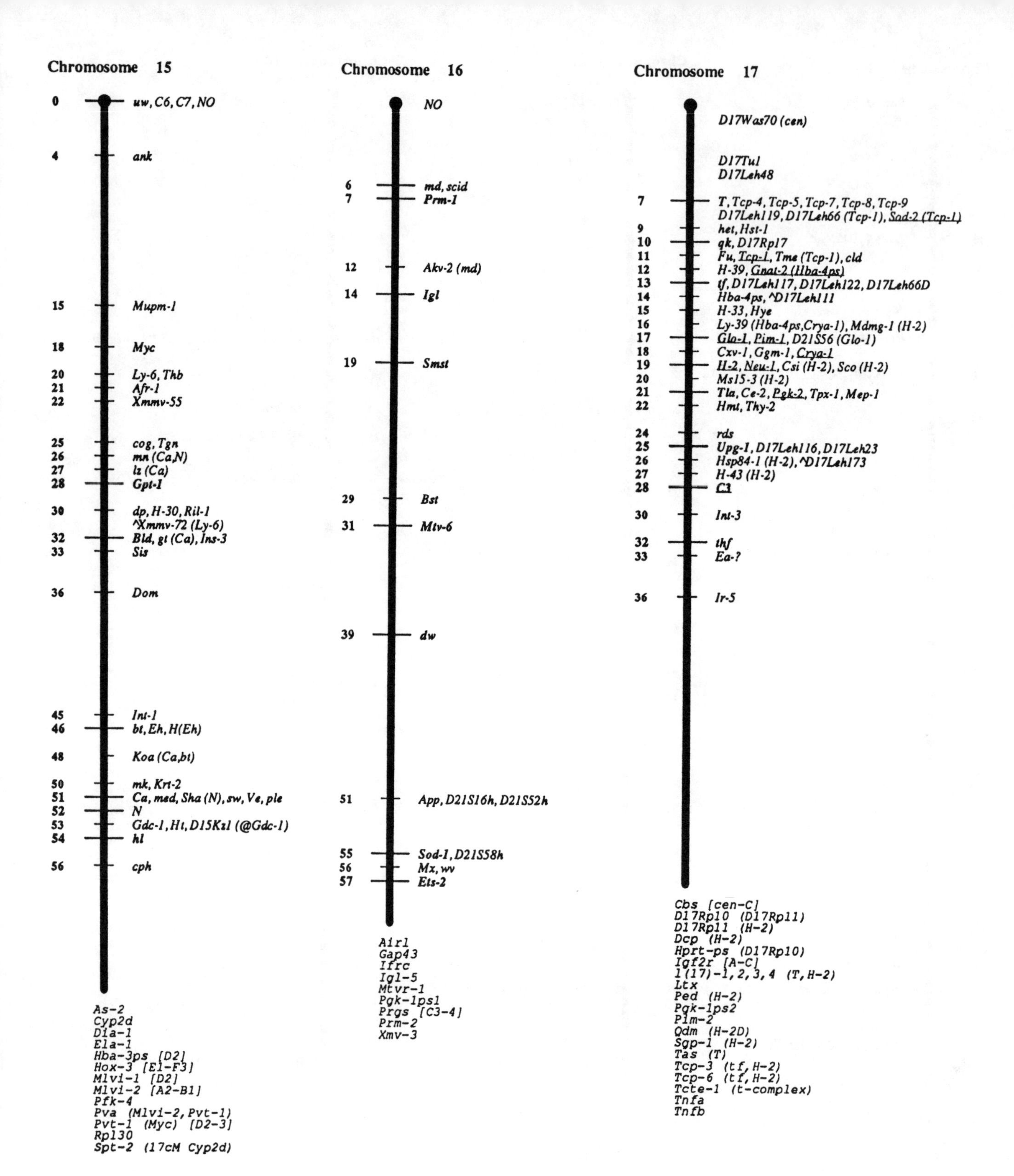
Chromosome 15
0 uw, C6, C7, NO
4 ank
15 Mupm-1
18 Myc
20 Ly-6, Thb
21 Afr-1
22 Xmmv-55
25 cog, Tgn
26 mn (Ca,N)
27 lz (Ca)
28 Gpt-1
30 dp, H-30, Ril-1
^Xmmv-72 (Ly-6)
32 Bld, gt (Ca), Ins-3
33 Sis
36 Dom
45 Int-1
46 bt, Eh, H(Eh)
48 Koa (Ca,bt)
50 mk, Krt-2
51 Ca, med, Sha (N), sw, Ve, ple
52 N
53 Gdc-1, Ht, D15Kz1 (@Gdc-1)
54 hl
56 cph
As-2
Cyp2d
Dia-1
Ela-1
Hba-3ps [D2]
Hox-3 [E1-F3]
Mlvi-1 [D2]
Mlvi-2 [A2-B1]
Pfk-4
Pva (Mlvi-2, Pvt-1)
Pvt-1 (Myc) [D2-3]
Rpl30
Spt-2 (17cM Cyp2d)
Chromosome 16
NO
6 md, scid
7 Prm-1
12 Akv-2 (md)
14 Igl
19 Smst
29 Bst
31 Mtv-6
39 dw
51 App, D21S16h, D21S52h
55 Sod-1, D21S58h
56 Mx, wv
57 Ets-2
Airl
Gap43
Ifrc
Igl-5
Mtvr-1
Pgk-1ps1
Prgs [C3-4]
Prm-2
Xmv-3
Chromosome 17
D17Was70 (cen)
D17Tu1
D17Leh48
7 T, Tcp-4, Tcp-5, Tcp-7, Tcp-8, Tcp-9
D17Leh119, D17Leh66 (Tcp-1), Sod-2 (Tcp-1)
9 het, Hst-1
10 qk, D17Rp17
11 Fu, Tcp-1, Tme (Tcp-1), cld
12 H-39, Gnat-2 (Hba-4ps)
13 tf, D17Leh117, D17Leh122, D17Leh66D
14 Hba-4ps, ^D17Leh111
15 H-33, Hye
16 Ly-39 (Hba-4ps, Crya-1), Mdmg-1 (H-2)
17 Glo-1, Pim-1, D21S56 (Glo-1)
18 Cxv-1, Ggm-1, Crya-1
19 H-2, Neu-1, Csi (H-2), Sco (H-2)
20 Ms15-3 (H-2)
21 Tla, Ce-2, Pgk-2, Tpx-1, Mep-1
22 Hmt, Thy-2
24 rds
25 Upg-1, D17Leh116, D17Leh23
26 Hsp84-1 (H-2), ^D17Leh173
27 H-43 (H-2)
28 C3
30 Int-3
32 thf
33 Ea-?
36 Ir-5
Cbs [cen-C]
D17Rp10 (D17Rp11)
D17Rp11 (H-2)
Dcp (H-2)
Hprt-ps (D17Rp10)
Igf2r [A-C]
l(17)-1, 2, 3, 4 (T, H-2)
Ltx
Ped (H-2)
Pgk-1ps2
Pim-2
Qdm (H-2D)
Sgp-1 (H-2)
Tas (T)
Tcp-3 (tf, H-2)
Tcp-6 (tf, H-2)
Tcte-1 (t-complex)
Tnfa
Tnfb

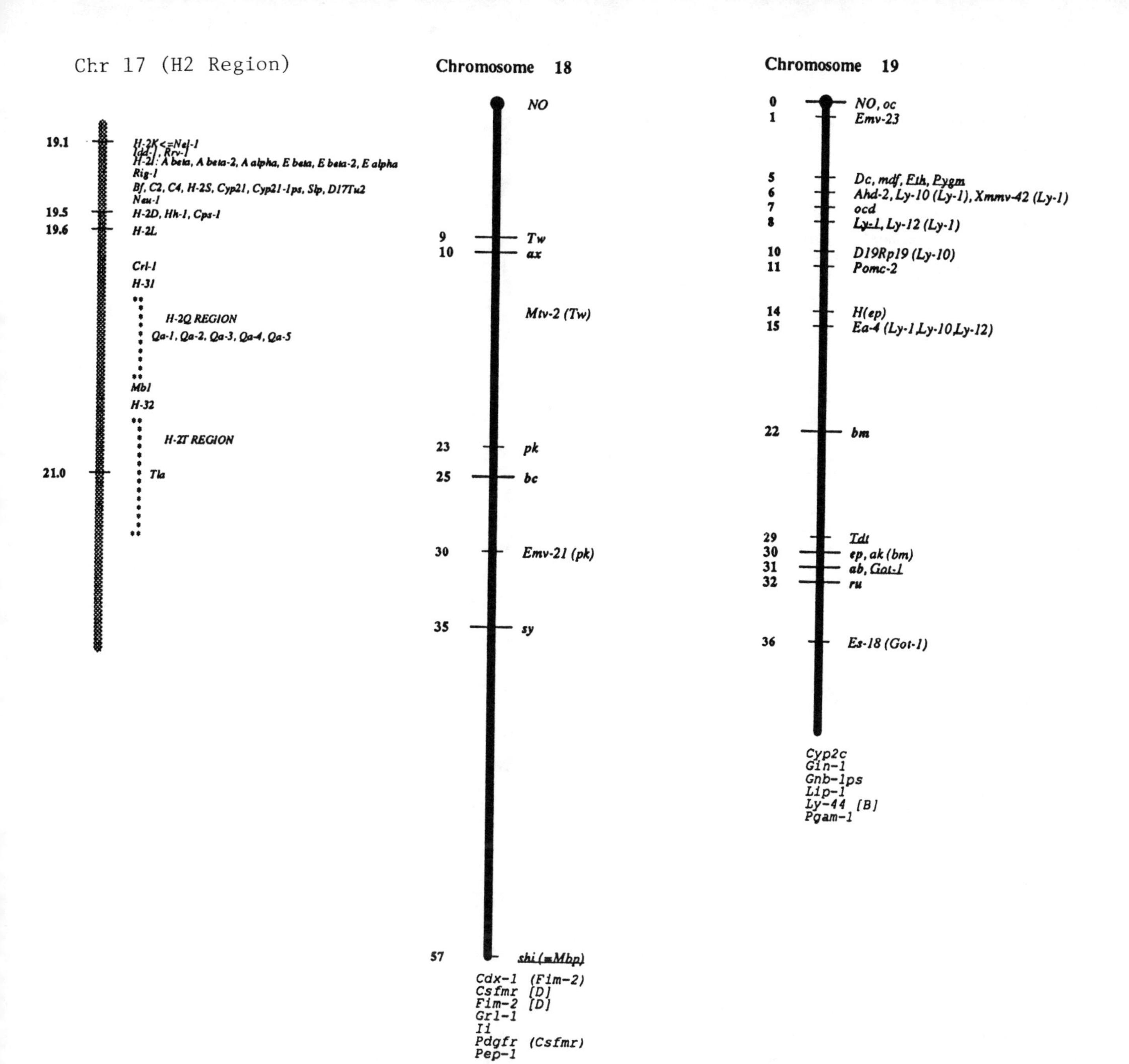

Chr 17 (H2 Region)
19.1
H-2K<=Nel-1
Idd-1, Rrv-1
H-2I: A beta, A beta-2, A alpha, E beta, E beta-2, E alpha
Rig-1
Bf, C2, C4, H-2S, Cyp21, Cyp21-1ps, Slp, D17Tu2
Neu-1
19.5
H-2D, Hh-1, Cps-1
19.6
H-2L
Crl-1
H-31
H-2Q REGION
Qa-1, Qa-2, Qa-3, Qa-4, Qa-5
Mb1
H-32
H-2T REGION
21.0
Tla
Chromosome 18
NO
9 Tw
10 ax
Mtv-2 (Tw)
23 pk
25 bc
30 Emv-21 (pk)
35 sy
57 shi (=Mbp)
Cdx-1 (Fim-2)
Csfmr [D]
Fim-2 [D]
Grl-1
Ii
Pdgfr (Csfmr)
Pep-1
Chromosome 19
0 NO, oc
1 Emv-23
5 Dc, mdf, Eth, Pygm
6 Ahd-2, Ly-10 (Ly-1), Xmmv-42 (Ly-1)
7 ocd
8 Ly-1, Ly-12 (Ly-1)
10 D19Rp19 (Ly-10)
11 Pomc-2
14 H(ep)
15 Ea-4 (Ly-1,Ly-10,Ly-12)
22 bm
29 Tdt
30 ep, ak (bm)
31 ab, Got-1
32 ru
36 Es-18 (Got-1)
Cyp2c
Gin-1
Gnb-1ps
Lip-1
Ly-44 [B]
Pgam-1

Chromosome X

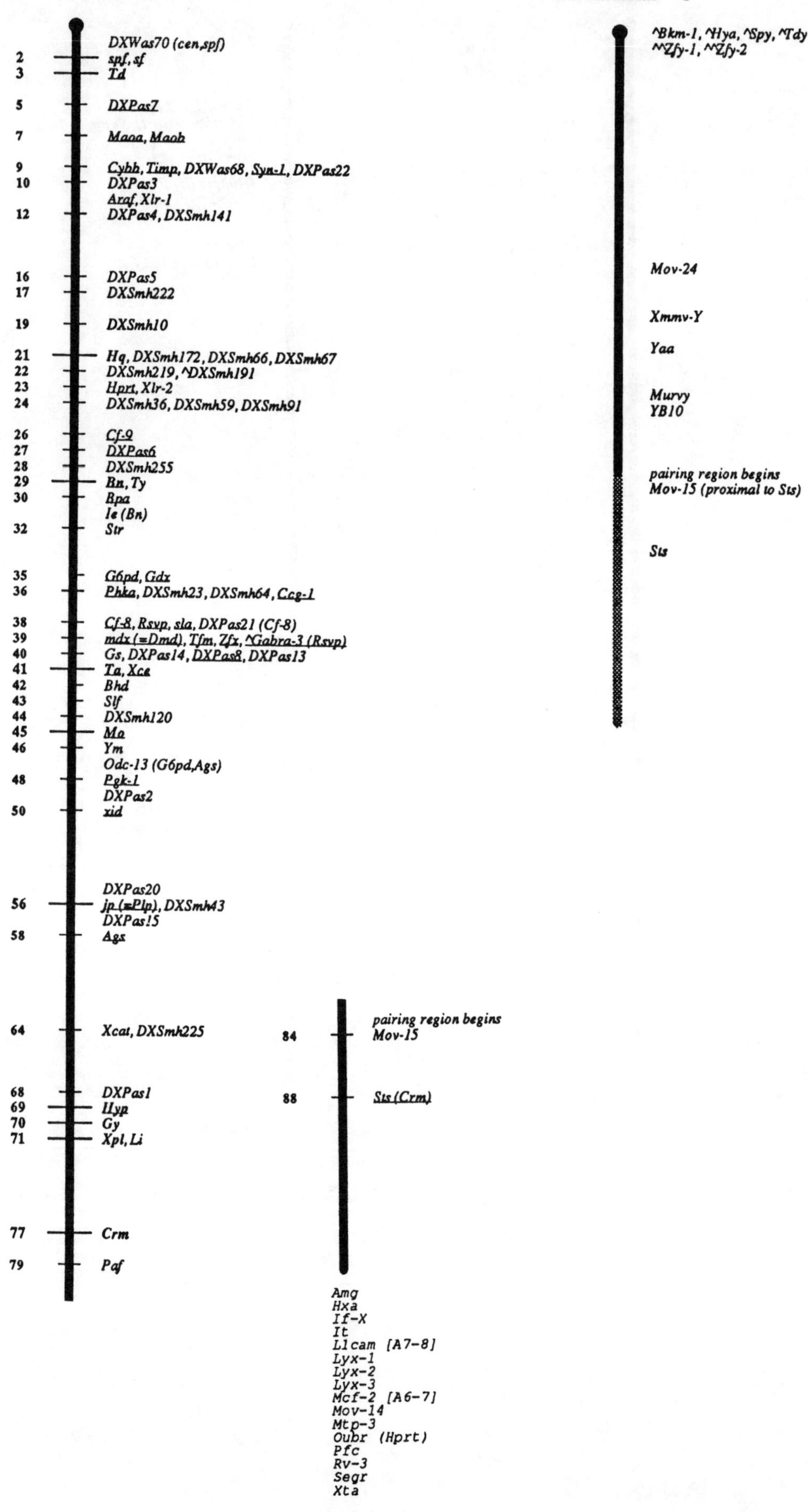

Chromosome Y

^Bkm-1, ^Hya, ^Spy, ^Tdy
^^Zfy-1, ^^Zfy-2

Mov-24

Xmmv-Y

Yaa

Murvy
YB10

pairing region begins
Mov-15 (proximal to Sts)

Sts

Table 1. GENETIC LOCI OF THE MOUSE as of August 1989

Explanation: Recessive locus symbols begin with a lower case letter. All other locus symbols begin with an upper case letter. Rules for genetic nomenclature in the mouse including DNA polymorphisms, can be found in Mouse News Letter 1985; 72:2-28. UN means the chromosomal location of the locus is unknown. RE means this locus symbol is reserved for a locus yet to be identified in the mouse, but for which a probable human homologue has been named.

SYMBOL	NAME OF LOCUS	CHROMOSOME
21OHA	withdrawn, = Oh21-1	17
21OHB	withdrawn, = Oh21-2	17
a	non-agouti	2
A<y>	yellow; pseudoallele of agouti locus	2
Aal	active avoidance learning	1
Aap	active avoidance performance(may not = single gene)	UN
Aat	alpha-1-antitrypsin (ex Pre-1)	12
ab	asebia	19
Abl	Abelson murine leukemia oncogene	2
Abll	Abelson related gene	1
abn	abnormal (provisional; extinct?)	UN
Abpa	androgen binding protein, alpha (ex Sal-1, Tcp, Abp)	7
Abpb	androgen binding protein, beta	UN
Abpg	androgen binding protein, gamma	7
ac	absent corpus callosum (prob. not a single gene)	UN
Acc	anterior capsular cataract (provisional)	UN
acd	adrenocortical dysplasia	UN
Acf-1	albumin conformation factor-1	1
Achr-1	withdrawn, = Acra	17
Achr-2	withdrawn, = Acrb	11
Achr-3	withdrawn, = Acrg	1
Achr-4	withdrawn, = Acrd	1
Ack	anti-C.kutscheri (ex ack)	UN
ack	withdrawn, = Ack	UN
acm	withdrawn, = Ack	UN
Aco-1	aconitase-1	4
Aco-2	aconitase-2 mitochondrial, monomorphic	RE
Acp-1	acid phosphatase-1	12
Acp-2	acid phosphatase-2	2
Acra	acetylcholine receptor alpha (ex Achr-1)	2
Acrb	acetylcholine receptor beta (ex Achr-2)	11
Acrd	acetylcholine receptor delta (ex Achr-4)	1
Acre	acetylcholine receptor epsilon	RE
Acrg	acetylcholine receptor gamma (ex Achr-3)	1
Acry-1	withdrawn, = Crya-1	17
act	adult cataract (provisional; extinct?)	UN
Actb	cytoplasmic beta-actin	5
Actc-1	cardiac alpha-actin	2
Acts	skeletal alpha-actin	3
Actx	melanoma X-actin	UN
Acy-1	aminoacylase-1	9
Acy-2	aminoacylase-2	UN
Ad	adult obesity and diabetes	7
Ada	adenosine deaminase	2
Adap	withdrawn, = App	16
Adh-1	alcohol dehydrogenase-1	3
Adh-1ps	alcohol dehydrogenase-1 pseudogene	3
Adh-1t	alcohol dehydrogenase-1-temporal	3
Adh-3e	alcohol dehydrogenase-3-electrophoretic	3
Adh-3t	alcohol dehydrogenase-3-temporal (ex Adt-1)	3
Adk	adenosine kinase	14
adr	arrested development of righting response	UN
Adre	age at disappearance of rooting response (prov.)	4
Ads-1	anti dsDNA antibody-1	UN
Ads-2	anti dsDNA antibody-2	UN
Ads-3	anti dsDNA antibody-3	17
Ads-4	anti dsDNA antibody-4	UN
Adt-1	withdrawn, = Adh-3t	3
Aem-1	antierythrocyte autoantibody modifier-1	4
Afp	alpha fetoprotein	5
Afr-1	alpha fetoprotein regulation-1(adult, ex Raf)	15
Afr-2	alpha fetoprotein regulation-2(inducibility, ex rif)	UN
Aft	abnormal feet and tail	9
Afuc	withdrawn, = Fuca	4
ag	agitans	14
Aglp	withdrawn, = Neu-1	17
Aglu	alpha-glucosidase	RE
Agp-1	anti gp70 immune complex-1	17
Agp-2	anti gp70 immune complex-2	17
Agp-3	anti gp70 immune complex-3	17
Ags	alpha-galactosidase	X
Agt	angiotensinogen	UN
Ah	aromatic hydrocarbon responsiveness (ex In, Ahh)	12
ah	withdrawn, allele of dt	1
aha	autoimmune hemolytic anemia (provisional)	UN
Ahd-1	aldehyde dehydrogenase-1, mitochondrial	4
Ahd-2	aldehyde dehydrogenase-2, cytoplasmic	19
Ahd-3	aldehyde dehydrogenase-3, microsomal	UN
Ahd-3r	aldehyde dehydrogenase-3 regulator	UN
Ahd-4	aldehyde dehydrogenase-4	UN
Ahd-5	aldehyde dehydrogenase-5(provisional)	UN
Ahd-6	aldehyde dehydrogenase-6	UN
Ahh	withdrawn, = Ah	12
Ahi-1	Abelson helper integration site	10
Ahr-1	aldehyde reductase-1	3
Ahr-2	aldehyde reductase-2 (= hexonate dehydrogenase)	UN
Ahr-3	aldehyde reductase-3, liver and kidney	7
Ahr-4	aldehyde reductase-4, liver and kidney	UN
Aia-1	autoimmune hemolytic anemia-1	4
Aia-2	autoimmune hemolytic anemia-2	UN
Airl	activator of immune response gene Ir-1	16
ak	aphakia	19
Ak-1	adenylate kinase-1	2
Ak-2	adenylate kinase-2	4
Akp-1	alkaline phosphatase-1,liver,bone,kidney,Mn req.	1
Akp-2	alkaline phosphatase-2,liver	4
Akp-3	alkaline phosphatase-3,intestine,not Mn req.	1
Akv-1	AKR leukemia virus inducer-1*(=Emv-11)	7
Akv-2	AKR leukemia virus inducer-2*(=Emv-12)	16
Akv-2J	withdrawn, = Akv-4	11
Akv-3	AKR leukemia virus inducer-3*(=Emv-13)	2
Akv-4	AKR leukemia virus inducer-4*(=Emv-14)	11
Akvp	AKR leukemia virus protein inducer*(may = Akv-3 = Emv-13)	UN
Akvr-1	withdrawn, = Fv-4	12
Al	alopecia	11
Ala-1	withdrawn, = Ly-6	15
Alb-1	serum albumin variant	5

ald	adrenocortical lipid depletion	1
Aldh	aldehyde dehydrogenase	UN
Aldo-1	aldolase-1	RE
Alf	albino lethal factor	7
Alm	anterior lenticonus with microphthalmia (provisional)	UN
alp	alopecie (provisional; extinct?)	UN
Alp-1	withdrawn, = Apoa-1 (ex Sep-1)	9
Alp-2	withdrawn, = Apoa-2	1
am	amputated	8
Amel	withdrawn, = Amg	X
Amg	amelogenin (ex Amel)	X
Ampd-1	AMP deaminase-1	3
Ampd-2	AMP deaminase-2	3
Amy-1	amylase-1, salivary	3
Amy-2	amylase-2, pancreatic	3
an	Hertwig's anemia	4
Anc	anisocoria (radiation induced eye mutation)	UN
Anf	withdrawn, = Pnd	4
ank	progressive ankylosis	15
Ank-1	ankyrin-1	RE
anx	anorexia	2
ao	apampischo	UN
Aox-1	aldehyde oxidase-1	1
Aox-2	aldehyde oxidase-2	1
ap	alopecia periodica (extinct)	UN
Ap2	adipocyte specific protein	3
Aph-1	alphaprotein-1	2
Aph-2	alphaprotein-2	2
Apk	acid phosphatase-kidney (ex Acp-2)	10
Apl	withdrawn, = Neu-1 (ex Acp-1)	17
Apnh	Na+/H+ antiporter	4
Apo	anterior polar opacity (provisional)	UN
Apoa-1	apolipoprotein A-I (ex Alp-1, Sep-1)	9
Apoa-2	apolipoprotein A-II	1
Apoa-4	apolipoprotein A-IV	9
Apob	apolipoprotein B	12
Apoc	anterior polar cataract (provisional)	UN
Apoe	apolipoprotein E	7
App	amyloid beta (A4) precursor protein (ex Cvap, Adap)	16
Aprt	adenine phosphoribosyl transferase	8
Aprt-ps	adenine phosphoribosyl transferase pseudogene	8
Apyc	anterior pyramidal cataract	UN
Ar	androgen receptor (= Tfm)	X
Araf	raf-related oncogene	X
Arp	Arp lymphoid/erythroid hyperplasia	5
Arsa	withdrawn, = As-2	15
As	withdrawn, allele of a	2
As-1	withdrawn, = As-1s	13
As-1r	arylsulfatase B regulation (ex Asr-1)	13
As-1s	arylsulfatase B structural	13
As-1t	arylsulfatase B temporal regulation (ex Ast-1)	13
As-2	arylsulfatase A (ex Arsa)	15
Asc-1	anterior suture cataract-1	UN
Asc-2	anterior suture cataract-2	UN
asd	withdrawn; allele at Gus locus	UN
Asgr-1	asialoglycoprotein receptor-1	11
Asgr-2	asialoglycoprotein receptor-2	11
ash	ashen	9
Asl	argininosuccinate lyase	5
asp	withdrawn, = asp-1	12
asp-1	audiogenic seizure prone-1 (ex asp, Ias)	12
asp-2	audiogenic seizure prone-2	4
asr	withdrawn, = ras	UN
Asr-1	withdrawn, = As-1r	13
Ass	arginosuccinate synthetase	2
Ass-ps1	arginosuccinate synthetase pseudogene-1	7
Ass-ps2	arginosuccinate synthetase pseudogene-2	10
Ast-1	withdrawn, = As-1t	13
at	atrichosis	10
At-3	anti-thrombin-3	1
Ath-1	atherosclerosis-1	1
Ath-2	atherosclerosis-2	UN
Atpa-1	Na,K-ATPase alpha-1	3
Atpa-2	Na,K-ATPase alpha-2	7
Atpa-3	Na,K-ATPase alpha-3	1
Atpb	Na,K-ATPase beta	1
av	Ames waltzer	10
Av-1	Abelson virus susceptibility-1	UN
Av-2	Abelson virus susceptibility-2	UN
ax	ataxia	18
azh	abnormal spermatazoon head shape	4
b	brown	4
B2m	beta-2 microglobulin (ex Ly-4, Ly-m11)	2
ba	bare	UN
Badm	B-adrenergic binding (provisional)	UN
bal	balding	UN
Bas	BALB murine sarcoma oncogene (may = Hras)	UN
Bbv	B10.BR/SgLi endogenous ecotropic virus*(B-tropic)	11
bc	bouncy	18
Bcd-1	butyryl CoA dehydrogenase-1	5
Bcg	withdrawn, = Lsh	1
Bcga	BCG-induced anergy	12
Bcl-1	B cell leukemia-1	RE
Bcl-2	B-cell leukemia/lymphoma-2	1
Bcm-1	B cell membrane protein 1	1
bd	bradypneic	12
Bda	bald-arthritic	11
Bdv-1	BALB/c defective provirus-1*	UN
bf	buff	5
Bf	complement component factor B (see H-2S)	17
Bfo	bell-flash ovulation	4
bg	beige	13
Bgb	beta-D-galactosidase in brain (provisional)	UN
Bge	withdrawn, = Bgl-e	9
Bgl-e	beta-galactosidase electrophoretic	9
Bgl-s	beta-galactosidase systemic regulator	9
Bgl-t	beta-galactosidase temporal	9
Bgs	withdrawn, = Bgl-s	9
Bgt	withdrawn, = Bgl-t	9
bh	brain-hernia	7
Bhd	broad headed	X
Bkm-1	banded krait minor satellite DNA-1	Y
Bkm-2	banded krait minor satellite DNA-2	X
Bkm-3	banded krait minor satellite DNA-3	X
Bkm-4	banded krait minor satellite DNA-4	17
Bkm-5	banded krait minor satellite DNA-5	4
bl	blebbed	5
Bld	blind	15
Blo	withdrawn, allele of Mo	X
bls	withdrawn, allele of uw	15

Symbol	Name	Chromosome
Blv-1	withdrawn, = Bv-1	8
Blvr	biliverdin reductase	2
bm	brachymorphic	19
bmf-1	withdrawn, = Bmfr-2	9
Bmfr-1	B-cell maturation factor responsiveness-1	4
Bmfr-2	B-cell maturation factor responsiveness-2 (ex bmf)	9
Bmk	B cell myeloid kinase	UN
Bmn	beta-mannosidase activity (liver, kidney)(provisional)	3
Bn	bent-tail	X
bp	brachypodism	2
Bpa	bare-patches	X
Bph	withdrawn, = Xt	13
Br	brachyrrhine	UN
brp	brachypod Japan	2
Brp-1	brain protein-1 (provisional)	6
Brp-10	brain protein-10 (provisional)	UN
Brp-11	brain protein-11 (provisional)	UN
Brp-12	withdrawn, = Ltw-4	1
Brp-13	brain protein-13 (provisional)	2
Brp-14	withdrawn, = Apoa-1	9
Brp-2	brain protein-2 (provisional)	UN
Brp-3	brain protein-3 (provisional)	UN
Brp-4	brain protein-4 (provisional)	UN
Brp-5	brain protein-5 (provisional)	UN
Brp-6	brain protein-6 (provisional)	UN
Brp-7	brain protein-7 (provisional)	UN
Brp-8	brain protein-8 (provisional)	UN
Brp-9	brain protein-9 (provisional)	UN
BRS-1	withdrawn, = Odc-1	1
BRS-10	withdrawn, = Odc-10	UN
BRS-11	withdrawn, = Odc-11	UN
BRS-2	withdrawn, = Odc-2	2
BRS-3	withdrawn, = Odc-3	3
BRS-4	withdrawn, = Odc-4	4
BRS-5	withdrawn, = Odc-5	6
BRS-6	withdrawn, = Odc-6	7
BRS-7	withdrawn, = Odc-7	7
BRS-8	withdrawn, = Odc-8	12
BRS-9	withdrawn, = Odc-9	14
Bru	bruised	UN
bs	blind-sterile	2
Bs	withdrawn, allele at W	5
Bsk	bare skin	11
Bsp	black spleen (splenic lipofuscinosis)	UN
Bst	belly spot and tail	16
bt	belted	15
bt-2	belted-2	UN
bus	bustling	UN
bv	Bronx waltzer	UN
Bv-1	C57BL/10 endogenous ecotropic virus*(N-tropic; ex Emv-2)	8
bw	withdrawn, = allele of mi	6
Bxv-1	B10 xenotropic virus-1*	1
c	albino	7
C2	complement component-2	17
C3	complement component-3	17
C4	complement component-4 (ex Ss)	17
C4bp	complement component-4 binding protein	1
C5	withdrawn, = Hc	2
C58v-1	C58 endogenous ecotropic virus-1* (= Emv-23)	8
C58v-2	C58 endogenous ecotropic virus-2*	UN
C58v-3	C58 endogenous ecotropic virus-3*	4
C58v-4	C58 endogenous ecotropic virus-4*	7
C6	complement component-6	15
C6r	complement component-6, regulatory	UN
C7	complement component-7	15
Ca	caracul	15
Caa	Ca ATPase activity	12
cab	cardiac abnormality	UN
cac	recessive cataract	UN
Cacy	calcyclin	3
Cad	congenital cataract (provisional)	UN
Caf	caffeine susceptibility	UN
Calh	calpactin I heavy chain	9
Call	calpactin I light chain	3
Calb	calbindin-D28K	4
Calc	calcitonin (ex Ct, Ctn)	7
Calm	calmodulin	RE
can	cartilage anomaly (extinct?)	UN
Capl	placental protein	3
Car-1	carbonic anhydrase-1 (ex Pro-1)	3
Car-2	carbonic anhydrase-2	3
Car-3	carbonic anhydrase-3 (prov.)	UN
Cas-1	catalase-1 (ex Cs-1)	2
Cat	dominant cataract	10
Cat-2	dominant cataract-2	UN
Cat-3	dominant cataract-3	UN
cb	cerebral degeneration	UN
Cbl-1	virally transduced oncogene-1	6
Cbl-2	virally transduced oncogene-2	9
Cbs	cystathionine beta-synthase	17
cby	chubby (provisional)	UN
Ccd	cleidocranial dysplasia (provisional)	UN
Ccg-1	cell cycle, G1 phase defect	X
Cck	cholecystokinin	9
Cd	crooked	6
Cd-18	CD18 antigen (provisional)	10
Cda	cytidine deaminase	RE
cdm	cadmium resistance	3
Cdx-1	caudal type homeo box-1	18
Ce-1	liver catalase activity	2
Ce-2	kidney catalase	17
Cf-5	coagulation factor 5	1
Cf-8	coagulation factor VIII	X
Cf-9	coagulation factor IX	X
Cfh	complement component factor h (prob. = Sas-1)	1
Cfhe	complement component factor h, electrophoretic variant	2
Cfp-1	cerebellar folial pattern-1	4
cg	circumgyrator (provisional)	UN
Cg	withdrawn, = Xce	X
Cgd	withdrawn, = Cybb	X
ch	congenital hydrocephalus	13
cho	chondrodysplasia	UN
Chrm-1	cholinergic receptor, muscarinic-1 (CNS)	UN
Chrm-2	cholinergic receptor, muscarinic-2 (cardiac)	RE
Chrm-3	cholinergic receptor, muscarinic-3 (cardiac)	RE
cht	chocolate	7
Chy	chylous ascites	11
ci	circler (provisional; extinct?)	UN
Cis	withdrawn, = Cs	10
ck	withdrawn, = cpk	UN

Symbol	Name	Chr
Ck-1	creatine kinase-1	RE
Ck-2	creatine kinase-2	RE
Ck-3	creatine kinase-3	RE
Ckmm	creatine kinase, muscle	7
cl	clubfoot (extinct)	UN
cla	clasper	4
cld	combined lipase deficiency	17
clf	cleft lip (provisional)	UN
Clo	withdrawn, allele of Sl	10
clp	claw paw	UN
cls	coatloss (provisional)	UN
Cm	coloboma	2
cmd	cartilage matrix deficiency	UN
Cms	resistance to Coccidioides immitis	UN
cn	achondroplasia	4
Cnp-1	cyclic nucleotide phosphodiesterase-1	11
Cnp-2	cyclic nucleotide phosphodiesterase-2	3
co	cocked	11
coa	cocoa	3
Coc	coralliform cataract	UN
cod	cerebellar outflow degeneration	UN
cog	congenital goiter	15
Coh	withdrawn, = Cyp2b	7
Col1a-1	withdrawn, = Cola-1	11
Col1a-2	withdrawn, = Cola-2	16
Col2a-	procollagen type II	RE
Col3a	procollagen type III, alpha 1	1
Col4a-1	procollagen type IV, alpha 1	UN
Col4a-2	procollagen type IV, alpha 2	UN
Col5a-	procollagen type V, alpha	RE
Col6a-1	procollagen type VI alpha 1	10
Col6a-2	procollagen type VI alpha 2	10
Cola-1	procollagen type I, alpha 1	11
Cola-2	procollagen type I, alpha 2	16
Con	withdrawn, allele of Sl	10
cou	withdrawn, = T	17
Cp	ceruloplasmin	9
Cpa	carboxypeptidase A	6
cph	congenital progressive hydronephrosis (ex jpk)	15
cpk	congenital polycystic kidneys (ex ck)	UN
Cpl-1	plasma corticosterone level-1	UN
Cpl-2	plasma corticosterone level-2	UN
cpr	withdrawn, allele of ho	6
Cps-1	cleft palate susceptibility-1	17
Cpz	chlorpromazine avoidance	UN
cr	crinkled	13
Cr-2	complement receptor-2	1
Crbp	cellular retinol binding protein	9
Crbp-2	cellular retinol binding protein II	9
crh	cryptothrix	UN
cri	cribriform degeneration	4
Crip	cysteine-rich intestinal protein	12
Crl-1	complement receptor lymphocyte-1	17
Crm	cream	X
crn	cranioschisis	UN
Crp	C-reactive protein	1
Crrx	withdrawn, = Crry-ps	8
Crry	complement receptor related protein Y	1
Crry-ps	complement receptor related protein pseudogene	8
Crya-1	alpha crystalline (ex Acry-1)	17

Symbol	Name	Chr
Cryb	beta crystalline (may = Len-2)	RE
Cryg-1	gamma crystalline-1 (may = Len-1)	1
Cryg-2	gamma crystalline-2	1
Cryg-3	gamma crystalline-3	1
Cryg-4	gamma crystalline-4	1
Crz	control of early proliferation of T. cruzi (provisional)	UN
Cs	citrate synthase (ex Cis, Cts)	10
Cs-1	withdrawn, = Cas-1	2
Csfg	colony stimulating factor, granulocyte	11
Csfgm	colony stimulating factor, granulocyte macrophage specif.	11
Csfm	colony stimulating factor, macrophage	3
Csfmr	colony stimulating factor 1 receptor(ex Fms, overlaps Fim-2)	18
Csfmu	colony stimulating factor, multi (contains Il-3)	11
Csi	corticosteroid side chain isomerase activity	1
Csn	casein gene family, -alpha, -beta, -gamma	5
Csna	alpha casein	5
Csnb	beta casein	5
ct	curly-tail	UN
Ct	withdrawn, = Calc	7
Ctg-3	withdrawn, = T3g	9
Ctl-1	cytotoxic T lymphocyte response-1	1
Ctla-1	cytotoxic T lymphocyte associated protein-1	14
Ctla-2	cytotoxic T lymphocyte associated protein-2	13
Ctla-3	cytotoxic T lymphocyte associated protein-3	13
Ctla-4	cytotoxic T lymphocyte associated protein-4	1
Ctn	withdrawn, = Calc	7
Ctrb	chymotrypsinogen B (ex Prt-2)	8
Cts	cataract and small eye (provisional)	UN
cu	clasper underweight (provisional;extinct?)	UN
Cu	withdrawn, = Ca	15
Cv-1	BALB/c virus inducibility (ex Inc-1, Emv-1)	5
Cvap	withdrawn, = App	16
cw	curly whiskers	9
cx	cerebellum ataxia (provisional; extinct?)	UN
Cxv-1	C xenotropic virus-1*	17
Cxv-2	C xenotropic virus-2*(ex XenCSA, provisional)	4
cy	crinkly-tail	4
Cybb	cytochrome b-245, beta polypeptide (ex Cgd)	X
Cyl	curly-tail lethal (provisional)	2
Cyp1a1	cytochrome P450, 1a1, aromatic carbon inducible (ex P450-1)	9
Cyp1a2	cytochrome P450, 1a2, aromatic carbon inducible (ex P450-3)	9
Cyp21	cytochrome P450, 21, steroid 21 hydroxylase (ex Oh21-1)	17
Cyp21-ps1	cytochrome P450, 21, pseudogene (ex Oh21-2)	17
Cyp2a	cytochrome P450, 2a, phenobarbitol inducible	7
Cyp2b	cytochrome P450, 2b, phenobarbital inducible (ex Coh)	7
Cyp2c	cytochrome P450, 2c, phenobarbital inducible	19
Cyp2d	cytochrome P450, 2d (ex P450IID)	15
Cyp2e	cytochrome P450, 2e, ethanol inducible	7
Cyp3	cytochrome P450, 3, steroid inducible (ex Pcn)	6
Cyp4	cytochrome P450, 4	4
Cyx	cycloheximide tasting	6
d	dilute (ex Emv-3)	9
D0Rp3	DNA segment, Roswell Park-3 (ex RP3)	UN
D0Rp5	DNA segment, Roswell Park-5 (ex RP5)	UN
D0Rp6	DNA segment, Roswell Park-6 (ex RP6)	UN
D11S16h	DNA segment, Chr 11, human D11S16	2
D11Was70	DNA segment, Chr 11, University of Washington-70	11
D12-1	withdrawn, = D12Nyu1	12
D12-2	withdrawn, = D12Nyu2	12
D12-3	withdrawn, = D12Nyu3	12

D12-4	withdrawn, = D12Nyu4	12
D12-5	withdrawn, = D12Nyu5	12
D12Mcg1	DNA segment, Chr 12, Med. Coll. of Georgia-1	12
D12Nyu1	DNA segment, Chr 12, NYU-1	12
D12Nyu10	DNA segment, Chr 12, NYU-10	12
D12Nyu2	DNA segment, Chr 12, NYU-2	12
D12Nyu3	DNA segment, Chr 12, NYU-3	12
D12Nyu4	DNA segment, Chr 12, NYU-4	12
D12Nyu5	DNA segment, Chr 12, NYU-5	12
D12Nyu6	DNA segment, Chr 12, NYU-6	12
D12Nyu7	DNA segment, Chr 12, NYU-7	12
D12Nyu8	DNA segment, Chr 12, NYU-8	12
D12Nyu9	DNA segment, Chr 12, NYU-9	12
D12Rp54	DNA segment, Chr 12, Roswell Park-54 (ex RP54)	12
D13Pas1	DNA segment, Chr 13, Pasteur Institute-1	13
D13Pas2	DNA segment, Chr 13, Pasteur Institute-2	13
D13Was70	DNA segment, Chr 13, University of Washington-70	13
D14Rp4	DNA segment, Chr 14, Roswell Park-4 (ex RP4)	14
D17Leh111	DNA segment, Chr 17, Lehrach-111	17
D17Leh116	DNA segment, Chr 17, Lehrach-116	17
D17Leh117	DNA segment, Chr 17, Lehrach-117	17
D17Leh119	DNA segment, Chr 17, Lehrach-119	17
D17Leh122	DNA segment, Chr 17, Lehrach-122	17
D17Leh173	DNA segment, Chr 17, Lehrach-173	17
D17Leh23	DNA segment, Chr 17, Lehrach-23	17
D17Leh48	DNA segment, Chr 17, Lehrach-48	17
D17Leh66	DNA segment, Chr 17, Lehrach-66	17
D17Leh66D	DNA segment, Chr 17, Lehrach-66D	17
D17Rp10	DNA segment, Chr 17, Roswell Park-10 (ex RP10)	17
D17Rp11	DNA segment, Chr 17, Roswell Park-11 (ex RP11)	17
D17Rp17	DNA segment, Chr 17, Roswell Park-17 (ex RP17)	17
D17Sil1	DNA segment, Chr 17, Silver-1	17
D17Tu1	DNA segment, Chr 17, Tubingen-1	17
D17Tu2	DNA segment, Chr 17, Tubingen-2	17
D17Was70	DNA segment, Chr 17, University of Washington-70	17
D19Rp19	DNA segment, Chr 19, Roswell Park-19	19
D19S19h	DNA segment, Chr 7, human D19S19	7
D1Pas1	DNA segment, Chr 1, Pasteur Institute-1	1
D21S16h	DNA segment, Chr 16, human D21S16	16
D21S52h	DNA segment, Chr 16, human D21S52	16
D21S56h	DNA segment, Chr 17, human D21S56	17
D21S58h	DNA segment, Chr 16, human D21S58	16
D2Mit1	DNA segment, Chr 2, MIT-1	2
D3S92h	DNA segment, Chr 9, human D3S92	9
D3S93h	DNA segment, Chr 9, human D3S93	9
D4Rp1	DNA segment, Chr 4, Roswell Park-1 (ex RP1)	4
D4S10h	DNA segment, Chr 11, human D4S10	11
D4S43h	DNA segment, Chr 5, human D4S43	5
D4S62h	DNA segment, Chr 5, human D4S62	5
D4S76h	DNA segment, Chr 5, human D4S76	5
D4S80h	DNA segment, Chr 5, human D4S80	5
D5Mit25	DNA segment, Chr 5, MIT-25	5
D5Pas1	DNA segment, Chr 5, Pasteur Institute-1	5
D6Mu1	DNA segment, Chr 6, Munich University-1	6
D7Ag2	DNA segment, Chr 7, Agouron Inst., brain specific	7
D7Rp2-r	DNA segment, Chr 7, Roswell Park regulator-2 (ex RP2-r)	7
D7Rp2-s	DNA segment, Chr 7, Roswell Park structural-2 (ex RP2-s)	7
D7Smh448	DNA segment, Chr 7, St. Mary's Hospital-448	7
D7Smh483	DNA segment, Chr 7, St. Mary's Hospital-483	7
D7Smh53	DNA segment, Chr 7, St.Mary's Hospital-53	7
D7Smh57	DNA segment, Chr 7, St.Mary's Hospital-57	7
D7Smh59	DNA segment, Chr 7, St. Mary's Hospital-59	7
D7Smh71	DNA segment, Chr 7, St.Mary's Hospital-71	7
D7Smh75	DNA segment, Chr 7, St. Mary's Hospital-75	7
D7Smh759	DNA segment, Chr 7, St. Mary's Hospital-759	7
D7Was12	DNA segment, Chr 7, University of Washington-12	7
D9Ag3	DNA segment, Chr 9, Agouron Inst., brain specific	9
da	dark	7
Dac	dactylaplasia	UN
DAcry-1	withdrawn, = Crya	17
Dao-1	D-amino acid oxidase	5
Das	dopamine agonist sensitivity (prov.)	UN
db	diabetes	4
Dbr	dominant brown (provisional)	UN
Dbv	withdrawn, = d	9
Dc	dancer	19
Dce	desmosterol-to-cholesterol enzyme	RE
Dcp-1	dexamethasone-induced cleft palate-1	17
Dcp-2	dexamethasone-induced cleft palate-2 (provisional)	17
dd	withdrawn, = sps	UN
de	droopy ear	3
Defcr	defensin-related cryptdin peptide	8
Den	withdrawn, allele of Re	11
dep	depilated	4
Dey	withdrawn, allele of Sey	2
df	Ames dwarf	11
Dfc	factor D, complement	RE
Dfp	dark foot pads	UN
dfw	deaf waddler	6
DGcry-1	withdrawn, = Cryg-1	1
DGcry-2	withdrawn, = Cryg-2	1
DGcry-3	withdrawn, = Cryg-3	1
DGcry-4	withdrawn, = Cryg-4	1
Dh	dominant hemimelia	1
Dhfr	dihydrofolate reductase	13
di	duplicate incisors	UN
Dia-1	diaphorase-1 (NADH)	15
Dia-2	diaphorase-2 (NADPH)	RE
Dip-1	withdrawn, = Pep-3	3
Dip-2	withdrawn, = Pep-1	18
Dipl1	diploid oocytes, metaphase I arrested(provisional)	UN
dl	downless	10
Dlb-1	dolichos lectin binding-1	11
dm	diminutive	2
Dmc	DMCM susceptibility (provisional)	UN
Dmd	Duchenne muscular dystrophy locus sequences(= mdx)	X
Dmm	disproportionate micromelia	UN
dn	deafness	UN
DNF15S2h	DNA segment, Chr 9, human DNF15S2	9
Do	disoriented	UN
Dom	dominant megacolon	15
dp	dilution-Peru	15
Dpgm	diphosphoglyceromutase	RE
dpy	dumpy	13
dr	dreher	1
Dre	dominant reduced ear	4
Ds	disorganization	14
Dsh	short digits	UN
Dsi-1	David Steffen integration-1	4
dsu	dilute suppressor	1

Symbol	Name	Chr
dt	dystonia musculorum	1
Dtc-1	detected by T-cells-1 (ex tra-1)	UN
du	ducky	9
dup	dumpy-Oda (provisional)	UN
dv	dervish	UN
dw	dwarf	16
DXPas1	DNA segment, Chr X, Pasteur Institute-1	X
DXPas10	DNA segment, Chr X, Pasteur Institute-10	X
DXPas13	DNA segment, Chr X, Pasteur Institute-13	X
DXPas14	DNA segment, Chr X, Pasteur Institute-14	X
DXPas15	DNA segment, Chr X, Pasteur Institute-15	X
DXPas2	DNA segment, Chr X, Pasteur Institute-2	X
DXPas20	DNA segment, Chr X, Pasteur Institute-20	X
DXPas21	DNA segment, Chr X, Pasteur Institute-21 (= Cf-8)	X
DXPas22	DNA segment, Chr X, Pasteur Institute-22	X
DXPas3	DNA segment, Chr X, Pastuer Institute-3	X
DXPas4	DNA segment, Chr X, Pasteur Institute-4	X
DXPas5	DNA segment, Chr X, Pasteur Institute-5	X
DXPas6	DNA segment, Chr X, Pasteur Institute-6(human DXS144)	X
DXPas7	DNA segment, Chr X, Pasteur Institute-7(human DXS32)	X
DXPas8	DNA segment, Chr X, Pasteur Institute-8(human DXS52)	X
DXPas9	DNA segment, Chr X, Pasteur Institute-9	X
DXS120h	DNA segment, Chr X, human MDXS120	X
DXSmh10	DNA segment, Chr X, St.Mary's Hospital-10	X
DXSmh117	DNA segment, Chr X, St. Mary's Hospital-114	X
DXSmh120	DNA segment, Chr X, St.Mary's Hospital-120	X
DXSmh14	DNA segment, Chr X, St. Mary's Hospital-14	X
DXSmh141	DNA segment, Chr X, St.Mary's Hospital-141	X
DXSmh172	DNA segment, Chr X, St.Mary's Hospital-172	X
DXSmh191	DNA segment, Chr X, St.Mary's Hospital-191	X
DXSmh219	DNA segment, Chr X, St.Mary's Hospital-219	X
DXSmh222	DNA segment, Chr X, St.Mary's Hospital-222	X
DXSmh225	DNA segment, Chr X, St.Mary's Hospital-225	X
DXSmh23	DNA segment, Chr X, St.Mary's Hospital-23	X
DXSmh255	DNA segment, Chr X, St.Mary's Hospital-255	X
DXSmh36	DNA segment, Chr X, St.Mary's Hospital-36	X
DXSmh43	DNA segment, Chr X, St.Mary's Hospital-43	X
DXSmh44	DNA segment, Chr X, St. Mary's Hospital-44	X
DXSmh59	DNA segment, Chr X, St.Mary's Hospital-59	X
DXSmh64	DNA segment, Chr X, St.Mary's Hospital-64	X
DXSmh66	DNA segment, Chr X, St.Mary's Hospital-66	X
DXSmh67	DNA segment, Chr X, St.Mary's Hospital-67	X
DXSmh91	DNA segment, Chr X, St.Mary's Hospital-91	X
DXSmh93	DNA segment, Chr X, St. Mary's Hospital-93	X
DXSmhG28	withdrawn, = Gdx	X
DXWas107	DNA segment, Chr X, University of Washington-107	X
DXWas17	DNA segment, Chr X, University of Washington-17	X
DXWas31	DNA segment, Chr X, University of Washington-31	X
DXWas68	DNA segment, Chr X, University of Washington-68	X
DXWas70	DNA segment, Chr X, University of Washington-70	X
dy	dystrophia muscularis	10
dyl	dysgenetic lens	4
dyn	neuraxonal dystrophy (provisional)	UN
e	recessive yellow (extension)	8
Ea-1	erythrocyte antigen-1	8
Ea-10	erythrocyte antigen-10	6
Ea-2	erythrocyte antigen-2 (ex H-14)	UN
Ea-3	erythrocyte antigen-3	UN
Ea-4	erythrocyte antigen-4	19
Ea-430	withdrawn; = Ea-9	4
Ea-5	erythrocyte antigen-5 (ex H-5)	UN
Ea-6	erythrocyte antigen-6 (ex H-6)	2
Ea-7	erythrocyte antigen-7	UN
Ea-8	erythrocyte antigen-8	UN
Ea-9	erythrocyte antigen-9 (ex Ea-430)	4
Ea-?	erythrocyte antigen-? (ex Ea-2)	17
Eam	ethanol activity modifier	4
eb	eye blebs	10
ec	ectopic	UN
ecl	epistatic circling C57L/J	4
ecs	epistatic circling SWR/J	UN
Ed	extra digit	UN
Ee-2	withdrawn, Es-3	11
Eftu	elongation factor Tu (prov.)	UN
Eg	endoplasmic beta-glucuronidase (prob. = Es-22)	8
Egf	epidermal growth factor	3
Egfbp-1	epidermal growth factor binding protein type A (prov.)	7
Egfbp-2	epidermal growth factor binding protein type B (prov.)	7
Egfbp-3	epidermal growth factor binding protein type C (prov.)	7
Egr-1	early growth response-1	UN
Eh	hairy ears	15
El	epilepsy (provisional)	UN
Ela-1	elastase-1 (structural or pseudogene)	15
Elo	eye lens obsolescence	1
Em	Emory cataract	UN
Ema	electrophoretic mobility and agglutinibility	UN
Empb3	erythrocyte membrane protein band 3	UN
Emv-1	withdrawn, = Cv-1	5
Emv-10	endogenous ecotropic MuLV-10*	UN
Emv-11	withdrawn, = Akv-1	7
Emv-12	withdrawn, = Akv-2	16
Emv-13	withdrawn, = Akv-3	2
Emv-14	withdrawn, = Akv-4	11
Emv-15	endogenous ecotropic MuLV-15*(A<y>-assoc.)	2
Emv-16	endogenous ecotropic MuLV-16*	1
Emv-17	endogenous ecotropic MuLV-17*	1
Emv-18	endogenous ecotropic MuLV-18*	9
Emv-19	endogenous ecotropic MuLV-19*	3
Emv-2	withdrawn, = Bv-1	8
Emv-20	endogenous ecotropic MuLV-20*	1
Emv-21	endogenous ecotropic MuLV-21*	18
Emv-22	withdrawn, = C58v-4	19
Emv-23	endogenous ecotropic MuLV-23	7
Emv-24	endogenous ecotropic MuLV-24*	5
Emv-25	endogenous ecotropic MuLV-25*	10
Emv-26	withdrawn, = C58v-1	8
Emv-27	endogenous ecotropic MuLV-27*	3
Emv-28	endogenous ecotropic MuLV-28*	11
Emv-29	endogenous ecotropic MuLV-29*	8
Emv-3	withdrawn, = d	9
Emv-4	endogenous ecotropic MuLV-4*	UN
Emv-5	endogenous ecotropic MuLV-5*	UN
Emv-6	endogenous ecotropic MuLV-6*	UN
Emv-7	endogenous ecotropic MuLV-7*	UN
Emv-8	endogenous ecotropic MuLV-8*	UN
Emv-9	endogenous ecotropic MuLV-9*	1
En-1	engrailed-1	1
En-2	engrailed-2	5
Enc	embryonic nucleus cataract	UN
Eno-1	enolase-1	4

Env-	withdrawn, = Xmmv-	2
Eo	eye-opacity	UN
Eox-1	ocular oxidase-1	1
ep	pale ear	19
Epa-1	epidermal antigen-1	UN
epf	epileptiform (proposed symbol)	UN
Eph-1	epoxide hydratase-1	1
epi	exocrine pancreatic insufficiency	UN
Epo	erythropoietin	5
Er	repeated epilation	4
Erba	avian erythroblastosis oncogene A	11
Erbb	avian erythroblastosis oncogene B	11
Erbb-2	avian erythroblastosis oncogene B-2	11
Ercc-2	excision repair-2	7
Erp-1	erythrocyte protein-1	8
Es-1	esterase-1	8
Es-10	esterase-10	14
Es-11	esterase-11	8
Es-12	esterase-12	6
Es-13	esterase-13	9
Es-14	esterase-14	9
Es-15	esterase-15	UN
Es-16	esterase-16	3
Es-17	esterase-17 (may = Ups)	9
Es-18	esterase-18	19
Es-19	esterase-19	UN
Es-2	esterase-2	8
Es-20	withdrawn, trimer of Es-9, Es-6	8
Es-21	esterase-21 (may = Es-13)	9
Es-22	esterase-22 (prob. = Eg)	8
Es-23e	esterase-23, electrophoretic	8
Es-23t	esterase-23, temporal	8
Es-24	esterase-24	8
Es-25	esterase-25	12
Es-26	esterase-26	3
Es-27	esterase-27, serum cholinesterase	3
Es-3	esterase-3	11
Es-4	withdrawn, = Es-1	8
Es-5	esterase-5	8
Es-6	esterase-6	8
Es-7	esterase-7	8
Es-8	esterase-8	7
Es-9	esterase-9	8
Esr	esterase-5 regulator	6
Ets-1	E26 avian leukemia oncogene-1, 5′ domain	9
Ets-2	E26 avian leukemia oncogene-2, 3′ domain	16
Evi-1	ecotropic viral integration site-1	3
Evi-2	ecotropic viral integration site-2	11
Ex	earlier X-zone degeneration	7
ex	withdrawn, = exf	UN
Exa	exploratory activity	4
exf	exfoliative (ex ex)	UN
Exnm	neuromuscular excitability (Russian, not available)	1
Exz-1	adrenal x-zone degeneration-1	UN
Exz-2	adrenal x-zone degeneration-2	UN
ey-1	eyeless-1	UN
ey-2	eyeless-2	UN
Ezg	extent of zona glomerulosa	UN
f	flexed-tail	13
fa	falter	UN

Fabph-1	fatty acid binding protein heart-1	4
Fabph-2	fatty acid binding protein heart-2	8
Fabph-3	fatty acid binding protein heart-3	UN
Fabph-4	fatty acid binding protein heart-4	4
Fabpi	fatty acid binding protein intestinal	3
Fabpl	fatty acid binding protein liver	6
far	first arch malformation	UN
fat	fat	UN
Fbp-1	fructose bisphosphatase-1	UN
Fbp-2	fructose bisphosphatase-2	UN
fc	flecking	2
Fcela	Fc epsilon low affinity receptor alpha	1
Fcr	Fc receptor	12
Fcr-3	Fc-gamma receptor	1
Fcr-5	Fc-epsilon receptor	1
fd	fur deficient	9
Fdc	withdrawn, allele of W	5
fe	faded	6
Fes	feline sarcoma oncogene	7
fg	withdrawn, = myd	8
Fgg	gamma-fibrinogen	3
Fgv-1	C3H/Fg virus-1*	7
Fgv-2	C3H/Fg virus-2*	UN
fh	fetal hematoma	UN
Fh-1	fumarate hydratase-1	RE
fhd	fathead	2
Fhe	Friend helper virus erythroblastosis susceptibility	UN
fi	fidget	2
Fim-1	Friend MuLV integration site-1	13
Fim-2	Friend MuLV integration site-2 (overlaps Csfmr)	18
Fim-3	Friend MuLV integration site-3	3
Fis-1	Friend virus integration site-1	7
Fk	fleck	UN
fl	flipper-arm	UN
Fla	F liver antigen	5
fld	fatty liver dystrophy	UN
Flp	withdrawn, = Fla	5
Flv	flavivirus resistance	UN
fm	foam-cell reticulosis	UN
fmd	withdrawn, = mdg	UN
Fms	withdrawn, = Csfmr	18
fmt	formentin	UN
Fn-1	fibronectin-1	1
Fnp-1	withdrawn, = Zfp-1	11
Fnp-2	withdrawn, = Zfp-2	8
For-1	formamidase-1, liver, kidney	UN
For-5	formamidase-5, brain	14
Fos	FBJ osteosarcoma oncogene	12
fp	withdrawn, = bv	UN
Fpgs	folylpolyglutamyl synthetase	2
fr	frizzy	7
fri	frissonant (provisional)	UN
Frl	furloss	UN
fro	fragilitas ossium	UN
fs	furless	13
Fshb	follicle stimulating hormone beta	2
ft	flaky tail	3
Fth	ferritin heavy chain	19
Fu	fused	17
Fuca	alpha-L-fucosidase (ex Afuc)	4

Symbol	Name	Chr
Fv-1	Friend virus susceptibility-1	4
Fv-2	Friend virus susceptibility-2	9
Fv-3	Friend virus susceptibility-3	UN
Fv-4	Friend virus susceptibility-4 (=Akvr-1)	12
Fv-5	Friend virus P-induced early anemia, polycythemia susc.	UN
Fv-6	Friend virus susceptibility-6	5
Fw	fawn (provisional)	UN
fy	withdrawn, allele of fz	1
fz	fuzzy	1
fzt	fuzzy tail (extinct?)	UN
G6pd	glucose-6-phosphate dehydrogenase-1, X-linked	X
Ga	greying with age(probably caused by milk-transmitted virus)	UN
Gabra-3	GABA-A receptor, subunit alpha-3	X
gad	gracile axonal dystrophy	5
Gad-1	glutamic acid decarboxylase-1 (provisional)	UN
Gad-2	glutamic acid decarboxylase-2 (provisional)	UN
Galt	galactose-1-phosphate uridyl transferase	4
Gap43	growth accentuating protein 43	16
Gapd	glyceraldehyde-3-phosphate dehydrogenase	6
Gaps	withdrawn, = Prgs	16
Gata-1	GATA repeat sequence-1	RE
Gb	withdrawn, allele of Sl	10
Gba	glucosidase, beta, acid	3
Gbp-1	guanine nucleotide-binding protein-1	3
gc	grey coat	5
Gcg	glucagon	2
Gda	guanine deaminase	UN
Gdc-1	glycerolphosphate dehydrogenase-1	15
Gdc-2	glycerolphosphate dehydrogenase-2	9
Gdcr-1	glycerolphosphate dehydrogenase regulator-1	UN
Gdcr-2	glycerolphosphate dehydrogenase regulator-2	UN
Gdh-X	withdrawn, = Gludl	14
gdn	golden	UN
Gdr-1	G6PD regulator-1	UN
Gdr-2	G6PD regulator-2	UN
Gdx	anonymous locus near G6pd	X
Gfap	glial fibrillary acidic protein	11
Gfrp	growth factor response protein	UN
Ggc	gamma-glutamyl cyclotransferase	6
Ggm-1	ganglioside expression-1	17
Ggm-2	ganglioside expression-2	UN
Ggtb	glycoprotein 4-beta-galactosyltransferase	4
Gh	growth hormone	11
Ghr	growth hormone receptor	15
Gia	Gi protein, alpha chain(inhib. reg. adenylate cyclase)	9
Gin-1	Gross passage A viral integration region-1	19
Gk	glucokinase activity	UN
gl	grey-lethal	10
Glb	glycine bitterness	6
gld	generalized lymphoproliferative disease	1
Glk	galactokinase	11
Glns	glutamine synthetase	UN
Glo-1	glyoxylase-1 (ex Qglo)	17
Gls	glutaminase	1
Gludl	glutamate dehydrogenase-like sequence (provisional)	14
gm	gunmetal	14
Gm-3	granulocyte-macrophage antigen-3	2
Gma	withdrawn, = Mag	9
Gnai-2	guanine nucleotide binding protein, alpha inhibiting-2(prov)	9
Gnas	guanine nucleotide binding protein, alpha stimulating (prov)	2
Gnat-1	guanine nucleotide binding protein, alpha transducing-1(prov	9
Gnat-2	guanine nucleotide binding protein, alpha transducing-2(prov	17
Gnb-1	guanine nucleotide binding protein, beta (prov)	4
Gnb-1ps?	guanine nucleotide binding protein, beta-1 pseudogene (?)	19
Gnb-2	guanine nucleotide binding protein, beta-2	5
Gnb-3	guanine nucleotide binding protein, beta-3	8
go	angora	5
Got-1	glutamate oxaloacetate transaminase-1, soluble	19
Got-2	glutamate oxaloacetate transaminase-2, mitochondrial	8
gp	gaping lids	UN
Gpd-1	glucose-6-phosphate dehydrogenase-1	4
Gpd-2	glucose-6-phosphate dehydrogenase-2	5
Gpdx	withdrawn, = G6pd	X
Gpi-1r	glucose phosphate isomerase-1, regulatory	7
Gpi-1s	glucose phosphate isomerase-1, structural	7
Gpi-1t	glucose phosphate isomerase-1, temporal (ex Org)	7
Gpt-1	glutamic-pyruvic transaminase-1, soluble	15
Gpx	glutathione peroxidase	RE
gr	grizzled	10
Gr-1	glutathione reductase-1	8
Grl-1	glucocorticoid receptor-1	18
Grmp	granulocyte membrane protein	1
gs	glabrous (provisional)	UN
Gs	greasy	X
Gsa	Gs protein, alpha chain(stim. reg. adenylate cyclase)	2
Gss	glutamate-gamma-semialdehyde synthetase	RE
Gsta	glutathione S-transferase Ya subunit	9
gt	gray tremor	15
Gt-1	withdrawn, not single locus	UN
Gt-2	withdrawn, not single locus	UN
Guk-1	guanylate kinase-1	RE
Gus-r	beta-glucuronidase-regulator	5
Gus-s	beta-glucuronidase-structural	5
Gus-t	beta-glucuronidase-temporal	5
Gus-u	beta-glucuronidase-systemic regulator	5
Gv-1	Gross virus antigen-1*	UN
Gv-2	Gross virus antigen-2*	7
Gvh	graft versus host regulation	RE
Gy	gyro	X
H(a<t>)	histocompatibility(a<t>)(provisional)	2
H(Eh)	histocompatibility(Eh)(provisional)	15
H(ep)	histocompatibility(ep)(provisional)	19
H(go)	histocompatibility(go)(provisional)	5
H(Igh)	withdrawn, = H-40	12
H(js)	histocompatibility(js)(provisional)	11
H(ln)	histocompatibility(ln)(provisional)	1
H(lt)	histocompatibility(lt)(provisional)	11
H(pi)	histocompatibility(pi) (provisional)	5
H(tn)	histocompatibility(tn) (provisional)	11
H-1	histocompatibility-1	7
H-10	histocompatibility-10	UN
H-11	histocompatibility-11	UN
H-12	histocompatibility-12	UN
H-13	histocompatibility-13	2
H-14	withdrawn, = Ea-2	17
H-15	histocompatibility-15	4
H-16	histocompatibility-16	4
H-17	histocompatibility-17	UN
H-18	histocompatibility-18	4
H-19	histocompatibility-19	8

Symbol	Name	Chr
H-2	histocompatibility-2 (major histocompatibility complex)	17
H-20	histocompatibility-20	4
H-21	histocompatibility-21	4
H-22	histocompatibility-22	7
H-23	histocompatibility-23	3
H-24	histocompatibility-24	7
H-25	histocompatibility-25	1
H-26	histocompatibility-26	UN
H-27	histocompatibility-27	5
H-28	histocompatibility-28	3
H-29	histocompatibility-29	8
H-2D	(controls antigenic specificities of D-region loci)	17
H-2G	(serologically detected antigen abundant on erythrocytes)	17
H-2I	(immune response region)	17
H-2K	(antigens for graft-versus-host mixed lymphocyte reaction)	17
H-2L	(controls antigenic specificities H-2.1, H-2.28)	17
H-2S	(region contains Ss=C4, Slp, Bf, C2)	17
H-2T	(target antigens reactive in cell-mediated reaction)	17
H-3	histocompatibility-3	2
H-30	histocompatibility-30	15
H-31	histocompatibility-31	17
H-32	histocompatibility-32	17
H-33	histocompatibility-33	17
H-34	histocompatibility-34	UN
H-35	histocompatibility-35 (12 cM from H-25)	1
H-36	histocompatibility-36	UN
H-37	histocompatibility-37	3
H-38	histocompatibility-38	UN
H-39	histocompatibility-39	17
H-4	histocompatibility-4	7
H-40	histocompatibility-40	12
H-41	histocompatibility-41	UN
H-42	histocompatibility-42	2
H-43	histocompatibility-43 (ex H-42)	17
H-44	histocompatibility-44	2
H-45	histocompatibility-45	2
H-5	withdrawn, = Ea-5	UN
H-6	withdrawn, = Ea-6	2
H-7	histocompatibility-7	9
H-8	histocompatibility-8	14
H-9	histocompatibility-9	UN
H-X	withdrawn, = Hxa	X
H-Y	withdrawn, = Hya	Y
H19	H19 fetal liver mRNA (provisional, non-standard symbol)	UN
ha	withdrawn, allele of sph	1
Hadh	hydroxylacyl-CoA-dehydrogenase	RE
Hao-1	hydroxyacid oxidase-1, liver	2
Hao-2	hydroxyacid oxidase-2, kidney	3
Hat-1	histidine aminotransferase-1	UN
Hba	hemoglobin alpha-chain complex	11
Hba-3ps	hemoglobin alpha-3, pseudogene	15
Hba-4ps	hemoglobin alpha-4, pseudogene	17
Hba-x	hemoglobin X (alpha-like embryonic chain in Hba complex)	11
Hbb	hemoglobin beta-chain complex	7
Hbb-y	hemoglobin Y (beta-like embryonic chain in Hbb complex)	7
hbd	hemoglobin deficient	UN
hbs	withdrawn, allele of un	2
Hc	hemolytic complement, (= C5)	2
Hck-1	hemopoietic cell kinase-1	2
Hck-2	hemopoietic cell kinase-2 (may = Lyn)	4
Hck-3	hemopoietic cell kinase-3 (may = Lck)	4
hcp	hypochondrodysplasia (provisional)	UN
Hcs	hepatocarcinogenesis susceptibility (provisional)	UN
Hct	hair constriction	UN
Hd	hypodactyly	6
Hdc-a	histidine decarboxylase testosterone reponse	2
Hdc-c	histidine decarboxylase concentration (kidney)	2
Hdc-e	histidine decarboxylase estrogen response	2
Hdc-s	histidine decarboxylase, structural (kidney)	2
Hdl-1	high density lipoprotein-1	1
He	withdrawn, = Hc	2
hea	hereditary erythroblastic anaemia	UN
heb	head blebs	4
het	head-tilt	17
Hex-1	withdrawn, = Hexa	RE
Hexa	hexosaminidase A	RE
Hexb	hexosaminidase B	13
Hf	hanukah factor (provisional, bad symbol, see hf)	UN
hf	hepatic fusion	7
hg	high growth	UN
Hgh-3	human growth hormone-3	UN
Hgh-4	human growth hormone-4	UN
Hh-1	hemopoietic histocompatibility (within H-2D)	17
Hi	hare tail (provisional)	UN
hid	hair interior defect	UN
Hiomt	hydroxyindole-O-methyltransferase	UN
his	histidinemia (probably mutation at Hsd locus)	10
Hist1	histone gene complex-1	13
Hist2	histone gene complex-2	3
Hk	hook	8
Hk-1	hexokinase-1	10
hl	hair-loss	15
Hld	hippocampal lamination defect	UN
Hm	hammer-toe	5
Hma	HPRT mobility alteration	7
Hmt	histocompatibility maternally transmitted	17
hn	withdrawn, = xn	UN
Hnl	hypothalamic norepinephrine level (near H-23)	3
ho	hotfoot	6
Hom-1	hormone metabolism-1 (possibly within H-2)	17
hop	hop-sterile	6
Hox-1	homeo box-1 cluster	6
Hox-1.1	homeo box-1 cluster, gene 1	6
Hox-1.2	homeo box-1 cluster, gene 2	6
Hox-1.3	homeo box-1 cluster, gene 3	6
Hox-1.4	homeo box-1 cluster, gene 4	6
Hox-1.5	homeo box-1 cluster, gene 5	6
Hox-1.6	homeo box-1 cluster, gene 6	6
Hox-1.7	homeo box-1 cluster, gene 7	6
Hox-2	homeo box-2 cluster	11
Hox-2.1	homeo box-2 cluster, gene 1	11
Hox-2.2	homeo box-2 cluster, gene 2	11
Hox-2.3	homeo box-2 cluster, gene 3	11
Hox-2.4	homeo box-2 cluster, gene 4	11
Hox-2.5	homeo box-2 cluster, gene 5	11
Hox-2.6	homeo box-2 cluster, gene 6	11
Hox-2.7	homeo box-2 cluster, gene 7	11
Hox-3	homeo box-3 cluster	15
Hox-3.1	homeo box-3 cluster, gene 1	15
Hox-3.2	homeo box-3 cluster, gene 2	15

Symbol	Name	Chr
Hox-3.3	homeo box-3 cluster, gene 3	15
Hox-3.4	homeo box-3 cluster, gene 4	15
Hox-4	homeo box-4 cluster	12
Hox-5	homeo box-5 cluster	2
Hox-6	homeo box-6 cluster (may = Hox-3)	14
Hox-6.1	homeo box-6 cluster, gene 1 (= Hox-3.3?)	15
Hox-6.2	homeo box-6 cluster, gene 2 (+ Hox-3.4?)	15
Hox-7	homeo box-7 cluster, msh like	5
Hp	haptoglobin	8
Hpa	withdrawn, = Hp	8
Hpb	withdrawn, = Hp	8
hpc	hyperspiny Purkinje cell	UN
hpg	hypogonadal	UN
hph-1	hyperphenylalaninemia-1	14
hph-2	hyperphenylalaninemia-2	UN
Hpl	hepatic lipase	UN
Hprt	hypoxanthine-guanine phosphoribosyl transferase	X
Hpt	hair patches	4
hpx	hypotransferrinemia	9
hpy	withdrawn, allele of hop	6
Hq	harlequin	X
hr	hairless	14
Hras-1	Harvey rat sarcoma virus oncogene	7
Hrt-1	heart protein-1	UN
hs	head spot	UN
Hsd	histidase synthetic rate	10
hsdr-1	hepatocyte specific developmental regulation-1	7
Hsp70	heat shock protein 70 (provisional)	RE
Hsp84-1	heat shock protein 84-1	17
Hsp84-2	heat shock protein 84-2	2
Hsp84-3	heat shock protein 84-3	12
Hsp86-1	heat shock protein 86-1	12
Hsp86-2	heat shock protein 86-2	11
Hsp86-3	heat shock protein 86-3	3
Hst-1	hybrid sterility-1	17
Hst-2	hybrid sterility-2	9
Hsv-1	withdrawn, = If-X	X
Ht	hightail	15
htd	hair-thyroid deficient	UN
hti	histidine decarboxylase thyroxin inducibility	UN
hub	hyper-unconjugated bilirubinemia (jaundice)	UN
hug	hugger	UN
Hv-1	hepatitis virus (MHV-2) susceptibility	UN
Hv-2	hepatitis virus (MHV-4) susceptibility	7
Hx	hemimelic extra toes	5
Hxa	histocompatibility-X, ex H-X	X
Hxd	hallux duplex	UN
hy-1	hydrocephalus-1 (probably extinct)	UN
hy-2	hydrocephalus-2 (probably extinct)	UN
hy-3	hydrocephalus-3	8
Hya	histocompatibility-Y, ex H-Y	Y
Hye	Hya expression	17
hyh	hydrocephaly with hop gait	7
Hyp	hypophosphatemia	X
hyt	hypothyroid (ex pet)	12
hz	withdrawn, allele of ru-2	7
Ia-1	I-region-associated-antigen-1	17
Ia-2	I-region-associated-antigen-2	17
Ia-3	I-region associated antigen-3	17
Ia-4	I-region-associated-antigen-4	17

Symbol	Name	Chr
Ia-5	I-region-associated-antigen-5	17
Iac	iris anomaly with cataract	UN
Iap	intracisternal A particles (provisional)	RE
Ias	withdrawn, = asp-1	4
ic	ichthyosis	1
Id-1	withdrawn, = Idh-1	1
Idc	iris dysplasia with cataract (provisional)	UN
Idd-1	insulin dependent diabetes susceptibility-1 (provisional)	17
Idd-2	insulin dependent diabetes susceptibility-2 (provisional)	9
Idd-3	insulin dependent diabetes susceptibility-3 (provisional)	UN
Idh-1	isocitrate dehydrogenase-1	1
Idh-2	isocitrate dehydrogenase-2	7
Idh-3	isocitrate dehydrogenase-3	RE
Ie	eye-ear reduction	X
If-1	NDV-induced circulating interferon	3
If-2	MTV-induced circulating interferon	UN
If-3	Sendai virus-induced interferon-3	UN
If-4	Sendai virus-induced interferon-4	UN
If-X	withdrawn, = Ifx	X
Ifa	interferon alpha gene family (leukocyte)	4
Ifb	interferon beta (fibroblast)	4
Ifbip-1	interferon beta induced protein	1
Ifg	interferon gamma	10
Ifgr	interferon gamma receptor	10
Ifrc	interferon receptor	16
Ifx	NDV-induced circulating interferon, X-linked (ex If-X)	X
Igf1r	insulin-like growth factor I receptor	7
Igf2r	insulin-like growth factor II receptor (prov.)	17
Igh	immunoglobulin heavy chain complex	12
Igh-1	immunoglobulin heavy chain-1 (serum IgG2a)	12
Igh-2	immunoglobulin heavy chain-2 (serum IgA)	12
Igh-3	immunoglobulin heavy chain-3 (serum IgG2b)	12
Igh-4	immunoglobulin heavy chain-4 (serum IgG1)	12
Igh-5	immunoglobulin heavy chain-5 (delta-like heavy chain)	12
Igh-6	immunoglobulin heavy chain-6 (heavy chain of IgM)	12
Igh-7	immunoglobulin heavy chain-7 (heavy chain of IgE)	12
Igh-Aa1	immunoglobulin heavy chain Aa1	12
Igh-Aa2	immunoglobulin heavy chain Aa2	12
Igh-Aa3	immunoglobulin heavy chain Aa3	12
Igh-Ars	immunoglobulin heavy chain Ars	12
Igh-Bgl	immunoglobulin heavy chain Bgl	12
Igh-C	immunoglobulin heavy-chain constant region	12
Igh-Dex	immunoglobulin heavy chain Dex	12
Igh-Gte	immunoglobulin heavy chain Gte	12
Igh-Inu	immunoglobulin heavy chain Inu	12
Igh-Lev	immunoglobulin heavy chain Lev	12
Igh-Nbp	immunoglobulin heavy chain Nbp	12
Igh-Np	immunoglobulin heavy chain Np	12
Igh-Ns1	immunoglobulin heavy chain Ns1	12
Igh-Ns2	immunoglobulin heavy chain Ns2	12
Igh-Ns3	immunoglobulin heavy chain Ns3	12
Igh-Ns4	immunoglobulin heavy chain Ns4	12
Igh-Ns5	immunoglobulin heavy chain Ns5	12
Igh-Pc	immunoglobulin heavy chain Pc	12
Igh-Sa1	immunoglobulin heavy chain Sa1	12
Igh-Sa2	immunoglobulin heavy chain Sa2	12
Igh-Sa3	immunoglobulin heavy chain Sa3	12
Igh-Sa4	immunoglobulin heavy chain Sa4	12
Igh-Sa5	immunoglobulin heavy chain Sa5	12
Igh-Src	immunoglobulin heavy chain Src	12

Symbol	Name	Chromosome
Igh-V	immunoglobulin heavy-chain variable region	12
Igj	immunoglobulin joining gene (not J region)	5
Igk	immunoglobulin kappa gene complex	6
Igk-C	immunoglobulin kappa chain constant region	6
Igk-Ef1	immunoglobulin kappa chain Ef1	6
Igk-Ef2	immunoglobulin kappa chain Ef2	6
Igk-J	immunoglobulin kappa chain, joining region	6
Igk-Pc	immunoglobulin kappa chain Pc	6
Igk-Trp	immunoglobulin kappa chain Trp	6
Igk-V	immunoglobulin kappa chain variable region	6
Igk-V1	immunoglobulin kappa chain V1	6
Igk-V10	immunoglobulin kappa chain V10	6
Igk-V11	immunoglobulin kappa chain V11	6
Igk-V12,13	immunoglobulin kappa chain V12,13	6
Igk-V19	immunoglobulin kappa chain V19	6
Igk-V21	immunoglobulin kappa chain V21	6
Igk-V23	immunoglobulin kappa chain V23	6
Igk-V24	immunoglobulin kappa chain V24	6
Igk-V28	immunoglobulin kappa chain V28	6
Igk-V4	immunoglobulin kappa chain V4	6
Igk-V8	immunoglobulin kappa chain V8	6
Igk-V9	immunoglobulin kappa chain V9	6
Igk-V9-26	immunoglobulin kappa chain V9-26	6
Igl	immunoglobulin lambda gene complex	16
Igl-1	immunoglobulin lambda chain, cluster 1	16
Igl-1r	immunoglobulin lambda chain regulator	16
Igl-2	immunoglobulin lambda chain, cluster-2	16
Igl-5	immunoglobulin lambda chain-5	16
Igl-Lo	withdrawn, = Igl-1r	16
Igl-V1	immunoglobulin lambda chain variable-1	16
Igl-x	immunoglobulin lambda chain-x	UN
Ihe-1	intestinal helminth expulsion-1 (provisional)	UN
Ii	Ia associated invariant chain	18
Il-1	interleuken-1 complex	2
Il-1a	interleukin-1 alpha	2
Il-1b	interleukin-1 beta	2
Il-2	interleukin-2	3
Il-3	interleukin-3 (contained in Csfmu)	11
Il-4	interleukin-4	11
Il-5	interleukin-5	11
In	withdrawn, = Ah	12
Inb-1	withdrawn, = Bv-1	8
Inc-1	withdrawn, = Cv-1	5
Inha	inhibin-alpha	1
Inhba	inhibin-beta-A	13
Inhbb	inhibin-beta-B	1
Ins-1	insulin I or insulin pseudogene	6
Ins-2	insulin II	7
Ins-3	insulin I or insulin pseudogene	15
Insr	insulin receptor	8
Int-1	mammary tumor integration site-1	15
Int-2	mammary tumor integration site-2	7
Int-3	mammary tumor integration site-3	17
Int-4	mammary tumor integration site-4	11
Ipo-1	withdrawn, = Sod-1	16
Ir-1	immune response-1	UN
Ir-2	immune response-2	2
Ir-3	immune response-3	UN
Ir-4	immune response-4	UN
Ir-5	immune response-5	17
Ir-6	withdrawn, = Ir-7	17
Ir-7	immune response-7	UN
It	irregular teeth	X
Itp	inosine triphosphatase	2
ity	withdrawn, = Ity	1
Ity	withdrawn, = Lsh	1
iv	situs inversus viscerum	UN
j	jaw lethal (probably extinct)	UN
ja	jaundiced	UN
jb	juvenile bare	7
jc	Jackson circler	10
je	jerker	4
jg	jagged-tail	5
ji	jittery	10
jp	jimpy (= Plp)	X
jpk	withdrawn, = cph	15
js	Jackson shaker	11
jsd	juvenile spermatogonial depletion	1
Jsr	jumbled spine and ribs	UN
jt	joined toes	UN
Jt-1	I-J antigen expression	4
Jun	Jun oncogene	UN
Junb	Jun-B oncogene	UN
Jund	Jun-D oncogene	UN
jv	Jackson waltzer	UN
Kal	kallikrein gene family	7
Kap	kidney androgen-regulated protein (provisional)	UN
kd	kidney disease	10
Kfo-1	kidney 40-50 thousand M.Wt. protein (provisional)	9
Kit	kit oncogene*	5
kr	kreisler	2
Kras-1	Kirsten rat sarcoma oncogene-1, pseudogene	RE
Kras-2	Kirsten rat sarcoma oncogene-2, expressed	6
Krt-1	keratin gene complex-1	11
Krt-2	keratin gene complex-2	15
Kth-1	kidney 30-40 thousand M.Wt. protein-1 (provisional)	17
Kth-2	kidney 20-40 thousand M.Wt. protein-2 (provisional)	UN
Kw	kinky-waltzer	UN
ky	kyphoscoliosis	UN
l(1)-1	chromosome 1 lethal-1 (provisional)	1
l(1)-2	chromosome 1 lethal-2 (provisional)	1
l(1)-4	chromosome 1 lethal-4 (provisional)	1
l(1)-5	chromosome 1 lethal-5 (provisional)	1
l(1)-6	chromosome 1 lethal-6 (provisional)	1
l(1)-7	chromosome 1 lethal-7 (provisional)	1
l(17)-1	chromosome 17 lethal-1 (Wis)(provisional)	17
l(17)-2	chromosome 17 lethal-2 (Pas)(provisional)	17
l(17)-3	chromosome 17 lethal-3 (provisional)	17
l(17)-4	chromosome 17 lethal-4 (provisional)	17
l(5)-1	chromosome 5 lethal-1 (provisional)	5
l(7)-1	chromosome 7 lethal-1 (provisional)	7
L1cam	L1 adhesion molecule	X
la	withdrawn, allele of tg	8
Laf	withdrawn, = Fla	5
Lamb-1	laminin, B1 subunit	12
Lamb-2	laminin, B2 subunit	1
Lamp-1	lysosomal membrane glycoprotein-1 (provisional)	UN
Lap-1	leucine arylaminopeptidase-1, intestinal	9
Lap-2	leucine aminopeptidase-2, serum	UN
Lbt-1	lupus band test-1 (provisional)	17

Symbol	Name	Chr
Lc	lurcher	6
Lcam	liver cell adhesion molecule	RE
Lcat	lecithin cholesterol acyltransferase	8
Lck	lymphocyte protein tyrosine kinase	4
lcsd	lysosomal cholesterol storage disease	UN
ld	limb deformity	2
Ldh-1	lactate dehydrogenase-1, A chain	7
Ldh-1ps	lactate dehydrogenase-1, A chain, pseudogene	UN
Ldh-2	lactate dehydrogenase-2, B-chain	6
Ldh-3	lactate dehydrogenase-3, C chain, sperm specific	UN
Ldlr	low density lipoprotein receptor	9
Ldr-1	lactate dehydrogenase regulator-1	6
Ldr-2	lactate dehydrogenase regulator-2	6
le	light ear	5
lec	laryngotracheo-esophageal cleft (provisional)	7
Len-1	eye lens protein-1 (gamma-crystallin)	1
Len-2	eye lens protein-2 (beta-crystallin-like)	1
Lfo-1	liver 40-50 thousand M.Wt. protein-1	7
Lfo-2	liver 40-50 thousand M.Wt. protein-2	1
lg	lid gap	UN
lgh	long hair	8
lgl	legless (provisional)	UN
lgr	London grey (provisional)	UN
lh	lethargic	2
Lhb	luteinizing hormone beta	7
Li	lined	X
Lif	leukemia inhibitory factor	UN
Lip-1	lysosomal acid lipase-1 (ex Lipa)	19
lit	little	6
lm	lethal milk	2
Lm-1	lymphomyeloid antigen-1	12
Lmp-2	low molecular weight polypeptide-2 (provisional)	17
Lmp-7	low molecular weight polypeptide-7 (provisional)	17
Lmyc-1	lung carcinoma myc-related oncogene-1	4
Lmyc-2	lung carcinoma myc-related oncogene-2 (pseudogene?)	12
ln	leaden	1
LN-1	withdrawn, = Lpn-1	17
LN-2	withdrawn, = Lpn-2	17
LN-3	withdrawn, = Lpn-3	UN
Lna-1	lymph node antigen-1	UN
Lop	withdrawn, allele of Cat	10
lop-2	lens opacity-2	UN
Lop-3	lens opacity-3 (provisional)	UN
Lop-4	lens opacity-4	2
Lop-5	lens opacity-5 (prov.)	UN
Lop-6	lens opacity-6 (prov.)	UN
Lop-7	lens opacity-7 (prov.)	UN
Lop-8	lens opacity-8 (prov.)	UN
Lop-9	lens opacity-9 (prov.)	UN
Low	withdrawn, allele at T	17
Lp	loop-tail	1
Lpl	lipoprotein lipase	8
Lpn-1	NZ lupus nephritis-1 (ex LN-1)	17
Lpn-2	NZ lupus nephritis-2 (ex LN-2)	17
Lpn-3	NZ lupus nephritis-3 (ex LN-3)	UN
lpr	lymphoproliferation	UN
Lps	lipopolysaccharide response	4
lr	lens rupture	UN
Lr	withdrawn, allele at T	17
Lr-1	withdrawn, = Lsr-1	2
ls	lethal spotting	2
Lsd	lymphocyte-stimulating determinant	1
Lse	low set ears	UN
Lsh	Leishmaniasis resistance (ex Ity, Bcg)	1
Lsr-1	listeria resistance	2
lst	Strong's luxoid	2
lt	lustrous	11
Ltf	lactotransferrin	9
Lth-1	liver 30-40 thousand M.Wt. protein-1	13
Lth-2	liver 30-40 thousand M.Wt. protein-2 (provisional)	UN
Ltk	leukocyte tyrosine kinase	UN
Ltn-1	withdrawn, = Mup-1(?)	4
Ltn-2	liver 10-20 thousand M.Wt. protein-2 (provisional)	UN
Ltw-1	withdrawn, = Apoa-1	9
Ltw-2	liver 20-30 thousand M.Wt. protein-2 (provisional)	12
Ltw-3	liver 20-30 thousand M.Wt. protein-3 (provisional)	9
Ltw-4	liver 20-30 thousand M.Wt. protein-4 (provisional)	1
Ltw-5	withdrawn, = Car-2	3
Ltw-6	liver 20-30 thousand M.Wt. protein-6 (provisional)	UN
Ltx	lymphotoxin (provisional)	17
lu	luxoid	9
Lus	lymphoid cytostasis suppressor	UN
Lv	delta-aminolevulinate dehydratase	4
Lvfo-1	withdrawn, = Lfo-1	7
Lvfo-2	withdrawn, = Lfo-2	1
Lvp-1	major liver protein-1	6
Lvth-1	withdrawn, = Lth-1	13
Lvth-2	withdrawn, = Lth-2	UN
Lvtn-1	withdrawn, = Mup-1	4
Lvtn-2	withdrawn, = Ltn-2	UN
Lvtw-1	withdrawn, = Apoa-1	9
Lvtw-2	withdrawn, = Ltw-2	12
Lvtw-3	withdrawn, = Ltw-3	9
Lvtw-4	withdrawn, = Ltw-4	1
Lvtw-5	withdrawn, = Car-2	3
lx	luxate	5
Ly-1	lymphocyte antigen-1 (ex Lyt-1)	19
Ly-10	lymphocyte antigen-10 (ex Ly-m10)	19
Ly-11	lymphocyte antigen-11	2
Ly-12	lymphocyte antigen-12	19
Ly-13	lymphocyte antigen-13	UN
Ly-14	lymphocyte antigen-14	7
Ly-15	lymphocyte antigen-15 (= LFA-1)	7
Ly-16	lymphocyte antigen-16 (ex Ly-18)	12
Ly-17	lymphocyte antigen-17 (= Ly-20, LyM-1, Ly-m1)	1
Ly-18	lymphocyte antigen-18 (ex Ly-m18, linked to Ltw-2)	12
Ly-19	lymphocyte antigen-19 (ex Ly-m19)	4
Ly-2	lymphocyte antigen-2 (ex Lyt-2)	6
Ly-20	lymphocyte antigen-20 (ex Ly-22a)	4
Ly-21	lymphocyte antigen-21	7
Ly-22	lymphocyte antigen-22 (ex Ly-m22)	1
Ly-23	lymphocyte antigen-23	2
Ly-24	lymphocyte antigen-24 (ex Pgp-1)	2
Ly-25	lymphocyte antigen-25	2
Ly-26	lymphocyte antigen-26	2
Ly-27	lymphocyte antigen-27	2
Ly-28	lymphocyte antigen-28	2
Ly-29	lymphocyte antigen-29	4
Ly-3	lymphocyte antigen-3 (ex Lyt-3)	6
Ly-30	lymphocyte antigen-30	4

Symbol	Name	Chr
Ly-31	lymphocyte antigen-31 (ex Ly-m31; provisional)	4
Ly-32	lymphocyte antigen-32	4
Ly-33	lymphocyte antigen-33	1
Ly-34	lymphocyte antigen-34	UN
Ly-35	lymphocyte antigen-35	6
Ly-36	lymphocyte antigen-36	6
Ly-37	lymphocyte antigen-37	3
Ly-38	lymphocyte antigen-38	3
Ly-39	lymphocyte antigen-39	17
Ly-4	L3T4 T cell differentiation antigen	6
Ly-40	lymphocyte antigen-40 (= Mac-1)	UN
Ly-41	lymphocyte antigen-41 (ex Pca-1)	UN
Ly-42	lymphocyte antigen-42	1
Ly-43	lymphocyte antigen-43	UN
Ly-44	lymphocyte antigen-44	19
Ly-5	lymphocyte antigen-5 (ex Lyt-4)	1
Ly-6	lymphocyte antigen-6 (ex Ly-8, Ala-1, DAG, H9/25, Ly-27)	15
Ly-7	lymphocyte antigen-7	UN
Ly-8	lymphocyte antigen-8 (ex Ly-11)	UN
Ly-9	lymphocyte antigen-9 (ex Lgp100, T100)	1
Ly-m20	withdrawn, = Ly-17	1
Lyb-2	B-lymphocyte antigen-2	4
Lyb-3	B-lymphocyte antigen-3	UN
Lyb-4	B-lymphocyte antigen-4	4
Lyb-5	B-lymphocyte antigen-5	UN
Lyb-6	B-lymphocyte antigen-6	4
Lyb-7	B-lymphocyte antigen-7	12
Lyb-8	B-lymphocyte antigen-8	7
LyM-1	withdrawn, = Ly-17	1
Lyt-1	withdrawn, = Ly-1	19
Lyt-2	withdrawn, = Ly-2	6
Lyt-3	withdrawn, = Ly-3	6
Lyx-1	lymphocyte antigen X-1	X
Lyx-2	lymphocyte antigen X-2	X
Lyx-3	lymphocyte antigen X-3	X
lz	lizard (extinct)	15
Lzm-s1	M lysozyme structural-1	UN
Lzm-s2	M lysozyme structural-2	UN
Lzp-r	P lysozyme regulator	UN
Lzp-s	P lysozyme structural	UN
m	misty	4
M(c<m>)	withdrawn, = Mcm-1	UN
ma	matted	3
Mag	myelin-associated glycoprotein	7
Mal-1	malaria resistance (may = Lsh, Ity)	1
Maoa	monoamine oxidase A	X
Maob	monoamine oxidase B	X
Map-1	mannosidase processing-1	5
Map-2	withdrawn, = Neu-1	17
Mb1	class I MHC subfamily gene (provisional)	17
Mbp	myelin basic protein (= shi)	18
Mbp*1	myelin basic protein transgene (provisional)	UN
mc	marcel	5
Mcf-2	mcf.2 transforming sequence	X
Mch	modifier of chinchilla (provisional)	UN
Mcm-1	modifier of c<m>-1	UN
Mcm-2	modifier of c<m>-2 (provisional)	UN
Mcr	mast cell regulator	UN
md	mahoganoid	16
mdac	modifier of Dac	UN
mdf	muscle deficient	19
mdg	muscle dysgenesis	UN
Mdh-1	withdrawn, Mod-1	9
Mdh-2	withdrawn, Mor-1	5
mdm	muscular dystrophy with myositis	2
Mdm-1	transformed mouse 3T3 cell double minute-1	10
Mdm-2	transformed mouse 3T3 cell double minute-2	10
Mdmg-1	mandibular morphogenesis-1	17
mdr	multiple drug resistance	UN
Mdr	withdrawn, = Mod-2r	7
Mdr-1	withdrawn, = Mod-2r	7
mdx	X-linked muscular dystrophy (ex pke; = Dmd)	X
me	motheaten	6
Mea	male-enhanced antigen (provisional)	UN
mea	meander tail	4
med	motor end-plate disease	15
Mel	cell line NK14 derived transforming oncogene	9
Melb	melanoblast-melanocyte population size	UN
Melc	melanocyte population size	UN
Mep-1	meprin-1	17
mes	mesenchymal dysplasia	13
Met	met proto-oncogene*	6
Met-1	methyltransferase-1	UN
Met-2	methyltransferase-2	12
Meta	meth A tumor antigen	12
mfd	withdrawn, = mdf	19
mg	mahogany	2
Mgpt-1	gpt-insertion-1	UN
Mgsa	melanoma growth stimulatory activity	5
mh	mocha	10
mi	microphthalmia	6
mic	microphthalmia Japan (provisional)	UN
Micrl	microwave-induced increase in complement-receptor B cells	5
Mis-1	withdrawn, = Pvt-1	15
mk	microcytic anemia	15
ml	myelin-less (ex uc; provisional)	UN
Ml	withdrawn, = Nil	7
mld	withdrawn, allele of shi	18
mlp	withdrawn, = Bmn	3
Mls-1	minor lymphocyte stimulator-1	1
Mls-2	minor lymphocyte stimulator-2	UN
Mlv-1	murine leukemia virus-1*	UN
Mlvi-1	Moloney-MuLV integration site-1	15
Mlvi-2	Moloney-MuLV integration site-2	15
Mlvi-3	Moloney-MuLV integration site-3	UN
Mmv-1	MCF endogenous virus-1*	5
Mmv-10	MCF endogenous virus-10*	1
Mmv-11	MCF endogenous virus-11*	11
Mmv-12	MCF endogenous virus-12*	3
Mmv-13	MCF endogenous virus-13*	11
Mmv-2	MCF endogenous virus-2*	3
Mmv-3	MCF endogenous virus-3*	7
Mmv-4	MCF endogenous virus-4*	7
Mmv-5	MCF endogenous virus-5*	1
Mmv-6	MCF endogenous virus-6*	12
Mmv-7	MCF endogenous virus-7*	12
Mmv-71	withdrawn, = Xmmv-71	2
Mmv-8	MCF endogenous virus-8*	11
Mmv-9	MCF endogenous virus-9*	1
mn	miniature	5

Symbol	Name	Chr
Mnd	motor neuron degeneration	UN
Mo	mottled	X
Mo-en.1	withdrawn, = En-1	1
Mod-1	malic enzyme, supernatant	9
Mod-2	withdrawn, = Mod-2s	7
Mod-2r	malic enzyme mitochondrial regulatory (ex Mdr, Mdr-1)	7
Mod-2s	malic enzyme, mitochondrial	7
Mor-1	malate dehydrogenase, mitochondrial	5
Mor-2	malate dehydrogenase, soluble	UN
Mos	Moloney sarcoma oncogene	4
Mov-1	Moloney leukemia virus-1*	6
Mov-10	Moloney leukemia virus-10*	3
Mov-11	Moloney leukemia virus-11*	UN
Mov-12	Moloney leukemia virus-12*	UN
Mov-13	Moloney leukemia virus-13*	11
Mov-14	Moloney leukemia virus-14*	X
Mov-15	Moloney leukemia virus-15*	XY
Mov-2	Moloney leukemia virus-2*	UN
Mov-24	Moloney leukemia virus-24*	Y
Mov-3	Moloney leukemia virus-3*	UN
Mov-34	Moloney leukemia virus-34*	UN
Mov-4	Moloney leukemia virus-4*	UN
Mov-5	Moloney leukemia virus-5*	UN
Mov-6	Moloney leukemia virus-6*	UN
Mov-7	Moloney leukemia virus-7*	1
Mov-8	Moloney leukemia virus-8*	UN
Mov-9	Moloney leukemia virus-9*	11
Mp	micropinna-microphthalmia	UN
Mph-1	macrophage antigen-1	7
Mpi-1	mannose phosphate isomerase-1	9
Mpo	myeloperoxidase	11
mr	withdrawn, allele of ru-2	7
Ms15-1	minisatellite 15-1	4
Ms15-2	minisatellite 15-2	UN
Ms15-3	minisatellite 15-3	17
Ms15-4	minisatellite 15-4	UN
Ms15-5	minisatellite 15-5	5
Ms15-6	minisatellite 15-6	5
Ms15-7	minisatellite 15-7	14
Ms15-8	minisatellite 15-8	UN
Ms6-1	minisatellite 6-1	UN
Ms6-2	minisatellite 6-2	4
Ms6-3	minisatellite 6-3	UN
Ms6-4	minisatellite 6-4	6
Ms6-5	minisatellite 6-5	14
Ms6hm	minisatellite 6 hypermutable	UN
msd	withdrawn, allele of jp	X
Msp-1	mouse salivary protein-1 (= Ssp)	UN
Msz	resistance to diabetes induction by MSz (prov.)	UN
Mt-1	metallothionein-1	8
Mt-2	metallothionein-2	8
mto	withdrawn, allele of adr	UN
Mtp-1	mouse tear protein-1	7
Mtp-3	mouse tear protein-3	X
Mtv-1	mammary tumor virus locus-1*	7
Mtv-10	withdrawn, = Mtv-7	1
Mtv-11	mammary tumor virus locus-11*(ex Mtv-7a)	14
Mtv-12	withdrawn, = Mtv-11	14
Mtv-13	mammary tumor virus locus-13*	4
Mtv-14	mammary tumor virus locus-14*	6
Mtv-15	withdrawn, = Mtv-17	4
Mtv-16	withdrawn, = Mtv-8	6
Mtv-17	mammary tumor virus locus-17*(ex Mtv-10, Mtv-15, Mtv-20)	4
Mtv-18	withdrawn, = Mtv-3	11
Mtv-19	withdrawn, = Mtv-7	16
Mtv-2	mammary tumor virus locus-2*	18
Mtv-20	mammary tumor virus locus-20*	4
Mtv-21	mammary tumor virus locus-21*	8
Mtv-22	mammary tumor virus locus-22*(prov., only in B6.C-KH-84)	UN
Mtv-23	mammary tumor virus locus-23*	UN
Mtv-24	mammary tumor virus locus-24*	UN
Mtv-25	mammary tumor virus locus-25*	UN
Mtv-26	mammary tumor virus locus-26*	UN
Mtv-3	mammary tumor virus locus-3*	11
Mtv-4	mammary tumor virus locus-4*	UN
Mtv-5	mammary tumor virus locus-5*	UN
Mtv-6	mammary tumor virus locus-6*	16
Mtv-7	mammary tumor virus locus-7*(ex Mtv-10, Mtv-19)	1
Mtv-7a	withdrawn, = Mtv-11	14
Mtv-8	mammary tumor virus locus-8*(ex Mtv-16)	6
Mtv-9	mammary tumor virus locus-9*	12
Mtvr-1	mammary tumor virus receptor-1	16
mu	muted	13
Mud-1	withdrawn, = Sas-1	1
Mup-1	major urinary protein-1	4
Mup-a	withdrawn, = Mup-1	4
Mupm-1	major urinary protein modifier-1	15
Murvy	murine repeated virus, Y-linked (provisional)	Y
mut-1	mutator-1 (provisional)	UN
mut-2	mutator-2 (provisional)	UN
Mv	malformed vertebrae	UN
Mx	withdrawn, = Mx-1	16
Mx-1	myxovirus (influenza virus) resistance-1 (ex Mx)	16
Mx-2	myxovirus (influenza virus) resistance-2 (prov.)	16
Mxv-1	MA/My xenotropic MuLV-1*	UN
my	blebs	3
Myb	myeloblastosis oncogene	10
Myc	myelocytomatosis oncogene	15
myd	myodystrophy	8
Myhc-a	myosin heavy chain, cardiac muscle, adult	14
Myhc-b	myosin heavy chain, cardiac muscle, foetal	14
Myhs	myosin heavy chain, skeletal muscle	11
Myhs-e	myosin heavy chain, skeletal muscle, embryonic	11
Myhs-f	myosin heavy chain, skeletal muscle, adult fast	11
Myhs-p	myosin heavy chain, skeletal muscle, perinatal	11
Myla	myosin light chain, alkali, cardiac atria	11
Mylc	myosin light chain, alkali, cardiac ventricles	9
Mylf	myosin light chain, alkali, fast skeletal muscle	1
Mylf-ps	myosin light chain, alkali, fast skeletal muscle, pseudo.	12
Mylpa	myosin light chain, phosphorylatable, cardiac atria	RE
Mylpc	myosin light chain, phosphorylatable, cardiac ventricles	RE
Mylpf	myosin light chain, phosphorylatable, fast skeletal muscle	7
N	naked	15
Nan	neonatal anemia	8
Nat-1	N-acetyl transferase-1, liver, blood	8
Nat-2	N-acetyl transferase-2, pineal gland	UN
nb	normoblastic anemia	UN
nc	withdrawn, allele of md	16
Ncam	neural cell adhesion molecule (may = sg)	9
nct	Nakano cataract	UN

Nel-1	CML-detected lymphocyte antigen	17
Neu-1	neuraminidase-1 (ex Apl, Map-2, Aglp)	17
Nfh	neurofilament, heavy polypeptide	UN
Nfl	neurofilament, light polypeptide	UN
Nfm	neurofilament, medium polypeptide	UN
Nfxv-1	NFS/N xenotropic virus-1	UN
Ng	nackig	UN
Ngcam	neural glial cell adhesion molecule	RE
Ngfa	nerve growth factor, alpha	7
Ngfb	nerve growth factor, beta	3
Ngfg	nerve growth factor, gamma	7
Ngfr	nerve growth factor receptor	11
Niard	non-immunoglobulin rearranging DNA sequence(ex Nird)	15
Nil	neonatal intestinal lipidosis (extinct)	7
Nird	withdrawn, = Niard	15
Nk-1	natural killer cell-associated antigen-1	UN
Nk-2	natural killer cell-associated antigen-2	2
Nmyc-1	neuroblasmtoma myc-related oncogene-1 (ex Nmyc)	12
Nmyc-2	neuroblastoma myc-related oncogene-2 (psuedogene?)	5
Nop	nuclear opacity	UN
Np-1	nucleoside phosphorylase-1	14
Np-2	nucleoside phosphorylase-2	14
nr	nervous	8
Nras	neuroblastoma ras oncogene	3
Nras-ps	neuroblastoma ras oncogene pseudogene	14
nu	nude	11
nuc	nuclear cataract	UN
Nuca	dominant nuclear cataract (provisional)	UN
nv	Nijmegen waltzer	7
Nxv-1	NZB/BlNJ xenotropic MuLV*(provisional)	UN
Nxv-2	NZB/BlNJ xenotropic MuLV*(provisional)	UN
Ny	progressive fur loss	UN
Nzc	nuclear and zonular cataract (provisional)	UN
Nzv-1	NZB virus-1*	UN
Nzv-2	NZB virus-2*	UN
O<hv>	organizer, high variegation (provisional)	X
o<s>G	withdrawn, mutation in Gus-s complex	5
Oat	ornithine aminotransferase	7
ob	obese	6
obl	withdrawn, allele at db	4
oc	osteosclerotic	19
ocd	osteochondrodystrophy	19
Och	ochre	4
Oct	withdrawn, = spf	X
Odc	ornithine decarboxylase structural gene	12
Odc-1	ornithine decarboxylase-1 (ex BRS-1)	1
Odc-10	ornithine decarboxylase-10 (ex BRS-10)	UN
Odc-11	ornithine decarboxylase-11 (ex BRS-11)	UN
Odc-12	ornithine decarboxylase-12	UN
Odc-13	ornithine decarboxylase-13	X
Odc-2	ornithine decarboxylase-2 (ex BRS-2)	2
Odc-3	ornithine decarboxylase-3 (ex BRS-3)	3
Odc-4	ornithine decarboxylase-4 (ex BRS-4)	4
Odc-5	ornithine decarboxylase-5 (ex BRS-5)	6
Odc-6	ornithine decarboxylase-6 (ex BRS-6)	7
Odc-7	ornithine decarboxylase-7 (ex BRS-7)	7
Odc-8	ornithine decarboxylase-8 (ex BRS-8)	12
Odc-9	ornithine decarboxylase-9 (ex BRS-9)	14
oe	open eyelids	11
oed	edematous	UN

oel	open eyelids with cleft palate	UN
oh	obstructive hydrocephalus	UN
Oh21-1	wuthdrawn, = Cyp21	17
Oh21-2	withdrawn, = Cyp21-1ps	17
ol	oligodactyly	7
olt	oligotriche	UN
om	ovum mutant	UN
op	osteopetrosis	3
opt	opisthotonus	6
or	ocular retardation	UN
Org	withdrawn, = Gpi-1t	7
Orm-1	orosomucoid-1 (ex Agp-1)	4
Orm-2	orosomucoid-2 (ex Agp-2)	4
Os	oligosyndactylism	8
ot	oscillator	UN
Otc	withdrawn, = spf	X
Otf-1	octamer-binding transcription factor-1	1
Otf-2	octamer-binding transcription factor-2	1
oto	otocephaly	1
Oua-1	ouabain resistance-1	3
Oubr	ouabain resistance (X-linked)	X
Ox-1	menadione oxidoreductase-1	UN
Ox-2	menadione oxidoreductase-2	UN
p	pink-eyed dilution	7
P450-1	withdrawn, = Cyp1a1	9
P450-2C	withdrawn, = Cyp2c	19
P450-2D	withdrawn, = Cyp2d	15
P450-3	withdrawn, = Cyp1a2	9
p63	withdrawn, = Tcp-1	17
pa	pallid	2
pad	paddle (provisional)	UN
Pad-1	MMTV LTR integration site	11
Pah	phenylalanine hydroxylase	10
Pan-1	pancreas protein-1	4
par	paralyse (provisional)	UN
Pas-1	pulmonary adenoma susceptibility-1	UN
Pas-2	pulmonary adenoma susceptibility-2	UN
Pas-3	pulmonary adenoma susceptibility-3	UN
Pax-1	paired box homeotic gene-1	2
Pax-2	paired box homeotic gene-2	7
Pax-3	paired box homeotic gene-3	1
Pbr	resistance to Paracoccidioides brasiliensis (prov.)	UN
pc	phocomelic	UN
Pca	withdrawn, = Ly-41	UN
Pca-1	withdrawn, = Ly-41	UN
pcd	Purkinje cell degeneration	13
Pck-1	phosphoenolpyruvate carboxykinase-1, cytosolic	2
Pcn	withdrawn, = Cyp3	6
pcp	polydactyly with cleft palate (provisional)	UN
Pcp-1	Purkinje cell protein-1	6
Pcp-2	Purkinje cell protein-2	8
Pcs-1	polar cataract and small eye-1	UN
Pcs-2	polar cataract and small eye-2	UN
pcy	polycystic kidney disease	UN
Pd	pyrimidine degrading	UN
pd	withdrawn, = pf	4
Pdeg	cGMP-phosphodiesterase gamma	11
Pdgfr	platelet derived growth factor receptor	18
Pdha-1	pyruvate dehydrogenase Elalpha subunit	X
Pdhal	pyruvate dehydrogenase Elalpha like	19

Symbol	Name	Chr
Pdn	polydactyly Nagoya	UN
pe	pearl	13
Ped	preimplantation embryo development (provisional)	17
Pep-1	peptidase-1 (ex Dip-2)	18
Pep-2	peptidase-2	10
Pep-3	peptidase-3 (ex Dip-1)	1
Pep-4	peptidase-4	7
Pep-7	peptidase-7	5
pet	withdrawn, = hyt	12
pf	pupoid fetus	4
Pfc	properdin factor, complement	X
Pfk-4	phosphofructokinase-4	15
pg	pygmy	10
Pgam-1	phosphoglyceromutase-1	19
Pgd	6-phosphogluconate dehydrogenase	4
pge	phosphoglyceromutase expression	UN
Pgk-1	phosphoglycerate kinase-1	X
Pgk-1ps1	phosphoglycerate kinase-1 pseudogene 1	16
Pgk-1ps2	phosphoglycerate kinase-1 pseudogene 2	17
Pgk-1ps3	phosphoglycerate kinase-1 pseudogene 3	3
Pgk-1ps4	phosphoglycerate kinase-1 pseudogene 4	4
Pgk-2	phosphoglycerate kinase-2 (= Tcp-2)	17
Pgm-1	phosphoglucomutase-1	5
Pgm-2	phosphoglucomutase-2	4
Pgm-3	phosphoglucomutase-3	9
Pgp-1	withdrawn, = Ly-24	2
Pgr	progesterone receptor	9
Pgy-1	P glycoprotein-1	5
Ph	patch	5
Phil	Philly cataract	UN
Phk	phosphorylase kinase (may = Phka)	X
Phka	phosphorylase kinase alpha (may = Phk)	X
Phkg	phosphorylase kinase gamma	UN
Phr	pheromonal response	UN
pi	pirouette	5
Pim-1	proviral integration, MCF	17
Pit-1	pituitary specific transcription factor-1	1
pk	plucked	18
Pk-1	pyruvate kinase-1	3
Pk-2	withdrawn, = Pk-3	9
Pk-3	pyruvate kinase-3	9
Pkca	protein kinase C alpha	11
Pkcc	protein kinase C gamma	7
pke	withdrawn, = mdx	X
pl	paraparesis-lethal	UN
Pl-1	placental lactogen-1	13
Pl-2	placental lactogen-2	13
Plat	plasminogen activator, tissue	8
Plau	plasminogen activator, urokinase	14
ple	perinatal lethality	15
Plf	proliferin	13
Plfr	proliferin-related protein	13
Plp	myelin proteolipid protein (= jp)	X
pma	peroneal muscular atrophy	UN
pn	pugnose	14
pnc	pancreas defect	UN
Pnd	pronatriodilatin (ex Anf)	4
Po	postaxial polydactyly	UN
Pol-20	viral polymerase gene-20 (provisional)	5
Pol-23	viral polymerase gene-23 (provisional)	2
Pol-5	viral polymerase gene-5 (wdrn = Xmmv-72, prob. = Xmmv-75)	15
Pol-6	viral polymerase gene-6	8
Pol-7	viral polymerase gene-7 (provisional)	12
Pomc-1	pro-opiomelanocortin-alpha	12
Pomc-2	pro-opiomelanocortin-beta	19
Por	NADPH-cytochrome P-450 oxidoreductase	6
Pp	passive performance	UN
pr	porcine tail	UN
Pre	withdrawn, = Pre-1	12
Pre-1	withdrawn, = Aat	12
Pre-2	prealbumin-2 (ex Pre-4)	12
Pre-4	withdrawn, = Pre-2	12
Prgs	phoshoribosyl glycinamide synthetase (ex Gaps)	16
Prl	prolactin	13
Prm-1	protamine-1	16
Prm-2	protamine-2	16
Prn-i	prion protein, inducible (may = Sinc)	2
Prn-p	prion protein, structural locus	2
Pro-1	proline oxidase-1	UN
Prp	proline-rich protein	6
Prps-1	phosphoribosyl pyrophosphate synthetase-1	UN
Prps-2	phosphoribosyl pyrophosphate synthetase-2	UN
Prt-1	pancreatic proteinase-1 (no recomb. with Prt-3)	UN
Prt-2	withdrawn, = Ctrb	8
Prt-3	pancreatic proteinase-3 (no recomb. with Prt-1)	UN
Prt-4	pancreatic proteinase-4 (ex Smg-1)	7
Prt-5	pancreatic proteinase-5 (ex Smg-2)	7
Prt-6	pancreatic proteinase-6	UN
Ps	polysyndactyly	4
Pscr-1	corneal resistance to Ps. aeruginosa-1 (may = Pas-1)	UN
Pscr-2	corneal resistance to Ps. aeruginosa-2 (may = Pas-2)	UN
Pscs	corneal suscept. to Ps. aeruginosa (may = Pas-3)	UN
Psp	parotid secretory protein	2
Psp-2	parotid secretory protein-2	14
Psph	phosphoserine phosphatase (ex Psp)	5
Pst	polydactyly short tail	UN
Pt	pintail	4
Pth	parathyroid hormone	7
Pthlh	parathyroid hormone-like peptide	6
ptr	pulmonary tumor resistance	UN
Ptv-1	polytropic virus-1	UN
pu	pudgy	7
Pv	pivoter	UN
Pva	parvalbumin	15
Pvt-1	plasmacytoma variant translocation-1 (ex Mis-1)	15
px	postaxial hemimelia	6
py	polydactyly	1
Pygb	brain glycogen phosphorylase	2
Pygl	liver glycogen phosphorylase	12
Pygm	muscle glycogen phosphorylase	19
Pyp	pyrophosphatase	10
Q	quinky	8
Q10	Q region gene 10	17
Qa-1	Qa lymphocyte antigen-1	17
Qa-11	Qa lymphocyte antigen-11	17
Qa-2	Qa lymphocyte antigen-2	17
Qa-3	Qa lymphocyte antigen-3	17
Qa-4	Qa lymphocyte antigen-4 (ex Qat-4)	17
Qa-5	Qa lymphocyte antigen-5 (ex Qat-5)	17
Qa-6	Qa lymphocyte antigen-6	17

Symbol	Name	Chr
Qa-7	Qa lymphocyte antigen-7	17
Qa-8	Qa lymphocyte antigen-8	17
Qa-9	Qa lymphocyte antigen-9	17
Qb-1	Qb lymphocyte antigen-1 (provisional)	17
Qd	queue deformee (provisional)	UN
Qdm	Qa-1 determinant modifier (provisional)	17
Qdpr	quininoid dihydropteridine reductase	5
Qed-1	Q lymphocyte antigen; prob. = Qa-1	17
Qglo	withdrawn, = Glo-1	17
qk	quaking	17
Qui	quinine sensitivity (taste)	6
qv	quivering	7
r	withdrawn, = rd	5
Ra	ragged	2
Raf	withdrawn, = Afr-1	UN
Raf-1	murine sarcoma 3611 oncogene-1	6
Ram-1	replication of amphotropic virus-1	8
ras	resistance to audiogenic seizures (provisional)	UN
Rasa	Ras p21 protein activator	13
Rb-1	retinoblastoma-1	14
Rbp-3	retinal binding protein-3, interstitial	11
rc	rough coat	9
Rcs-1	reticular cell sarcoma suppression-1	UN
rd	retinal degeneration (ex r)	5
rds	retinal degeneration-slow	17
Re	rex	11
Rec-1	ecotropic MuLV receptor-1 (ex Rev-1)	5
Rec-2	ecotropic MuLV M813 receptor	2
Rel	reticuloendotheliosis oncogene	11
Ren	withdrawn, = Ren-1, Ren-2	1
Ren-1	renin-1 structural (ex Rnr, Rn-1, Ren-A)	1
Ren-2	renin-2 tandem dup. of Ren-1 (ex Rnr, Rn-2, Ren-B)	1
Ren-A	withdrawn, = Ren-1	1
Ren-B	withdrawn, = Ren-2	1
Rev-1	withdrawn, = Rec-1	5
Rf	rib fusion	UN
Rfv-1	recovery from Friend virus-1	17
Rfv-2	recovery from Friend virus-2	17
Rfv-3	recovery from Friend virus-3	UN
rg	rotating	UN
Rgv-1	resistance to Gross virus-1	17
Rgv-2	resistance to Gross virus-2	UN
rh	rachiterata	2
rhg	retarded hair growth	UN
Rho	rhodopsin	6
Rhv-1	resistance to hepatitis virus-1	UN
Rhv-2	resistance to hepatitis virus-2 (may = Hv-2)	UN
Ri	withdrawn, allele of Re	11
Ri-1	recognition of identity-1	17
Ri-2	recognition of identity-2	17
Rib-1	ribonuclease-1, pancreatic	14
Ric	rickettsia tsutsugamushi resistance	5
rif	withdrawn, = Afr-2	UN
Rig-1	regulation of Igh-1b-1	17
Rig-2	regulation of Igh-1b-2 (provisional)	UN
Ril-1	radiation-induced leukemia sensitivity-1	15
Ril-2	radiation-induced leukemia sensitivity-2	4
Ril-3	radiation-induced leukemia sensitivity-3	1
Rip	regulation of phenobarbitol-inducible P450	7
Ripr	repression of phenobarbitol-inducible P450	7

Symbol	Name	Chr
rl	reeler	5
Rmc-1	receptor for MCF virus-1	1
Rmcf	resistance to MCF virus	5
Rmp-1	resistance to mousepox (provisional)	UN
Rmrpr	RNAase mitochondrial RNA processing RNA	4
Rmv-1	resistance to Moloney virus-1 (prov.; prob. = H-2I)	17
Rmv-2	resistance to Moloney virus-2 (prov.; prob. = H-2I)	17
Rmv-3	resistance to Moloney virus-3 (prov.; prob. = H-2I)	17
Rn	roan	14
Rn-1	withdrawn, = Ren-1	1
Rn-2	withdrawn, = Ren-2	1
Rn7s-2	7s RNA-2	2
Rn7s-5	7s RNA-5	5
Rn7s-6	7s RNA-6	6
Rn7s-9	7s RNA-9	9
Rnr	withdrawn, = Ren-1, Ren-2	1
Rnr12	ribosomal RNA-12	12
Rnu1-1	U1 small nuclear RNA-1 (provisional)	1
Rnu1-2	U1 small nuclear RNA-2 (provisional)	9
Rnu1-ps1	U1 small nuclear RNA psuedogene-1	1
Rnu1-ps2	U1 small nuclear RNA pseudogene-2	UN
Rnu1a-1	U1a1 small nuclear RNA (provisional)	11
Rnu1a-2	U1a2 small nuclear RNA	12
Rnu1b-1	U1b1 small nuclear RNA	3
Rnu1b-3	U1b3 small nuclear RNA	3
Rnu2	U2 small nuclear RNA sequences	RE
ro	rough	2
rol	resistance to osmotic lysis	UN
rp	reduced pigmentation	7
RP1	withdrawn, = D4Rp1	4
RP1 series	withdrawn, = D4Rp1 series	4
RP10	withdrawn, = D17Rp10	17
RP11	withdrawn, = D17Rp11	17
RP17	withdrawn, = D17Rp17	17
RP2-r	withdrawn, = D7Rp2-r	7
RP2-s	withdrawn, = D7Rp2-s	7
RP3	withdrawn, = D0Rp3	UN
RP4	withdrawn, = D14Rp4	14
RP5	withdrawn, = D0Rp5	UN
RP54	withdrawn, = D12Rp54	12
RP6	withdrawn, = D0Rp6	UN
Rpl18	ribosomal protein L18 (provisional)	RE
Rpl30	ribosomal protein L30	15
Rpl32	ribosomal protein L32	6
Rpl7	ribosomal protein L7 (provisional)	RE
Rpn-1	ribophorin I	6
Rpn-2	ribophorin II	2
Rpo2-1	RNA polymerase II-1	11
Rpo3-1	RNA polymerase III	UN
Rps16	ribosomal protein S16	7
Rras	Harvey rat sarcoma oncogene, subgroup R	7
Rrm1	ribonucleotide reductase M1	7
Rrm2	ribonucleotide reductase M2	12
Rrm2-ps1	ribonucleotide reductase M2 pseudogene 1	4
Rrm2-ps2	ribonucleotide reductase M2 pseudogene 2	7
Rrm2-ps3	ribonucleotide reductase M2 pseudogene 3	13
Rrs	resistance to Rous sarcoma	17
Rrv-1	resistance to RadLV-1	17
rs	recessive spotting (may be allele at W)	5
rsh	rump shaker	UN

Symbol	Name	Chr
Rsm-1	resistance to Schistosoma mansoni (provisional)	UN
rst	rosette	UN
Rsvp	red sensitive visual pigment	X
Rtp	resistance to transplantable plasmacytoma MPC-11 (prov.)	UN
ru	ruby-eye	19
ru-2	ruby-eye-2	7
Rua	raffinose acetate tasting (provisional)	6
ruf	rough fur	9
rv	rib-vertebrae	UN
Rv-1	Rauscher leukemia virus susceptibility-1	UN
Rv-2	Rauscher leukemia virus susceptibility-2	9
Rv-3	Rauscher leukemia virus susceptibility-3	X
Rvil-1	radiation-induced leukemia virus susceptibility (ex RLVil)	2
Rw	rump white	5
s	piebald	14
S100b	S100 protein, beta polypeptide (neural)	10
sa	satin	13
Saa-1	serum amyloid A-1	7
Saa-2	serum amyloid A-2	7
Saa-3	serum amyloid A-3	7
Saa-4	serum amyloid A-4	7
Saac	sialic acid O-acetylation	9
Sac	saccharin preference	UN
Sag	retinal S-antigen	1
Sal	satin-like (provisional)	UN
Sal-1	withdrawn, = Abpa	7
Sap	serum amyloid P-component (provisional)	1
Sas-1	serum antigenic substance-1 (NOM, probably = Cfh)	1
Sas-2	serum antigenic substance-2	1
sb	stub (extinct)	UN
sc	screwtail (extinct)	UN
scb	scabby	8
Scd	stearoyl-CoA desaturase	UN
sch	scant hair	9
scid	severe combined immunodeficiency	16
Scl-1	susceptibility to cutaneous leishmaniasis-1	8
Scl-2	susceptibility to cutaneous leishmaniasis-2	UN
Sco	scopolamine modification of exploratory activity	17
Sd	Danforth's short tail	2
Sdh-1	sorbitol dehydrogenase-1	2
Sdr-1	serine dehydratase regulator-1	UN
Sdr-2	serine dehydratase regulator-2	UN
sdy	sandy	13
se	short ear	9
sea	sepia	1
seb	seborrhoic dermatitis	UN
Segr	segregation reversal	X
Sep-1	withdrawn, = Apoa-1	9
Sep-2	serum protein-2	9
Sev-1	withdrawn, = d	9
Sey	small eye	2
sf	scurfy	X
Sf	withdrawn, = spf	X
Sfpi-1	SFFV proviral integration-1	UN
sg	staggerer (may = Ncam)	9
Sgp-1	serum gp70 production-1 (provisional)	17
Sgp-2	serum gp70 production-2 (provisional; may = Gv-2)	7
sh-1	shaker-1	7
sh-2	shaker-2	11
Sha	shaven	15

Symbol	Name	Chr
shi	shiverer (= Mbp)	18
shm	shambling	11
Shmt	serine hydroxymethyl transferase	RE
sho	shorthead	UN
si	silver	10
Sig	sightless	6
Sigje	small inducible gene JE (provisional)	11
Sinc	scrapie incubation period (may = Prn-i)	UN
Sip	schedule-induced polydipsia	UN
Sis	simian sarcoma oncogene	15
sj	withdrawn, = fmd	UN
Sk	scaly	UN
Skn-1	skin antigen-1 (ex Sk-1)	UN
Skn-2	skin antigen-2 (ex Sk-2)	UN
sks	skeletal fusions with sterility	4
skt	small with kinky tail	7
Sl	steel	10
sla	sex-linked anemia	X
sld	sublingual gland differentiation arrest (provisional)	UN
Slf	sex-linked fidget	X
sll-1	sex-linked lethal-1 (provisional)	X
sll-2	sex-linked lethal-2 (provisional)	X
Slp	sex-limited protein (within H-2S)	17
slt	slaty	14
sm	syndactylism	12
smc	spondylo-metaphyseal chondrodysplasia	UN
Sme	small ear	UN
Smg-1	withdrawn, = Prt-4	7
Smg-2	withdrawn, = Prt-5	7
smk	smoky	UN
Smst	somatostatin (SST in human)	16
sno	snubnose	4
Soa	sucrose octaacetate aversion	6
soc	soft coat	3
Sod-1	superoxide dismutase-1, soluble	16
Sod-2	superoxide dismutase-2, mitochondrial	17
Sp	splotch	1
spa	spastic	3
Sparc	secreted acidic cysteine rich glycoprotein	11
spc	sparse coat	14
spd	spasmodic	11
spf	sparse-fur	X
sph	spherocytosis (prob. = Spna-1)	1
sph-(H)	sphaerozytose (provisional)	UN
Spi-1	serine protease inhibitor-1 (prob. = Pre-1 or -2)	12
Spi-2	serine protease inhibitor-2 (prob. = Pre-1 or -2)	12
Spi-2r	Spi-2 transcription regulator	12
Spl	plasma serotonin level	1
Spl-1	spleen antigen-1	UN
spm	sphingomyelinosis	UN
Spn	sialophorin	RE
Spna-1	alpha-spectrin-1, erythroid (prob. = sph)	1
Spna-2	alpha-spectrin-2, brain	2
Spnb-1	beta-spectrin-1	12
Spp	secreted phosphoprotein	RE
Spp-1	secreted phosphoprotein-1	5
sps	spontaneous seizure (ex dd; prov.)	UN
Spy	spermatogenesis Y	Y
sr	spinner	9
Sr-1	spectrotype regulation	UN

Symbol	Name	Chromosome
Src	Rous sarcoma oncogene	2
Src-2	Rous sarcoma oncogene	RE
Srch	Src-homologous sequence	2
Srlv-1	susceptibility to RadLV-1	UN
srn	siren	UN
Srv	sensitivity to RADLV	UN
Ss	withdrawn, = C4	17
Ssm	withdrawn, allele of W	17
Ssp	sex-limited saliva pattern (= Msp-1)	UN
sst	withdrawn, allele of dr	1
st	shaker-short (probably extinct)	UN
Sta	autosomal striping	UN
stb	stubby	2
stg	stargazer	RE
Stk	stem cell kinetics (provisional)	UN
stm	stumpy	UN
Sto-1	16 steroid alpha-hydroxylase-1	UN
Sto-2	16 steroid alpha-hydroxylase-2 (provisional)	UN
Stp-1	withdrawn, = Tp-1	1
Str	striated	X
Sts	steroid sulfatase	XY
stu	stumbler	UN
su	surdescens	UN
Suc-1	sucrase-1	UN
Surf-1	surfeit gene-1 (prov.)	UN
Surf-2	surfeit gene-2 (prov.)	UN
Surf-3	surfeit gene-3 (prov.)	UN
Surf-4	surfeit gene-4 (prov.)	UN
Surf-5	surfeit gene-5 (prov.)	UN
Surf-6	surfeit gene-6 (prov.)	UN
sut	subtle gray	3
sv	Snell's waltzer	9
Svl	withdrawn, allele of Cat	10
Svp-1	seminal vesicle protein-1	2
Svp-2	seminal vesicle protein-2	7
Svp-3	seminal vesicle protein-3	2
svs	seminal vesicle shape	UN
sw	swaying	15
Swl	sprawling	UN
Sxa	sex chromosome association (provisional)	X
Sxr	sex reversed, chromosomal aberration, Tp(Y)1Ct	XY
Sxv	susceptibility to xenotropic virus (prob. = Rmc-1)	1
sy	shaker-with-syndactylism	18
syn	withdrawn, = jt	UN
Syn-1	synapsin I	X
T	brachyury	17
T3d	T3-delta chain	9
T3e	T3-epsilon chain	9
T3g	T3-gamma chain	9
T3z	T3-zeta chain	1
Ta	tabby	X
Tag	temporal alpha-galactosidase	UN
Tal	talipes	UN
Tam-1	tosyl argnine methylesterase-1	7
Tas	T-associated sex reversal	17
Tat	tyrosine aminotransferase	8
Tb	tibialess (provisional)	UN
tb	tumbler	1
Tbg	thyroxin binding globulin	RE
tc	truncate	6
Tcd-1	t-complex distorter-1	17
Tcd-2	t-complex distorter-2	17
Tcd-3	t-complex distorter-3	17
Tcd-4	t-complex distorter-4	17
Tcm	total cataract with microphthalmia	UN
Tcn-2	transcobalamin-2	11
Tcp	withdrawn, = Abpa	7
Tcp-1	t-complex protein-1 (ex Tp63, p63/6.9)	17
Tcp-10a	t-complex protein-10a (provisional)	17
Tcp-10b	t-complex protein-10b (provisional)	17
Tcp-10c	t-complex protein-10c (provisional)	17
Tcp-2	withdrawn, = Pgk-2	17
Tcp-3	t-complex protein-3	17
Tcp-4	t-complex protein-4	17
Tcp-5	t-complex protein-5	17
Tcp-6	t-complex protein-6	17
Tcp-7	t-complex protein-7	17
Tcp-8	t-complex protein-8	17
Tcp-9	t-complex protein-9	17
Tcr-1	transmission distortion, responder locus (in t complex)	17
Tcra	T cell antigen receptor alpha chain	14
Tcrb	T cell receptor beta chain	6
Tcrd	T cell receptor delta chain	14
Tcrg	T cell receptor gamma chain	13
tcs-1	t-complex sterility-1	17
tcs-2	t-complex sterility-2	17
tcs-3	t complex sterility-3	17
tct	withdrawn, = allele of T	17
Tcte-1	t-complex associated testis expressed-1	17
Td	tattered	X
td	withdrawn, allele of du	9
Tda-1	testis-determining autosomal-1 (provisional)	UN
Tda-2	testis-determining autosomal-2 (provisional)	RE
Tdf	withdrawn, = Tdy	Y
Tdt	terminal deoxynucleotidyl transferase	19
Tdy	testis-determining-Y (= Zfy-1, Zfy-2?)	Y
te	lighthead (extinct)	UN
ter	teratoma	UN
tf	tufted	17
Tfm	testicular feminization (= Ar)	X
tg	tottering	8
Tgfb	transforming growth factor, beta	7
Tgfb-2	tumor growth factor, beta-2	1
Tgn	thyroglobulin	15
Th	tyrosine hydroxylase	7
th	withdrawn, = thd	1
Thb	ThB cell surface antigen	15
thd	tilted head (ex th)	1
thf	thin fur	17
Tht	thick tail	5
Thy-1	thymus cell antigen-1(theta)	9
Thy-2	thymus cell antigen-2	17
ti	tipsy	11
Tib	withdrawn, = Tcrb	6
Timp	tissue inhibitor of metalloproteinase	X
Tind	T-peripheral cell antigen	12
tint	interaction with T complex	UN
tip	tippy	9
Tis	TPA-induced sequences (provisional)	UN
tk	tail-kinks	9

Symbol	Name	Chr
Tk-1	thymidine kinase-1	11
tl	nonerupted teeth	UN
Tla	thymus leukemia antigen	17
tlt	tilted	5
tm	tremulous (extinct)	UN
Tme	T-associated maternal effect	17
Tmevd-1	TMEV-induced demyelinating disease susceptibility(prov.)	6
Tmtr	tracheal mucociliary transport rate	UN
tn	teetering	11
Tnfa	tumor necrosis factor, alpha	17
Tnfb	tumor necrosis factor, beta (lymphotoxin)	17
Tob	withdrawn, allele of e	8
Tol-1	tolerance to BCG	UN
tor	tortured	UN
tp	taupe	7
Tp-1	transition protein-1 (prov.)	1
Tp-2	transition protein-2 (prov.)	UN
Tp63	withdrawn, = Tcp-1	17
Tph	tryptophane hydrolase (provisional)	UN
Tpi-1	triosephosphate isomerase-1	6
Tpmt	thiopurine methyltransferase	UN
Tpre	pre-T-cell alloantigen	12
Tpx-1	testis specific gene-1	17
Tr	trembler	11
Tra-1	tumor rejection antigen gp96	10
tra-1	withdrawn, = Dtc-1	UN
Tre	tRNA glutamic acid	RE
Trel-1	tRNA glutamic acid like-1	RE
Trf	transferrin	9
Trip-1	withdrawn, = Pep-2	10
Trk-1	tRNA lysine-1	RE
Trl-1	tRNA leucine-1	RE
Trl-2	tRNA leucine-2	RE
trm	tremor	UN
Trm-1	tRNA methionine initiator-1	RE
Trm-2	tRNA methionine initiator-2	RE
Trp-1	tRNA proline-1	RE
Trp-2	tRNA proline-2	RE
Trp53	transformation-related protein 53	11
Trp53-ps	transformation-related protein 53, pseudogene	14
Trq-1	tRNA glutamine-1	RE
Trsp	tRNA phosphoserine (opal suppressor)	RE
Trsp-ps1	tRNA phosphoserine pseudogene-1	RE
Try-1	trypsin-1 (may = Prt-1 or Prt-3)	6
Ts	tail-short	11
Ts-1	Trichinella spiralis resistance-1	17
Ts-2	Trichinella spiralis resistance-2	17
Ts-3	Trichinella spiralis resistance-3	UN
Ts-4	Trichinella spiralis resistance-4	UN
Tse-1	tissue specific extinction-1, of TAT	11
Tse-2	tissue specific extinction-2, of L-ADH	UN
Tse-3	tissue specific extinction-3, of albumin	UN
Tsha	thyrotropin-stimulating hormone, alpha subunit	4
Tshb	thyrotropin-stimulating hormone, beta subunit	3
Tsk	tight-skin	2
Tsk-2	tight skin-2	1
Tsu	T-suppressor cell alloantigen	12
Tsv	variable short tail	5
Tsz-1	thymus size-1	5
Tthy	thymocyte alloantigen	12

Symbol	Name	Chr
tu	toe-ulnar	UN
tub	tubby	7
Tubb	tubulin, beta	UN
Tw	twirler	18
twi	twitcher	12
twt	twister	7
twy	tiptoe walking-Yoshimura (provisional)	UN
tx	toxic milk	UN
ty	trembly	X
Tyr	tyrosinase	7
Tyrp	tyrosinase related protein (prov.)	4
U	umbrous	UN
ub	withdrawn, = th	1
uc	withdrawn, = ml	UN
Ucp	mitochondrial uncoupling protein	8
Udpgt-3	UDP glucuronosyltransferase-3	5
Udpk	uridine diphosphate kinase	RE
Ugpp	uridyl diphosphate glucose pyrophosphorylase	RE
Ul	ulnaless	2
Um	uvomorulin	8
Umph-1	uridine monophosphatase-1	UN
Umph-2	uridine monophosphatase-2	11
Umpk	uridine monophosphate kinase	RE
un	undulated	2
uns	unsteady	UN
Up	umbrous-patterned (provisional)	UN
Up-1	withdrawn, = Mup-1	4
Upg-1	urinary pepsinogen-1	17
Upg-2	urinary pepsinogen-2	1
Ups	uroporphyrinogen I synthetase	9
ur	urogenital (extinct)	UN
us	urogenital syndrome	2
uw	underwhite	15
v	waltzer	10
Va	varitint-waddler	3
van	withdrawn, allele of mk	15
vb	vibrator	11
vc	vacillans	4
Ve	velvet coat	15
vi	visceral inversion	UN
Vil	villin	1
vit	vitiligo (may = Ga; provisional)	UN
vl	vacuolated lens	1
Vlm	vacuolated lens with microphthalmia (provisional)	UN
Vpreb-1	pre-B lymphocyte gene-1 (prov., nonstandard symbol)	16
Vpreb-2	pre-B lymphocyte gene-2 (prov., nonstandard symbol)	16
vs	variable spotting	9
vt	vestigial-tail	11
W	dominant spotting	5
wa-1	waved-1	6
wa-2	waved-2	11
wal	waved alopecia	14
Wap	whey acidic protein	UN
War	warfarin resistance	7
Wc	waved coat	14
wd	waddler (probably extinct)	4
we	wellhaarig	2
Wf	wavy fur (provisional)	UN
wh	writher	UN
wi	whirler	4

Symbol	Name	Chr
wl	wabbler-lethal	14
wms	wriggle mouse sagami (provisional)	UN
wr	wobbler	UN
wst	wasted	2
Wt	waltzer-type	UN
wt	withdrawn, allele of un	2
wuf	white underfur (provisional, extinct?)	UN
wv	weaver	16
Xcat	X-linked cataract	X
Xce	X-chromosome controlling element (ex Cg)	X
xid	X-linked immune deficiency	X
Xld-1	xylose dehydrogenase-1	7
Xlr	X-linked gene family of B cell surface antigens	X
Xlr-1	X linked B cell surface antigen-1	X
Xlr-2	X linked B cell surfact antigen-2	X
Xmmv-15	xenotropic-MCF leukemia virus-15*(ex Env-15)	7
Xmmv-2	xenotropic-MCF leukemia virus-2*(ex Env-2)	9
Xmmv-21	xenotropic-MCF leukemia virus-21*(ex Env-21)	12
Xmmv-22	xenotropic-MCF leukemia virus-22*	3
Xmmv-23	xenotropic-MCF leukemia virus-23*(ex Env-23)	4
Xmmv-25	xenotropic-MCF leukemia virus-25*(ex Env-25)	12
Xmmv-27	xenotropic-MCF leukemia virus-27*(ex Env-27)	6
Xmmv-29	xenotropic-MCF leukemia virus-29*(ex Env-29)	8
Xmmv-3	xenotropic-MCF leukemia virus-3*(ex Env-3)	UN
Xmmv-31	xenotropic-MCF leukemia virus-31*(ex Env-31)	7
Xmmv-34	xenotropic-MCF leukemia virus-34*(ex Env-34, Env-39)	12
Xmmv-35	xenotropic-MCF leukemia virus-35*(ex Env-35)	7
Xmmv-36	xenotropic-MCF leukemia virus-36*(ex Env-36)	1
Xmmv-42	xenotropic-MCF leukemia virus-42*(ex Xp-1)	19
Xmmv-43	xenotropic-MCF leukemia virus-43*(ex Xp-2)	UN
Xmmv-44	xenotropic-MCF leukemia virus-44*(ex Xp-3)	UN
Xmmv-45	xenotropic-MCF leukemia virus-45*(ex Xp-4)	UN
Xmmv-46	xenotropic-MCF leukemia virus-46*(ex Xp-5)	UN
Xmmv-47	xenotropic-MCF leukemia virus-47*(ex Xp-6)	UN
Xmmv-48	xenotropic-MCF leukemia virus-48*(ex Xp-7)	UN
Xmmv-49	xenotropic-MCF leukemia virus-49*(ex Xp-8)	UN
Xmmv-5	xenotropic-MCF leukemia virus-5*(ex Env-5)	5
Xmmv-50	xenotropic-MCF leukemia virus-50*(ex Xp-9)	12
Xmmv-51	xenotropic-MCF leukemia virus-51*(ex Xp-10)	UN
Xmmv-52	xenotropic-MCF leukemia virus-52*(ex Xp-11)	5
Xmmv-53	xenotropic-MCF leukemia virus-53*(ex Xp-12)	UN
Xmmv-54	xenotropic-MCF leukemia virus-54*(ex Xp-13)	UN
Xmmv-55	xenotropic-MCF leukemia virus-55*(ex Xp-14)	15
Xmmv-56	xenotropic-MCF leukemia virus-56*(ex Xp-15)	UN
Xmmv-57	xenotropic-MCF leukemia virus-57*(ex Xp-16)	UN
Xmmv-58	xenotropic-MCF leukemia virus-58*(ex Xp-17)	UN
Xmmv-59	xenotropic-MCF leukemia virus-59*(ex Xp-18)	UN
Xmmv-6	xenotropic-MCF leukemia virus-6*(ex Env-6)	1
Xmmv-60	xenotropic-MCF leukemia virus-60*(ex Xp-19)	UN
Xmmv-61	xenotropic-MCF leukemia virus-61*(ex Xp-20, may = Xmmv-9)	1
Xmmv-62	xenotropic-MCF leukemia virus-62*(ex Xp-21)	4
Xmmv-63	xenotropic-MCF leukemia virus-63*(ex Xp-22)	UN
Xmmv-64	xenotropic-MCF leukemia virus-64*(ex Xp-23)	UN
Xmmv-65	xenotropic-MCF leukemia virus-65*(ex Xp-24)	3
Xmmv-66	xenotropic-MCF leukemia virus-66*(ex Xp-25)	UN
Xmmv-67	xenotropic-MCF leukemia virus-67*(ex Xp-26)	UN
Xmmv-68	xenotropic-MCF leukemia virus-68*(ex Xp-27)	UN
Xmmv-69	xenotropic-MCF leukemia virus-69*(ex Xp-28)	UN
Xmmv-70	xenotropic-MCF leukemia virus-70*(ex Xp-29)	UN
Xmmv-71	xenotropic-MCF leukemia virus-71*(ex Mmv-71)	2
Xmmv-72	xenotropic-MCF leukemia virus-72*	15
Xmmv-73	xenotropic-MCF leukemia virus-73*	7
Xmmv-74	xenotropic-MCF leukemia virus-74*	1
Xmmv-75	xenotropic-MCF leukemia virus-75*	15
Xmmv-76	xenotropic-MCF leukemia virus-76	7
Xmmv-8	xenotropic-MCF leukemia virus-8*(ex Env-8)	4
Xmmv-9	xenotropic-MCF leukemia virus-9*(ex Env-9)	1
Xmmv-Y	xenotropic-MCF leukemia virus-Y*(ex Env-Y)	Y
Xmv-1	xenotropic murine leukemia virus-1*	4
Xmv-2	xenotropic murine leukemia virus-2*	4
Xmv-3	xenotropic murine leukemia virus-3*	16
Xmv-4	xenotropic murine leukemia virus-4*	11
Xmv-5	xenotropic murine leukemia virus-5*	11
xn	exencephaly	UN
Xox-1	xanthine oxidase (provisional)	UN
Xp-1	withdrawn, = Xmmv-42 (and Xp-2 to -29 = Xmmv-43 to -70)	19
Xpa	DNA repair gene (complements XP group A)(provisional)	4
Xpl	X-linked polydactyly	X
Xs	extra-toes spotting	7
Xt	extra toes	13
Xta	testis ascorbic acid	X
Yaa	accelerated autoimmunity and lymphoproliferation	Y
YB10	Y-specific B10 derived sequences	Y
Ym	yellow mottled	X
z	lethal (provisional)	2
Zfa	zinc finger protein, autosomal	10
Zfp-1	zinc finger protein-1 (ex Fnp-1)	11
Zfp-10	zinc finger protein-10	UN
Zfp-11	zinc finger protein-11	UN
Zfp-12	zinc finger protein-12	UN
Zfp-13	zinc finger protein-13	UN
Zfp-14	zinc finger protein-14	UN
Zfp-15	zinc finger protein-15	UN
Zfp-16	zinc finger protein-16	UN
Zfp-17	zinc finger protein-17	UN
Zfp-18	zinc finger protein-18	UN
Zfp-19	zinc finger protein-19	UN
Zfp-2	zinc finger protein-2 (ex Fnp-2)	8
Zfp-20	zinc finger protein-20	UN
Zfp-21	zinc finger protein-21	UN
Zfp-22	zinc finger protein-22	UN
Zfp-23	zinc finger protein-23	UN
Zfp-24	zinc finger protein-24	UN
Zfp-25	zinc finger protein-25	UN
Zfp-3	zinc finger protein-3	11
Zfp-4	zinc finger protein-4	8
Zfp-5	zinc finger protein-5	UN
Zfp-6	zinc finger protein-6	UN
Zfp-7	zinc finger protein-7	UN
Zfp-8	zinc finger protein-8	UN
Zfp-9	zinc finger protein-9	UN
Zfx	X linked zinc finger protein	X
Zfy-1	Y linked zinc finger protein-1	Y
Zfy-2	Y linked zinc finger protein-2	Y
Zp-3	zona pellucida glycoprotein-3	UN

Mus musculus
Mouse DNA Clones
RFLP List
March 1989

Janan T. Eppig
The Jackson Laboratory
Bar Harbor, ME 04609
USA
phone: (207) 288-3371

List of Mouse DNA Clones and Probes

This booklet contains

1. **List of Mouse DNA Clones and Probes.** For each entry, the list contains the Chromosome assignment, Gene Symbol, Gene name, Probe name, RFLV (restriction fragment length variant, "+" indicates known RFLV), Holder or first author, and Reference. The list is arranged numerically by chromosome and the genes are listed alphabetically within each chromosome list. Genes and probes that have not been assigned to a specific chromosome are listed at the end.

2. **Map of Mouse Genes for which probes/clones are listed.** This linkage map of the mouse contains only those genes that appear in the list of probes and clones. Comments about the map appear following the map.

The database which forms the foundation for this booklet is continually being updated. A mailing list is maintained of those interested in receiving updates of the List of Mouse DNA Clones and Probes (provided at approximately six month intervals). Please be sure you have signed up to receive updates.

I would appreciate receiving contibuted information for these databases. I also encourage you to communicate any errors or omissions you observe. Your assistance will help keep this list current and complete.

Notes on Map of Mouse Genes for which probes/clones are listed

Genes may have multiple probes listed in the Probe/Clone List.

Probes for which gene location is unknown or which identify sequences on multiple chromosomes are not included.

Chromosome 17. Many probes designated by D17Lehxxx are located in the *t* region. These have been grouped as a single entry on Chromosome 17. The loci are as follows: D17Leh48, D17Leh54, D17Leh66, D17Leh80, D17Leh89, D17Leh94, D17Leh108, D17Leh111, D17Leh116, D17Leh119, D17Leh122, D17Leh173, D17Leh180, D17Leh443, D17Leh467, D17Leh508, D17Leh525, D17Leh550.

X Chromosome. There are too many genes to fit on the map of the X Chromosome. Locations listed as DX[1] - DX[6] correspond to the following:

DX[1] = DXPas3, DXWas68, DXSmh172, DXPas22, DXPas4, DXPas7, DXSmh141
DX[2] = DXPas5, DXSmh222, DXSmh10, DXSmh66, DXSmh67, DXSmh219, DXSmh191
DX[3] = DXSmh36, DXSmh59, DXSmh91
DX[4] = DXSmh23, DXSmh64, DXPas21
DX[5] = DXPas8, DXPas14, DXPas13
DX[6] = DXSmh43, DXPas15, DXPas20

Y Chromosome. No genes have been localized on the Y chromosome map. Several probes are available that detect genes syntenic with the Y chromosome (see Probe List).

Chr	Symbol&	Gene Name	Probe@	RFLV#	Holder‡	Reference∞

& Gene symbols enclosed in quotes " " are not standard gene symbols and will be replaced with appreoved symbols as they are assigned. Gene symbols followed by (x), x indicates parts of the gene detected by the probe. Symbols in () are not part of the gene symbol.
@ Probes and clones are from mouse except where preceded by [B]=bovine; [C]=chicken; [CH]=Chinese hamster; [H]=human; [Ha]=hamster; [Mp]=Mus pahari; [R]=rat; [SH]=Syriam hamster; [V]=virus
# RFLV (restriction fragment length variant). Known variants for a given probe indicated by "+" in RFLV column.
‡ Holders/Contributors preceded by * indicate that probe information was obtained from the literature only (Name is first author). Codes preceding holder indicate: 1-available; 2-collaborators only; 3-available in future; 4-not available; 5-not known; fr:"obtained from"
∞ If article is preceded by "ref:" the "Holder" is not the first author or the probe was obtained from the "Holder".

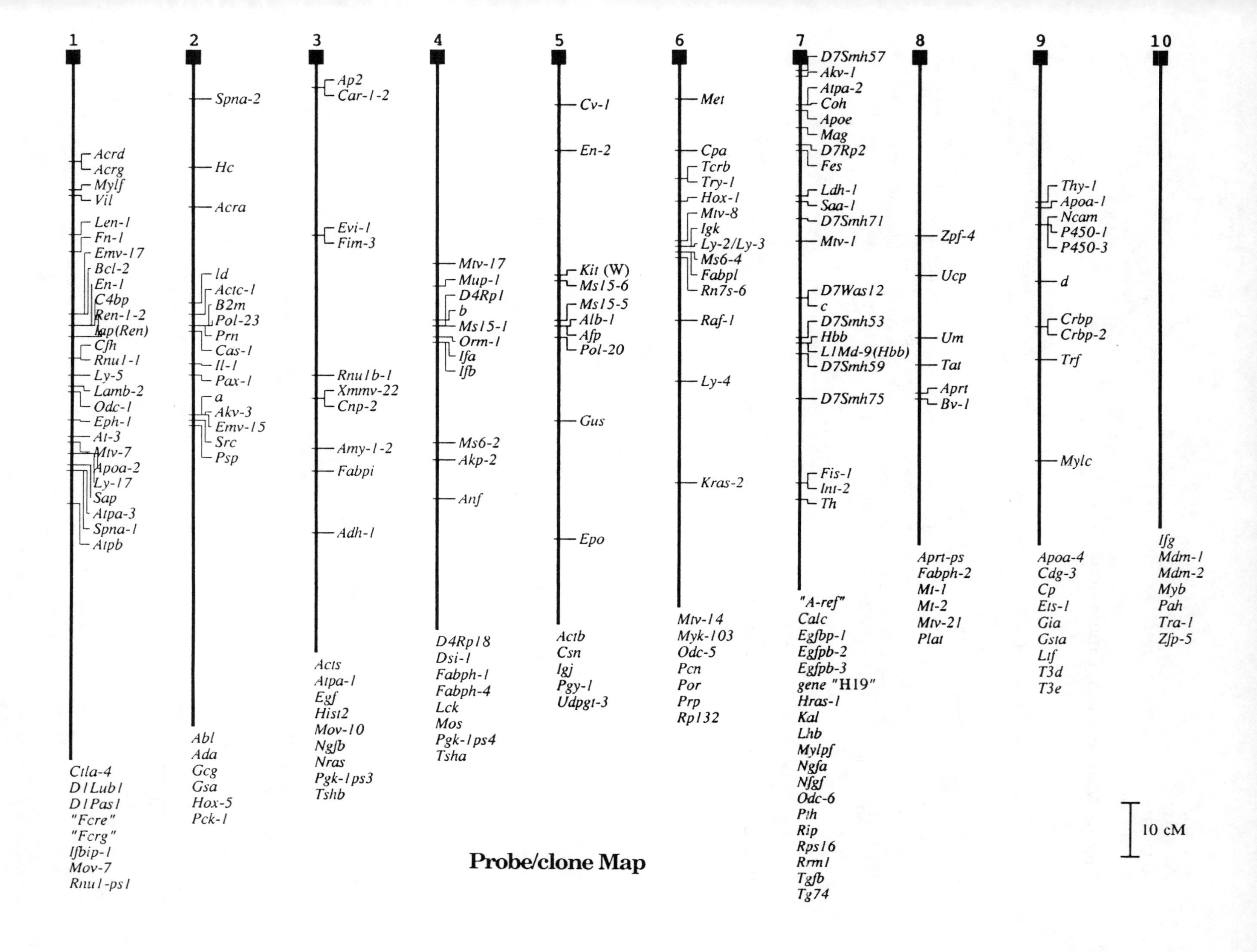

Probe/clone Map

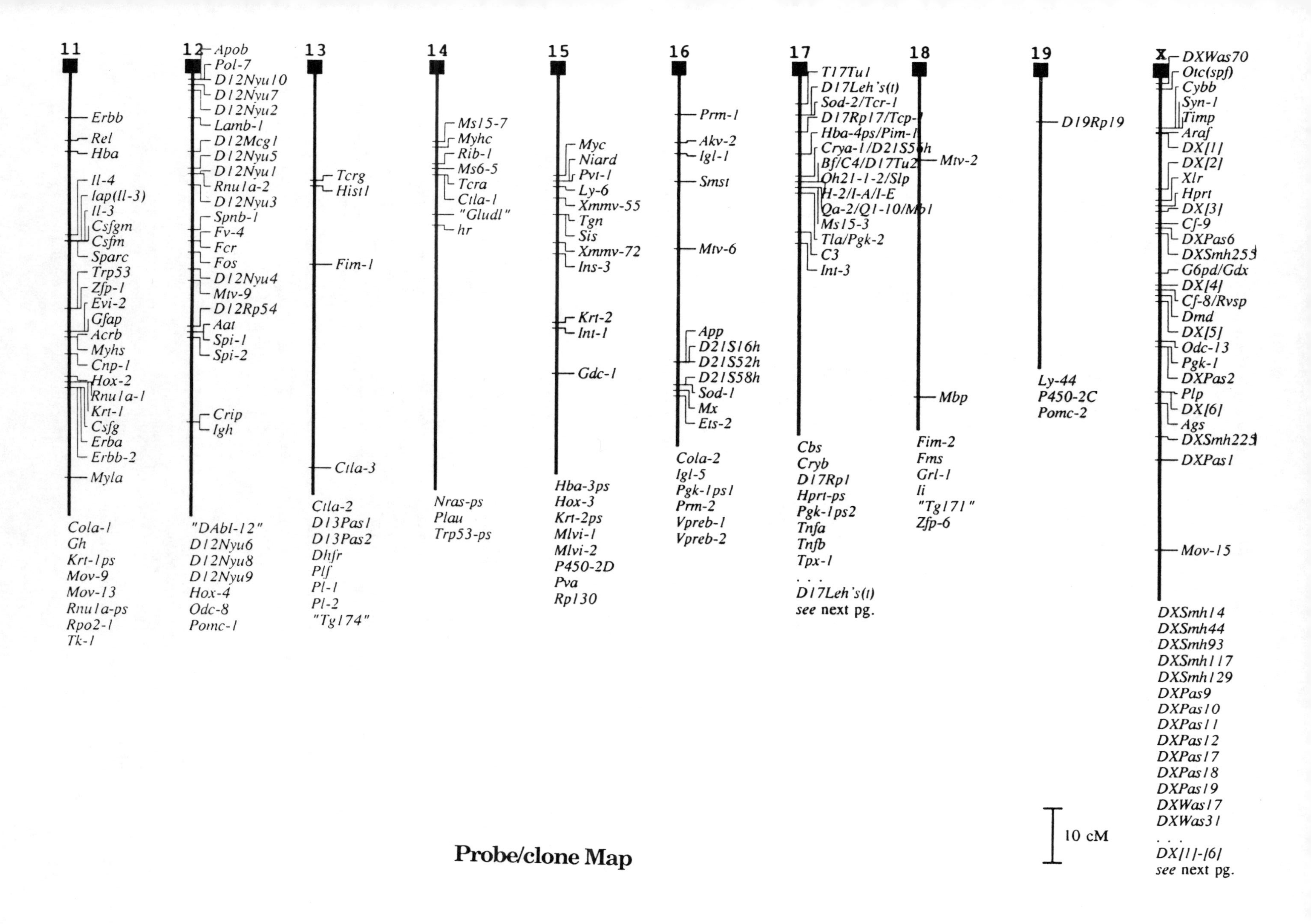

Probe/clone Map

List of Mouse DNA Clones and Probes

Chr	Symbol&	Gene Name	Probe@	RFLV#	Holder‡	Reference∞
1	Acrd	acetylcholine receptor δ	dAChr	+	5-Heidmann	Science 1986; 234:866
			p6H		*Crowder	PNAS 1986; 83:8405
			λ46, λ53, λ58, λ60, λ61		*LaPolla	PNAS 1984; 81:7970
1	Acrg	acetylcholine receptor γ	gAChr	+	5-Heidmann	Science 1986; 234:866
			γ18		*Buonanno	J Biol Chem 1986; 261:16451
			M169, M160		*Yu	Nuc Acid Res 1986; 14:3539
1	Apoa-2	apolipoprotein A-II	pSPA2-16, pSPA2-3,		*Kunisada	Nuc Acid Res 1986; 14:5729
			pSPA2-13, pSPA2-1		*Kunisada	Nuc Acid Res 1986; 14:5729
			pSPA2-15		*Kunisada	Nuc Acid Res 1986; 14:5729
			mouse cDNA	+	*fr:Lusis	ref: Seldin, Genomics 1988; 2:48
1	At-3	anti-thrombin-3	[H] pAt3		*Bock	Cytogenet Cell Genet 1985; 39:67
			"	+	*Seldin	J Exp Med 1988; 167:688
			"	+	*Seldin	Genomics 1988; 2:48
1	Atpa-3	Na, K-ATPase α-3	[R] rb13c	+	5-Herrera	ref: Kent, PNAS 1987; 84:5369
			"	+	*Kent	PNAS 1987; 84:5369
1	Atpb	Na, K-ATPase β	[R] rb19		*Mercer	Mol Cell Biol 1986; 6:3884
			"	+	*Kent	PNAS 1987; 84:5369
1	Bcl-2	B-cell leukemia/lymphoma-2	Ncm4		*Gurfinkel	Eur J Immunol 1987; 17:567
			"	+	*Mock	Cytogenet Cell Genet 1988; 47:11
			[H] pB4, pB16		*Tsujimoto	PNAS 1986; 83:5214
			λ114.2.1, λ114.5.1		*Negriniet	Cell 1987; 49:455
			λ114.11.1, λ114.11 2		*Negriniet	Cell 1987; 49:455
			λ113.15.5, λ114.11.3		*Negriniet	Cell 1987; 49:455
			pMBCL5.4, pMBCL3.3		*Negriniet	Cell 1987; 49:455
1	C4bp	complement component-4 binding protein	pMBP.15		*Kristensen	Biochemistry 1987; 26:4668
			"	+	*Seldin	J Exp Med 1988; 167:688
1	Cfh	complement component factor h	Mu23IV, pMH8	+	2-Tack	ref: D'Eustachio, J Immunol 1986; 137:3990
			Mu23IV		*Kristensen	PNAS 1986; 83:3963
			pMH8	+	*Seldin	Genomics 1988; 2:48
			"		*D'Eustachio	J Immunol 1986; 137:3990
1	Ctla-4	cytotoxic T lymphocyte associated protein-4	M17G7		*Brunet	Nature 1987; 328:267
			F41F4, F51G9		*Brunet	Nature 1987; 328:267
1	D1Lub1	DNA segment (homogeneous staining region)	λMmHSR1009	+	*Boldyreff	Cytogenet Cell Genet 1988; 47:84
			"		*Weith	EMBO J 1987; 6:1295
1	D1Pas1	DNA segment, testis specific transcript	PL10		*Leroy	Devel 1987; 101 (supp): 177
1	"D1?"	spleen gene	MScR1.0		1-Weis	J Immunol 1987; 138:3488
1	Emv-17	endogenous ecotropic MuLV-17	pPS1.25	+	1-Copeland	ref: Buchberg, Virology 1986; 60:1175
			"	+	"	ref: Moore, Genetics 1988; 119:933
1	En-1	engrailed-1 (formerly Mo-en.1)	pI4.U6	+	3-Hill	Cytogenet Cell Genet 1987; 44:171
			pλ4, pλ4U6		*Davidson	Development 1988; 104:305
			λMo-en.1		*Joyner	Cell 1985; 43:29
1	Eph-1	epoxide hydratase-1	[R] pEH-1		*Gonzales	J Biol Chem 1981; 256:4697
			"		*Simmons	J Biol Chem 1985; 260:515
1	"Fcre"	Fc receptor ε	[R] FcεRIα		*Kinet	Biochemistry 1987; 26:4605
			"	+	*Huppi	J Immunol 1988; 141:2807
1	"Fcrg"	Fc receptor γ	Fcγ2b/γ1 Rα		*Ravetch	Science 1986; 234:718
			"	+	*Huppi	J Immunol 1988; 141:2807
1	Fn-1	fibronectin-1	[H] unnamed		*Kornblihtt	PNAS 1983; 80:3218
			"	+	*Skow	Genomics 1987; 1:283
1	Iap	intercisternal A particle (Ren-assoc)	pDBRn3, MIARN		*Burt	Nuc Acid Res 1984; 12:8579
1	Ifbip-1	interferon β induced protein (gene 202)	clone 2, 3, 4, 5, p922		*Samanta	J Biol Chem 1986; 261:11849
1	Lamb-2	laminin-β2 subunit	pPE9	+	1-Elliott	In Vitro Cell Devel Biol 1985; 21:477
			p16, p57, p54, λ94, λ573		*Durkin	Biochemistry 1988; 27:5198
			λ281, λ34, λ6		*Durkin	Biochemistry 1988; 27:5198
1	Len-1 (or Cryg-1)	eye-lens protein-1 (γ-crystallin)	pMγ1, pMγ2		*Shinohara	PNAS 1982; 79:2783
			"		*Breitman	PNAS 1984; 81:7762
			"	+	*Quinlan	Genes & Dev 1987; 1:637
			pMγ3, pMγ4	+	*Shinohara,Breitman, Quinlan (see pMγ1, pMγ2 above)	
			γ2 probe		*Lok	Mol Cell Biol 1985; 5:2221
			"	+	*Quinlan	Genes & Dev 1987; 1:637
			λML1, γ6 clone		*Lok	Nuc Acid Res 1987; 12:4517
			"	+	*Quinlan	Genes & Dev 1987; 1:637
			pMg1CR1	+	*Skow	Biochem Genet 1988; 26:557
1	Ly-5	lymphocyte antigen-5	MT4-T200		*Raschke	PNAS 1987; 84:161
			pLy-5-68	+	*Shen	PNAS 1985; 82:7360
			"	+	*Seldin	J Exp Med 1988; 167:688
			"	+	*Seldin	Genomics 1988; 2:48
			pLy-5-B11, pLy-5-B15		*Saga	PNAS 1987; 84:5364
			pLy-T4		*Saga	PNAS 1987; 84:5364
			mLC-1		*Thomas	PNAS 1987; 84:5360
			probe A (from clone 70)	+	*Saga	Mol Cell Biol 1988; 8:4889
			probe B (from clone 1)	+	*Saga	Mol Cell Biol 1988; 8:4889
			clones: 1,2,6,17,27,34,35		*Saga	Mol Cell Biol 1988; 8:4889

Chr	Symbol&	Gene Name	Probe@	RFLV#	Holder‡	Reference∞
	Emv-15(3'unique)		p15.4	+	1-Copeland	ref: Siracusa, Genetics 1987; 117:85
2	Emv-15 flank		pRI		*Lovett	PNAS 1987; 84:2853
			"	+	*Lovett	Genetics 1987; 115:747
	Emv-15 (junct)		pRI pLTR		*Lovett	PNAS 1987; 84:2853
2	Gcg	glucagon	[SH] pshil		*Bell	Nature 1983; 302:716
			"		*Lalley	Cytogenet Cell Genet 1987; 44:92
2	Gsa	Gs protein, α chain	unnamed (several)		*Sullivan	PNAS 1986; 83:6687
			pGM13.1		*Ashley	J Biol Chem 1987; 262:15299
2	Hc	hemolytic complement (C5)	pC5	+	2-Tack	ref: D'Eustachio J, Immunol 1986; 137:3990
			"		*Wetsel	Complement 1985; 2:86
			pMC5.04		*Wetsel	Biochemistry 1987; 26:737
			"	+	*Birkenmeier	PNAS 1988; 85:8121
2	Hox-5	homeo box-5	unnamed genomic		*Featherstone	PNAS 1988; 85:4760
			probes a, b, c		*Featherstone	PNAS 1988; 85:4760
2	Il-1	interleukin-1	pIL1-1301		*Lomedico	Nature 1984; 312:458
			"	+	*fr:Lomedico	ref: Carlson, Mol Cell Biol 1988, 8:5528
			pIL1-31		*Lomedico	Nature 1984; 312:458
2	Il-1a	interleukin-1α	pMuIL-1α,	+	*Gray	J Immunol 1986; 137:3644
			"	+	*D'Eustachio	Immunogenetics 1987; 26:339
			cosmid α-1.3.2		*D'Eustachio	Immunogenetics 1987; 26:339
			cosmid α-RCF1		*D'Eustachio	Immunogenetics 1987; 26:339
2	Il-1b	interleukin-1β	pMuIL-1β	+	*Gray	J Immunol 1986; 137:3644
			"	+	*D'Eustachio	Immunogenetics 1987; 26:339
			unnamed		*Telford	Nuc Acid Res 1986; 14:9955
			cosmid b-7.1		*D'Eustachio	Immunogenet 1987; 26:339
2	ld	limb deformity (wild type)	probes A,B,C		*Woychik	Nature 1985; 318:36
2	Pax-1	paired box homeotic gene-1	unnamed		*Deutsch	Cell 1988; 53:617
			" (RFLV for *un* vs +)		*Balling	Cell 1988; 55:531
2	Pck-1	phosphoenolpyruvate caroxykinase-1	pPCK10		1-Lem	Somatic Cell Mol Genet 1985; 11:633
2	Pol-23	viral polymerase gene-23	B65.5PV2 derivative	+	*Rossomando	Immunogenetics 1986; 23:233
2	Prn-p	prion protein, structural	PRP-cDNA		5-Cheseboro	Nature 1985; 315:331
			PRP-cDNA subclone 13	+	*fr:Cheseboro	ref: Carlson, Mol Cell Biol 1988, 8:5528
			PRP-cDNA subclone 9	+	*fr:Cheseboro	ref: Carlson, Mol Cell Biol 1988, 8:5528
			[Ha] pHaPrP-cDNA-1	+	*fr:Oesch	ref: Carlson, Mol Cell Biol 1988, 8:5528
			"	+	*fr:Oesch	ref: Carlson, Cell 1986; 46:503
			"		*Oesch	Cell 1985; 40:735
			unnamed (several)	+	*Westaway	Cell 1987; 51:651
2	Psp	parotid secretory protein	pMPd16, pMpd39, pMPd12		*Madsen	Nuc Acid Res 1985; 13:1
2	Spna-2	α-spectrin, brain (fodrin)	[C] pUC8-13a		*Birkenmeier	PNAS 1985; 82:5671
			"	+	*Birkenmeier	PNAS 1988; 85:8121
2	Src	Rous sarcoma oncogene	v-sarc	+	1-Simon	ref: Blatt, Mol Cell Biol 1984; 4:978
			pSRA-2		*Delorbe	J Virol 1980; 36:50
			"		*fr:Delorbe	ref: Sakaguchi, PNAS 1984; 81:525
3	Acts	actin, skeletal α-actin	pmS1, pmS2, pmS4		*Leader	DNA 1986; 5:235
			pAM91	+	1-Buckingham	ref: Minty, J Mol Biol 1983; 167:77
			"		*Alonso	J Mol Evol 1986; 23:11
			[R] p749		*Katcoff	PNAS 1980; 77:960
			"		*Czosnek	EMBO J 1982; 1:1299
			[R] pAc15-2		*Nudel	PNAS 1982; 79:2763
			"		*Czosnek	EMBO J 1982; 1:1299
			pMαVSM-2		*Min	Nuc Acid Res 1988; 16:10374
3	Adh-1	alcohol dehydrogenase-1	pHJE-mYA1		*Zhang	Gene 1987; 57:27
			pHJE-mYA2	+	*Zhang	Gene 1987; 57:27
			pADHm9, pADHm16		*Ceci	Gene 1987; 59:171
			λADH22, λADH18	+	*Ceci	Gene 1987; 59:171
			pZK105-36,	+	*Edenberg	PNAS 1985; 82:2262
			pZK6-6, pZK7		*Edenberg	PNAS 1985; 82:2262
3	Amy-1	amylase-1, salivary	pMSa104	+	1-Paul,Elliott	Biochem Genet 1987; 25:569
			"	+	*Hagenbuchle	Cell 1980; 21:179
			"	+	*Blatt	Somatic Cell Mol Genet 1988; 14:133
3	Amy-1(p)	amylase-1, 5' flank promoter,noncoding	cSamD4		*Wiebauer	PNAS 1985; 82:5446
			"	+	*Meisler	Genetics 19867; 113:713
3	Amy-2	amylase-2, pancreatic	pMPa21	+	1-Paul,Elliott	Biochem Genet 1987; 25:569
			"		*Hagenbuchle	Cell 1980; 21:179
			"		1-Wellauer	ref: Schibler, J Mol Biol 1980; 142:93
3	Amy-2.1	amylase-2, pancreatic	cPamE1		*Meisler	ref: Osborn, Mol Cell Bio 1987; 7:326
3	Amy-2.2	amylase-2 pancreatic	cPam23, cPam24		*Meisler	ref: Osborn, Mol Cell Bio 1987; 7:326
			cADPb1		*Osborn	Mol Cell Biol 1987; 7:326
3	Ap2	adipocyte specific protein	paP2		*Hunt	PNAS 1986; 83:3786
3	Atpa-1	Na, K-ATPase α-1	[R] rb5	+	5-Levenson	ref: Kent, PNAS 1987; 84:5369
3	Car-1	carbonic anhydrase-1	pMCAI		*Fraser	J Mol Evol 1986; 23:294
3	Car-2	carbonic anhydrase-2	pMCAII	+	1-Venta	ref: Curtis, Gene 1983; 25:325
			"		*Fraser	J Mol Evol 1986; 23:294

Chr	Symbol&	Gene Name	Probe@	RFLV#	Holder‡	Reference∞
			p6-69		*Bishop	PNAS 1986; 83:5568
			"		*Curtis	J Biol Chem 1983 258:4459
3	Cnp-2	cyclic nucleotide phosphodiesterase-2	[R] pCNP-2	+	*Bernier	J Neurosci 1987; 7:2703
			"	+	*Bernier	J Neurosci Res 1988; 20:497
3	Egf	epidermal growth factor	pmegf10		*Zabel	PNAS 1985; 82:469
			"		*Scott	Science 1983; 221:236
			"	+	5-Rutter	ref: Kent, PNAS 1987; 84:5369
3	Evi-1	ecotropic viral integration site-1	pUC1.0M3	+	1-Copeland	ref: Mucenski, Mol Cell Biol 1988; 8:301
3	Evi-1(f)	5' to Evi-1	pUCO.5RP		*Mucenski	Mol Cell Biol 1988; 8:301
3	Evi-1(f)	3' to Evi-1	pUC1.OHP		*Mucenski	Mol Cell Biol 1988; 8:301
			"	+	*Sola	J Virol 1988; 62:3973
3	Fabpi	fatty acid binding protein intestinal	[R] unnamed	+	*Sweetser	J Biol Chem 1987; 262:16060
					*Alpers	PNAS 1984; 81:313
3	Fim-3	Friend MuLV integration site-3	206-PP20		4-Gisselbrecht	ref: Bordereaux, J Virol 1987; 61:4043
			"	+	*Sola	J Virol 1988; 62:3973
3	Hist2	histone gene complex-2 (H2a,H3)	MM614		*Graves	J Mol Biol 1985; 183:179
	Hist2	histone gene complex-2 (H3)	H3.2-614		*Taylor	J Mol Evol 1986; 23:242
3	Mov-10	Moloney leukemia virus-10	p10-2	+	*Munke	Cytogenet Cell Genet 1986; 43:140
			pMov-10		*Harbers	Nuc Acid Res 1982; 10:2521
3	Ngfb	nerve growth factor β	pmngf6	+	*Mucenski	Mol Cell Biol 1988; 8:301
			"		*Zabel	PNAS 1985; 82:469
			"		*Scott	Nature 1983; 302:538
3	Nras	neuroblastoma ras oncogene	λ3.2N-rasT		*Guerrero	Science 1984; 225:1041
			clones: 95a-4, 92b-4,		*Ryan	Nuc Acid Res 1984; 12:6063
			clones: 15a-2-4, 17b-4		*Ryan	Nuc Acid Res 1984; 12:6063
			probes: N3.8,N2.7,N8.5		*Ryan	Nuc Acid Res 1984; 12:6063
			unnamed		*Chang	Oncogene Res 1987; 1:129
3	Pgk-1ps3	phosphglycerate kinase-1 pseudogene3	B10 (pMPGK5-b)	+	*Adra	Somatic Cell Mol Genet 1988; 14:69
3	Rnulb-1	U1b small nuclear RNA	pUP136, pUP453,		*Michael	Somatic Cell Mol Genet 1986; 12:215
			pUP550		*Michael	Somatic Cell Mol Genet 1986; 12:215
			pUE236		*Michael	Somatic Cell Mol Genet 1986; 12:215
			pUP550, U1T-13	+	*Blatt	Somatic Cell Mol Genet 1988; 14:133
			mU1b3	+	*Lund	Somatic Cell Mol Genet 1988; 14:143
			pU1.1, pU1.2, pU1.3		*Marzluff	Nuc Acid Res 1983; 11:6255
			pU1b-453, pU1b-550		*Howard	Nuc Acid Res 1986; 14:9811
3	Tshb	thyrotropin stimulating hormone β subunit	pmTSH$_{\beta}$		*Pravtcheva	Somatic Cell Mol Genet 1986; 12:307
			"		*Chin	Endocrinol 1985; 116:873
			pTSH-β_H		*Gurr	PNAS 1983; 80:2122
			TSHβ1, TSHβ2,		*Gordon	DNA 1988; 7:17
3	Tshb	thyrotropin stimulating hormone β subunit	TSHβ3, TSHβ4		*Gordon	DNA 1988; 7:17
3	Xmmv-22	xenotropic-MCF leukemia virus-22	pEnv	+	*Blatt	Somatic Cell Mol Genet 1988; 14:133
			"		*Jenkins	Nature 1981; 293:370
4	Akp-2	alkaline phosphatase-2, liver	λmpA		*Terao	Somatic Cell Mol Genet 1988; 14:211
			λmpA, λmpB,		*Terao	PNAS 1987; 84:7051
			λmpC, λmpD		*Terao	PNAS 1987; 84:7051
4	Anf	atrial natriuretic factor	pCAR60	+	5-Mullins	Hypertension 1987; 9:518
			"	+	*fr:Mullins	ref:Huppi,Curr Top Micro Immun 1988;137:276
4	b	brown (tyrosinase-like)	pMT4, pMT5		*Shibahara	Nuc Acid Res 1986; 14:2413
			pMT4	+	*Jackson	PNAS 1988; 85:4392
4	b	brown (tyrosinase-like)	clone 5A		*Jackson	PNAS 1988; 85:4392
4	D4Rp1	DNA segment, Roswell Park-1	pMK1440, pMKA11		1-Berger	J Biol Chem 1984; 259:7941
			"	+	*Elliott	Mouse News Letter 1981; 64:87
			pODC1440	+	*fr:Berger	ref:Huppi,Curr Top Micro Immun 1988;137:276
4	D4Rp18	DNA segment, Roswell Park-18	pMK174	+	1-Mann	Genetics 1986; 114:993
4	Dsi-1	David Steffen integration-1	[R] dsi-HK, dsi-SR		*Vijaya	J Virol 1987; 61:1164
4	Fabph-1	fatty acid binding protein, heart-1	[R] H-FABP cDNA		*Heuckeroth	J Biol Chem 1987; 262:9709
4	Fabph-4	fatty acid binding protein, heart-4	[R] H-FABP cDNA		*Heuckeroth	J Biol Chem 1987; 262:9709
4	Ifa	interferon α	pBD2		*Daugherty	J Interfer Res 1984; 4:635
			λMIF-1,λMIF-2,λMIF-3,		*Kelley	Nuc Acid Res 1985; 13:805
			λMIF-4,λMIF-5,λMIF-6		*Kelley	Nuc Acid Res 1985; 13:805
			λMIF-7,λMIF-8,λMIF-9		*Kelley	Nuc Acid Res 1985; 13:805
		interferon α (α2, α6)	phage 19		*Zwarthoff	Nuc Acid Res 1985; 13:791
		interferon α (α5, α4, α1)	phage 10		*Zwarthoff	Nuc Acid Res 1985; 13:791
		interferon α (α1)	probe 10EF		*Zwarthoff	Nuc Acid Res 1985; 13:791
		interferon α (α_2)	pMuIFN-α_2		*Shaw	Nuc Acid Res 1983; 11:555
			"		5-Chang	Cytogenet Cell Genet 1986; 41:101
			pSV-19EE-M		*VanderKorput	J Gen Virol 1985; 66:493
			pMF1204		*Kelley	Gene 1983; 26:181
			"	+	*Nadeau	Genetics 1986; 104:1239
			"	+	*fr:Kelley	ref:Huppi,Curr Top Micro Immun 1988;137:276

Chr	Symbol&	Gene Name	Probe@	RFLV#	Holder‡	Reference∞
		interferon α (α4)	probe 10EC		*Zwarthoff	Nuc Acid Res 1985; 13:791
		interferon α (α5)	probe 10EE		*Zwarthoff	Nuc Acid Res 1985; 13:791
		interferon α (α6)	pSV10EF		*VanderKorput	J Gen Virol 1985; 66:493
		interferon α (αI9)	λMIF-2	+	*Seif	Gene 1986; 43:111
		interferon α (αI9-coding)	unnamed	+	1-DeMaeyer	J Hered 1987; 78:143
4	Ifa	interferon α (α10)	MuIFN-α10		*Trapman	J Gen Virol 1988; 69:67
4	Ifb	interferon β	pMβ3	+	*Nadeau	Genetics 1986; 104:1239
			pMβ3, IFN-β		3-Kawade	J Biol Chem 1983; 258:9522
			pMIF3/10, pMIF20/11		*Skup	Nuc Acid Res 1982; 10:3069
4	Lck	lymphocyte protein tyrosine kinase	NT18		*Marth	PNAS 1986; 83:7400
			"	+	*fr:Marth	ref:Huppi,Curr Top Micro Immun 1988;137:276
			T9, NT5, NT9		*Marth	Cell 1985; 43:393
			clones 1, 2, 42, 65		*Voronova	Nature 1986; 319:682
			clones 41.1, 22.2, 6.1, 13.1		*Garvin	Mol Cell Biol 1988; 8:3058
4	Mos	Moloney sarcoma oncogene	5'rc probe		*Canaani	PNAS 1983; 80:7118
			3'rcprobe		*Canaani	PNAS 1983; 80:7118
			MSH	+	2-VandeWoude	ref: Swan, J Virol 1982; 44:752
			"	+	*Propst	Genomics 1989
			pMS1,pMB7,pMB30,pHT13		*Blair	Science 1981; 212:941
			"		*McClements	Cold Spr Hbr Sympos 1981; 45:699
			"		*Blair	PNAS 1980; 77:3504
			"		*Osakarsson	Science 1980; 207:1222
			"		*Wood	PNAS 1984; 81:7817
			UMS		*Wood	PNAS 1984; 81:7817
4	Ms6-2	minisatellite 6-2	[H] probes 33.6, 33.15		*Jeffreys	Nuc Acid Res 1987; 15:2823
4	Ms15-1	minisatellite 15-1	[H] probes 33.6, 33.15		*Jeffreys	Nuc Acid Res 1987; 15:2823
4	Mtv-17	mammary tumor virus locus-17	Mtv17-3'C, -5'C		*Peters	J Virol 1986; 59:535
			pMMTV	+	*Callahan	J Virol 1984; 49:1005
			"	+	*fr:Callahan	ref:Huppi,Curr Top Micro Immun 1988;137:276
4	Mup-1	major urinary protein-1	p19		*Krauter	J Cell Biol 1982; 94:414
			p1057	+	1-Bennett	PNAS 1982; 79:1220
			BS-6		*Clark	EMBO J 1985; 4:3159
			BJ-31		*Hastie	ref: Held, Mol Cell Biol 1987; 7:3705
			p499,p199		*fr:Held	ref: Kuhn, Nuc Acid Res 1984; 12:6073
			"		*Sampsell	Genetics 1985; 109:549
			p499	+	*fr:Held	ref: Duncan, Mol Cell Biol 1988; 8:2705
			"	+	*fr:Held	ref:Huppi,Curr Top Micro Immun 1988;137:276
			BL-7,BL-14,BL-1,BL-15		*Bishop	EMBO J 1982; 1:615
			"		*Clissold	Gene 1981;15:225
			BS-5,BS-1,BS-2,BS-3,BS-4		*Bishop	EMBO J 1982; 1:615
			"		*Clissold	Gene 1981;15:225
			pBS6-2,pBS6-5,pBS6-1-1		*Clarke	EMBO J 1985; 4:3159
4	Mup	major urinary protein (Mup2)	BL6-25,BL6-51		*Held	Mol Cell Biol 1987; 7:3705
			pBS2-2,pBS3B-3		*Clarke	EMBO J 1985; 4:3159
		(Mup3)	BL6-3,BL6-11		*Held	Mol Cell Biol 1987; 7:3705
		(Mup4)	BL6-42		*Held	Mol Cell Biol 1987; 7:3705
4	Orm-1	orosomucoid-1	[R] p10-14		*Baumann	J Biol Chem 1983; 258:563
			pIRL-10	+	*Bauman	Mol Gen Genet 1985; 201:505
4	Pgk-1ps4	phosphoglycerate kinase-1 pseudogene 4	B15(probe pMPGK5-b)	+	*Adra	Somatic Cell Mol Genet 1988; 14:69
4	Tsha	thyrotopin stimulating hormone α subunit	pTSHα-1		*Chin	Endocrinology 1985; 78:5329
			"		*Naylor	Somatic Cell Genet 1983; 9:757
5	Actb	actin, cytoplasmic β-actin	pAL41	+	1-Minty	J Mol Biol 1983; 167:77
			"		*Alonso	J Mol Evol 1986; 23:11
			β28, pβ5'-Gem4		*Elder	Mol Cell Biol 1988; 8:480
			[R] pAc18-1		*Nudel	PNAS 1982; 79:2763
			"		*Czosnek	EMBO J 1982; 1:1299
5	Afp	alpha fetoprotein	unnamed		*Gorin	PNAS 1980; 77:1351
			unnamed		*Law	Gene 1980; 10:53
			MAF1,MAF2		*Bonner	Nuc Acid Res 1984; 12:6575
			HcH440	+	1-Tilghmann	ref: Belayew, Mol Cell Biol 1982; 2:1427
			λAFP14, λAFP15		*Ingram	PNAS 1981; 78:4694
			λAFP7, λAFP8, λAFP15		*Ingram	PNAS 1981; 78:4694
5	Alb-1	serum albumin	MSA1,MSA3		*Bonner	Nuc Acid Res 1984; 12:6575
			pMSA433	+	*Bellis	Mol Biol Evol 1987; 4:351
			λalb6, λalb5		*Ingram	PNAS 1981; 78:4694
5	Csn	cassein gene family (α,β,γ)	pCMα11, pCMβ13,		*Gupta	J Cell Biol 1982; 93:199
			pCMγ19, pCMδ40,		*Gupta	J Cell Biol 1982; 93:199
			pXMI-14		*Gupta	J Cell Biol 1982; 93:199
		(β-casein)	pA200		*fr:Henninghausen	ref: Schoenenberger, EMBO J 1988; 7:169

Chr	Symbol&	Gene Name	Probe@	RFLV#	Holder‡	Reference∞
5	Cv-1	BALB/c virus inducibility-1	unnamed(U3LTR -MuLV)		*Khan	Nuc Acid Res 1987; 15:7640
5	En-2	engrailed-2	clone En-2c		*Joyner	Genes & Dev 1987; 1:29
			pλ8, pλ8SR		*Davidson	Development 1988; 104:305
5	Epo	erythropoietin	clone12.a		*McDonald	Mol Cell Biol 1987; 7:365
			"		*McDonald	Mol Cell Biol 1986; 6:842
			λMG1a		*Shoemaker	Mol Cell Biol 1986; 6:849
		(rearranged)	clone12.c, clone 18.c		*McDonald	Mol Cell Biol 1987; 7:365
			"		*McDonald	Mol Cell Biol 1986; 6:842
5	Gus	β-glucoronidase	pGUS2		*fr:Granschow	ref: Bevilacqua, PNAS 1988; 85:831
			cosB1		*fr:Meisler	ref: Bevilacqua, PNAS 1988; 85:831
5	Gus-s	β-glucoronidase-structural	pGA-1	+	1-Paigen	ref: Watson, Gene 1985; 36:15
			PPv318		*Wang	J Biol Chem 1988; 263:15841
	Gus-s^{a}	β-glucoronidase-structural, a allele	pGUS7,pGUS5,		*Catterall	Biochem 1983; 22:6049
			"		*Watson	Ann NY Acad Sci 1984; 438:101
			"		*Funkenstein	Mol Cell Biol 1988; 8:1160
			pGUS48,pGUS22		*Funkenstein	Mol Cell Biol 1988; 8:1160
			pGUS142,pGUS121		*Funkenstein	Mol Cell Biol 1988; 8:1160
			GusA21,GusA12,GusA17		*Funkenstein	Mol Cell Biol 1988; 8:1160
5	Igj	immunoglobulin joining gene	unnamed		*Koshland	J Exp Med 1982; 155:647
			pJc3, pJc4, pJc5, pJc6,		*Mather	Cell 1981; 23:369
			pJc9, pJc10, pJc11		*Mather	Cell 1981; 23:369
			MJ-cDNA		*Cann	PNAS 1982; 72:6656
5	Kit (W)	kit oncogene	v-kit (feline)	+	*Chabot	Nature 1988; 335:88
					*Besmer	Nature 1986; 320:415
			pK3A	+	*Geissler	Cell 1988; 55:185
5	Ms15-5	minisatellite 15-5	[H] probes 33.6, 33.15		*Jeffreys	Nuc Acid Res 1987; 15:2823
5	Ms15-6	minisatellite 15-6	[H] probes 33.6, 33.15		*Jeffreys	Nuc Acid Res 1987; 15:2823
5	Pgy-1	P glycoprotein-1	[CH] unnamed		*VanderBliek	Mol Cell Biol 1986; 6:1671
			"		*Jakobsson	Anticanc Res 1988; 8:307
5	Pol-20	viral polymerase gene-20	B65.5PV2 derivative	+	*Rossomando	Immunogenetics 1986; 23:233
5	Udpgt-3	UDP glucuronosyltransferase-3	[R] pUDPGT$_{r}$3		*Krasnewich	Somatic Cell Mol Genet 1987; 13:179
			UDPGT$_{m-1}$, UDPGT$_{m-1a}$		*Kimura	Eur J Biochem 1987; 168:515
5	W	see Kit (above)				
6	Cpa	carboxypeptidase A	[R] clone 11-3	+	5-Rutter	ref: Bucan, EMBO J 1986; 5:2899
			"		*Quinto	PNAS 1982; 79:31
6	Fabpl	fatty acid binding protein, liver	[R] unnamed	+	*Sweetser	J Biol Chem 1987; 262:16060
6	Hox-1	homeo box-1	MH-3		*Rubin	Science 1986; 233:663
			pMo-10-1.4		*McGinnis	Cell 1984; 38:657
		5' to Hox-1	pG10-UPA		*Rabin	PNAS 1986; 83:9104
6	Hox-1.1	homeo box-1.1	cloneB21		*Kessel	PNAS 1987; 84:5306
			p577	+	*Bucan	EMBO J 1986; 5:2899
6	Hox-1.2	homeo box-1.2	λm6, λm20, λm1		*Colberg-Poley	Cell 1985; 43:39
			λm5, λm12		*Colberg-Poley	Cell 1985; 43:39
6	Hox-1.3	homeo box-1.3	λm5, λm2, λm12, λm1		*Colberg-Poley	Cell 1985; 43:39
6	Hox-1.4	homeo box-1.4	pHBT-1		*Wolgemuth	EMBO J 1986; 5:1229
			"	+	*Bucan	EMBO J 1986; 5:2899
			p2181-B2-4a		*Wolgemuth	PNAS 1987; 84:5813
6	Hox-1.4 & Hox-1-y	homeo box-1.4 & 1-y	cos2,cos3		*Duboule	EMBO J 1986; 5:1973
6	Hox-1.2,-1.3, -1.4,1-x	homeo box-1.2, 1.3, 1.4, 1-x	cos1		*Duboule	EMBO J 1986; 5:1973
6	Hox-1.5	homeo box-1.5	λMo-10		*McGinnis	Cell 1984; 38:675
			"	+	*Mock	Genetics 1987; 116:607
6	Hox-1.7	homeo box-1.7	MH-1G5, MH-1	+	*Rubin	Mol Cell Biol 1987; 7:3836
			MH-1G33b	+	*Rubin	Mol Cell Biol 1987; 7:3836
6	Igk-V	immunoglobulin κ variable chain (Vk1)	T105	+	*Gibson	ref: Moynet, J Immunol 1985; 135:727
			"	+	*D'Hoosterlaere	Curr Top Microbiol Immunol 1988;137:116
6	Igk-V	immunoglobulin κ variable chain (Vk4)	H76K10	+	*D'Hoostelaere	Immunogenetics 1986; 23:260
			"		*fr:Cory	ref: Gough, Biochem 1980; 19:2072
			"		*fr:Cory	ref: Cory, J Mol App Genet 1981 1:103
			"	+	*D'Hoosterlaere	Curr Top Microbiol Immunol 1988; 137:116
			"	+	*D'Hoosterlaere	Curr Top Microbiol Immunol 1988;137:116
6	Igk-V	immunoglobulin κ variable chain (Vk8)	M603K2	+	*D'Hoostelaere	Immunogenetics 1986; 23:260
			"		*fr:Cory	ref: Gough, Biochemistry 1980; 19:2072
			"		*Cory	J Mol App Genet 1981 1:103
6	Igk-V	immunoglobulin κ variable chain (Vk9)	MOPC41 5'FL		*fr:Qwen	ref: Stafford, Nature 1983; 306:77
			"	+	*D'Hoostelaere	Immunogenetics 1986; 23:260
			M41	+	*fr:Queen	ref: D'Hoosterlaere, Curr Top Microbiol Immunol 1988; 137:116
6	Igk-V	immunoglobulin κ variable chain (Vk10)	pC3386 (Vk10)	+	*fr:Shapiro	ref:D'Hoostelaere,Immunogen 1986;23:260
			"	+	*D'Hoosterlaere	Curr Top Microbiol Immunol 1988;137:116

Chr	Symbol&	Gene Name	Probe@	RFLV#	Holder‡	Reference∞
6	Igk-V	immunoglobulin κ variable chain (Vk11)	p6684^{K-}		*fr:Perry	ref: Kelley, Mol Cell Biol 1985; 5:1660
			"	+	*D'Hoostelaere	Immunogenetics 1986; 23:260
			"	+	*D'Hoosterlaere	Curr Top Microbiol Immunol 1988;137:116
6	Igk-V	immunoglobulin κ variable chain (Vk12)	pBC2-5(4.8)	+	*Lawler	ref: D'Hoosterlaere, Curr Top Microbiol Immunol 1988; 137:116
6	Igk-V	immunoglobulin κ variable chain (Vk19)	pK(11)^{24}-S	+	*Lawler	ref: D'Hoosterlaere, Curr Top Microbiol Immunol 1988; 137:116
6	Igk-V	immunoglobulin κ variable chain (Vk21)	B61K16	+	*D'Hoostelaere	Immunogenetics 1986; 23:260
			"		*fr:Cory	ref: Gough, Biochemistry 1980; 19:2072
			"		* Cory	J Mol App Genet 1981 1:103
			"	+	*D'Hoosterlaere	Curr Top Microbio Immunol 1988; 137:116
6	Igk-V	immunoglobulin κ variable chain (Vk22)	pVT15-S	+	*Lawler	ref: D'Hoosterlaere, Curr Top Microbiol Immunol 1988; 137:116
6	Igk-V	immunoglobulin κ variable chain (Vk23)	HyHEL10		*Lavoie	ref: Smith-Gill, J Mol Biol 1987; 194:713
			"	+	*D'Hoosterlaere	Curr Top Microbio Immunol 1988; 137:116
6	Igk-V	immunoglobulin κ variable chain (Vk24)	M167V_L		*fr:Storb	ref: Selsing, Cell 1981; 25:47
			"	+	*D'Hoostelaere	Immunogenetics 1986; 23:260
			"	+	*D'Hoosterlaere	Curr Top Microbiol Immunol 1988;137:116
6	Igk-V	immunoglobulin κ variable chain (Vk9-26)	pC9-26	+	*Shapiro	ref: D'Hoosterlaere, Curr Top Microbiol Immunol 1988; 137:116
6	Igk-V	immunoglobulin κ variable chain	EM3.C58-7, EM3.BALB-1		*Boyd	PNAS 1986; 83:9134
			EM3.C58-5, EM3.C58-10		*Boyd	PNAS 1986; 83:9134
			p9(35)		*Goldrick	J Exp Med 1985; 162:713
			Ch30.2	+	*Boyd	Immunogenetics 1986; 24:150
			pT1		4-Zachau	Nuc Acid Res 1980; 8:1693
			"	+	*Bucan	EMBO J 1986; 5:2899
			pECk	+	*D'Hoosterlaere	Immunogenetics 1985; 22:277
			"	+	*fr:Weigert	ref:D'Hoosterlaere, Curr Top Microbiol Immunol 1988; 137:116
6	Igk-C	immunoglobulin κ constant chain	pHBC$_κ$		*Lewis	Cell 1982; 30:807
			"		*Vasmel	Leukemia 1987; 1:155
6	Igk	immunoglobulin κ chain	$J_κ$1-$J_κ$5		*Muenski	Mol Cell Biol 1986; 6:4236
			"		*Dildrop	Eur J Immunol 1987; 17:731
			"		*Tonegawa	Nature 1983; 313:575
			[R] pKB/94		*fr:Schechter	ref: Breiner, Gene 1982; 18:165
			"		*Banergee	EMBO J 1985; 4:3183
			p$κ^o$		*Schuler	EMBO J 1988; 5:2139
6	Kras-2	Kirsten rat sarcoma oncogene-2	Ki-ras, KBE-2		1-Muller,Scolnick	ref: Ellis, Nature 1981; 292:506
			pKMSV	+	5-Robbins	J Virol 1986; 57:709
			pHiHi3	+	*Ryan	Nuc Acid Res 1986; 14:9222
			λMKN-142,λMKT-142		*Guerrero	Science 1984; 225:1159
			pY413		*Cahilly	Cytogenet Cell Genet 1985; 39:140
			[H]pSW-11-1		*Martin-DeLeon	Somatic Cell Mol Genet 1988; 14:205
6	Ly-2	lymphocyte antigen-2 (Ly-2.2)	λLy2.2		*Liaw	J Immunol 1986; 137:1037
			"		*Zamoyska	Cell 1985; 43:153
6	Ly-2	lymphocyte antigen-2 (Ly-2.1)	λLy2.1		*same as λLy-2.2	
			CAK411, CAK292		*Youn	Immunmogenet 1988; 28:345
6	Ly-2	lymphocyte antigen-2	λLy-2-B, λLy-2-D, λLy-2-F		*Gorman	J Immunol 1988; 140:3646
6	Ly-3	lymphocyte antigen-3 (Ly-3.2)	λLY3C-23, λLY3C-24		*Nakauchi	PNAS 1987; 84:4210
6	Ly-3	lymphocyte antigen-3 (Ly-3.1)	CAK812, CAK361, CAK302		*Youn	Immunogenetics 1988; 28:353
6	Ly-3	lymphocyte antigen-3	cosLy-3-17, λcLy-3-3		*Gorman	J Immunol 1988; 140:3646
			λcLy-3-5, λcLy-3-6	+	*Gorman	J Immunol 1988; 140:3646
			[R] pX9.5		*Johnson	Nature 1986; 323:74
6	Ly-4	L3T4 T cell differentiation antigen	p3C, L3T4.25		*Littman	Nature 1987; 325:453
6	Met	met protooncogene	pMet-1		2-Dean	Mol Cell Biol 1987; 7:921
			"	+	2-Dean	Genomics 1987; 1:167
			[H] pMet G PstI	+	*fr: Dean	ref: Sweet, Nuc Acid Res 1988; 16:8745
6	Ms6-4	minisatellite 6-4	[H] probes 33.6, 33.15		*Jeffreys	Nuc Acid Res 1987; 15:2823
6	Mtv-8	mammary tumor virus locus-8	pUII3P	+	5-Robbins	J Virol 1986; 57:709
			Mtv8-5'C, Mtv8-3'C		*Peters	J Virol 1986; 59:535
6	Mtv-14	mammary tumor virus locus-14	[V] MMTV probe	+	*Traina	J Virol 1981; 40:735
6	"Myk-103"	wildtype sequences (unnamed)	λ35	+	*Wilkie	Mol Cell Biol 1987; 7:1646
6	Odc-5	ornithine decarboxylase related seq-5	pMK217	+	1-Elliott,Berger	ref: Richards-Smith, MNL 1984; 71:46
6	Pcn	cytochrome P-450, pregnenolone carvonitrile inducible	[R] pP-450PCN-10		*Hardwick	J Biol Chem 1983; 258:10182
			"		*Simmons	J Biol Chem 1985; 260:515
6	Por	NADPH-cytochrome P-450 oxidoreductase	[R] pOR-7		*Gonzales	J Biol Chem 1982; 257:5962
			"		*Simmons	J Biol Chem 1985; 260:515
6	Prp	proline-rich protein	M14, M56, MP2		*Ann	J Biol Chem 1988; 263:10887
			MC16, MC22		*Ann	J Biol Chem 1988; 263:10887

Chr	Symbol&	Gene Name	Probe@	RFLV#	Holder‡	Reference∞
6	Raf-1	murine sarcoma 3611 oncogene-1	p171	+	5-Robbins	J Virol 1986; 57:709
			v-raf		1-Muller	PNAS 1983; 80:4218
			3611-MSV	+	*Bucan	EMBO J 1986; 5:2899
			"		*Kozak	J Virol 1984; 49:297
			"		*Rapp	PNAS 1983; 80:4218
6	Rn7s-6	7s RNA-6	pA6	+	*Taylor	Immunogenetics 1985; 22:471
6	Rpl32	ribosomal protein L32	p3AH3.4		*Dudov	Cell 1984; 37:457
			L32-I3		*Wiedemann	Somatic Cell Mol Genet 1987;13:77
6	"RS"	recombining sequence for κ	M315-rs1		*Siminovitch	Nuc Acid Res 1987; 15:2699
6	Tcrb	T cell receptor β chain	p4.1	+	*Bucan	EMBO J 1986; 5:2899
			"		*Snodgrass	Nature 1985; 313:592
			pCb2		*Schuler	EMBO J 1988; 5:2139
			p86T5	+	*Epstein	J Exp Med 1985; 161:1219
			"		*Vasmel	Leukemia 1987; 1:155
			pUCJ1 (Jβ1)		*Mucenski	Mol Cell Biol 1986; 6:4236
			"		*Kronenberg	Nature 1985; 313:647
			"		*Mucenski	J Virol 1988; 62:839
6	Tcrb	T cell receptor β chain	pUCJ2A (Jβ2)		*Mucenski	Mol Cell Biol 1986; 6:4236
			"		*Kronenberg	Nature 1985; 313:647
			"		*Mucenski	J Virol 1988; 62:839
			J15		*fr: Snodgrass	ref: Hedrick, Nature 1984; 308:153
			"		*Vasmel	Leukemia 1987;1:155
			pHDS11, pHDS4/203		*Saito	Nature 1984; 307:757
			"		*Saito	Nature 1984; 312:36
			cosmid D1A		*Lai	PNAS 1987; 84:3846
			pDOβ2		*Kotzin	Science 1985; 229:167
			"		*Born	PNAS 1985; 82:2925
			Dβ1		*Siu	Nature 1984; 311:344
			"		*Lindsten	PNAS 1987; 84:7639
			$V_{\beta}8$ probe		*Barth	Nature 1985; 316:517
			"		*Sim	Cell 1985; 42:89
			VβE1(Vβ2), VβC5(Vβ8.1)		*Patten	Nature 1984; 312:40
			"		*Lindsten	PNAS 1987; 84:7639
			VβLB2(Vβ6)		*Lindsten	PNAS 1987; 84:7639
			VβSJL73(Vβ15)		*Behlke	PNAS 1986; 83:767
			"		ref:Lindsten	PNAS 1987; 84:7639
			Vβ86T1(Vβ1)		*Hedrick	Nature 1984; 308:153
			"		*Lindsten	PNAS 1987; 84:7639
			"	+	*D'Hoostelaere	Immunogenetics 1985; 22:277
			Vβ2B4(Vβ3)		*Chien	Nature 1984; 309:322
			"		*Lindsten	PNAS 1987; 84:7639
			Vβ14(VβJ6.19)		*Malissen	Nature 1986; 319:28
			"		*Lindsten	PNAS 1987; 84:7639
			vB11,vB9,vB3,vB8,vB5		*Behlke	Science 1985; 229:566
			"		*Bougueleret	Immunogenetics 1987; 26:304
			vB6,vB2,mvB9,vB4,vB1		*Behlke	Science 1985; 229:566
			"		*Bougueleret	Immunogenetics 1987; 26:304
			vB17		*Kappler	Cell 1987; 49:263
			"		*Bougueleret	Immunogenetics 1987; 26:304
			TB3,TB2.1,TB12		*fr:Barth	ref:Bougueleret,Immunogenet 1987;26:304
			TB2,TB23,AR5		*fr:Barth	ref:Bougueleret,Immunogenet 1987;26:304
			2B4#7		*Chien	Nature 1984; 309:322
			"		*Bougueleret	Immunogenetics 1987; 26:304
6	Tcrb	T cell receptor β chain	C5,LB2,E1		*Patten	Nature 1984; 312:40
			"		*Bougueleret	Immunogenetics 1987; 26:304
6	Tcrb-V	T cell receptor β -variable	cosmids A5,B8,B9,B16,B22		*Chou	Science 1987; 238:545
			cosmids B29,B33,B39,B41,B47		*Chou	Science 1987; 238:545
			cosmids C1,C3,C10,C16,C42		*Chou	Science 1987; 238:545
			cosmids C47,C55,D5,D6,D7,D9		*Chou	Science 1987; 238:545
			cosmids D10,D13,D15,D18		*Chou	Science 1987; 238:545
			cosmids E19,E21,E40,F10,G6		*Chou	Science 1987; 238:545
6	Try-1	trypsin-1	[R] clone 4-79		*Craik	J Biol Chem 1984; 259:14255
			"	+	*Bucan	EMBO J 1986; 5:2899
			pMPt9		*Stevenson	Nuc Acid Res 1986; 14:8307
7	"A-ref"	unnamed	unnamed		*Huleihel	Mol Cell Biol 1986; 6:2655
7	Akv-1	AKR leukemia virus inducer-1	[V] pEnv	+	1-Jenkins	J Virol 1982; 43:26
7	Apoe	apolipoprotein E	p2C1	+	*Lusis	J Biol Chem 1987; 262:7594
			"		*Rajavashisth	PNAS 1985; 82:8085
7	Atpa-2	Na, K ATPase α-2	[R] rb-2	+	*Kent	PNAS 1987; 84:5369
7	c	albino (see tyrosinase below)				
7	Calc	calcitonin	[R] CT-89		*Jacobs	Science 1981; 213:457
			"		*Lalley	Cytogenet Cell Genet 1987; 44:92

Chr	Symbol&	Gene Name	Probe@	RFLV#	Holder‡	Reference∞
7	Coh	coumarin hydroxylase activity	[R] p450b-5	+	*Gonzalez	J Biol Chem 1982; 257:5962
			"	+	*Simmons	J Biol Chem 1983; 258:9585
			[R]R17 (P450-e)		fr:Adesnik	Noshiro, Arch Bioch Biophs 1986; 244:857
			pf26		*Noshiro	Biochemistry 1988; 27:6434
7	D7Rp2	DNA segment, Roswell Park-2	pMK908		*Berger	J Biol Chem 1981; 256:7006
			"	+	*Elliot	PNAS 1983; 80:501
			pMAK-1		*Snider	J Biol Chem 1985; 260:9884
			pMAK-1E	+	*King	Mol Cell Biol 1986; 6:209
			pMAK-5H, pMAK-4	+	*King	Mol Cell Biol 1986; 6:209
7	D7Smh57	DNA segment, St. Mary's Hospital-57	D7Smh57A	+	1-Brown	ref: Greenfield, Genomics 1987; 1:153
7	D7Smh71	DNA segment, St. Mary's Hospital-71	D7Smh71	+	1-Brown	ref: Greenfield, Genomics 1987; 1:153
7	D7Smh53	DNA segment, St. Mary's Hospital-53	D7Smh53	+	1-Brown	ref: Greenfield, Genomics 1987; 1:153
7	D7Smh59	DNA segment, St. Mary's Hospital-59	D7Smh59	+	1-Brown	ref: Greenfield, Genomics 1987; 1:153
7	D7Smh75	DNA segment, St. Mary's Hospital-75	D7Smh75	+	1-Brown	ref: Greenfield, Genomics 1987; 1:153
7	D7Was12	DNA segment, Univ. Wash.-12	clone 12A	+	*Disteche	Somatic Cell Mol Genet 1984; 10:211
7	Egfbp	epidermal growth factor binding protein	MB1-73, MB2-20A		*Blaber	Biochemistry 1987; 26:6742
7	Egfbp-1	EGF binding protein (type A)	mGK-22		*Evans	J Biol Chem 1987; 262:8027
			"		*Drinkwater	Biochemistry 1987; 26:6750
7	Egfbp-2	EGF binding protein (type B)	mGK-13		*Evans	J Biol Chem 1987; 262:8027
			"		*Drinkwater	Biochemistry 1987; 26:6750
7	Egfbp-3	EGF binding protein (type C)	mGK-9		*Evans	J Biol Chem 1987; 262:8027
			"		*Drinkwater	Biochemistry 1987; 26:6750
7	Fes	feline sarcoma oncogene	pN26	+	5-Blatt	Mol Cell Biol 1984; 4:978
			"		*Franchini	Mol Cell Biol 1982; 2:1014
			pC.1, λF18.1, λβ3.2		*Wilks	Oncogene 1988; 3:289
7	Fis-1	Friend virus integrations site-1	p1.8	+	1-Silver	J Virol 1986; 60:1156
			"		1-Silver	J Virol 1986; 57:526
7	"H19"	H19, fetal liver	H19	+	1-Pachnis	PNAS 1984; 81:5523
			λH19.5,λH19.6,λH19.10		*Pachnis	EMBO J 1988; 7:673
			λH19.12,λH19.13		*Pachnis	EMBO J 1988; 7:673
7	Hbb	hemoglobin β chain	pCR1βMG9	+	1-Edgell,Rougeon	ref: Rougeon, Gene 1977; 1:229
			λgtWES.MβG1		*Tilghman	PNAS 1977; 74:4406
			λgtWES.MβG2		*Tiemier	Cell 1978; 14:237
			"		*Tilghman	PNAS 1977; 74:4406
			λgtWES.MβG3		*Tiemier	Cell 1978; 14:237
			pMBd2, pMBJ		*fr:Leder	ref: Tilghman, PNAS 1977; 74:4406
			"		*Lo	Differentiation 1987; 35:37
7	Hbb	hemoglobin β chain (M. caroli)	λ83-4, λ83-2, λ83-12, λ83-10		*Casavant	Mol Cell Biol 1988; 8:4669
			λ83-25, λ83-5, λ83-8, λ83-30,		*Casavant	Mol Cell Biol 1988; 8:4669
			λ83-11, λ83-22, λ83-13		*Casavant	Mol Cell Biol 1988; 8:4669
7	Hbb-d	hemoglobin β chain (diffuse allele)	pHE100		*Holdener-Kenny	PNAS 1986; 83:4374
7	Hbb-s	hemoglobin β chain (single allele)	pβs11R3		*Todokoro	BBRC 1986; 135:1112
7	$Hbb\text{-}b1^{s}$	hemoglobin β chain (single, chain 1)	BA2,BA4	+	*Weaver	Cell 1981; 24:403
7	Hbb-b2	hemoglobin β chain (d-minor)	CA4		*Jahn	Cell 1980; 21:159
7	$Hbb\text{-}b2^{s}$	hemoglobin β chain (single, chain 2)	BA1,BA3	+	*Weaver	Cell 1981; 24:403
7	$Hbb^{th\text{-}1}$	hemoglobin β chain deleted (DA-I)	plaque5156		*Goldberg	J Biol Chem 1986; 261:12368
7	Hbb	hemoglobin β chain (β major)	pCR1β		*Rougeon	Gene 1977; 1:229
7	Hbb-y1	hemoglobin Y (β-like embryonic)	pYS1E/B		*Todokoro	BBRC 1986; 135:1112
			CE17, CE18, CE14		*Jahn	Cell 1980; 21:159
7	Hbb-y, -bh0, -bh1	hemoglobin Y + pseudogenes	BA11		*Weaver	Cell 1981; 24:403
			"		*Holdener-Kenny	PNAS 1986; 83:4374
7	Hbb-bh1, -bh2, -bh3	hemoglobin β pseudogenes	BA12		*Weaver	Cell 1981; 24:403
			"		*Holdener-Kenny	PNAS 1986; 83:4374
7	Hbb(bh1+bh3)	hemoglobin β pseudogenes	CA11		*Jahn	Cell 1980; 21:159
7	Hbb-bh3	hemoglobin β pseudo-bh3	bh3	+	*Bellis	Mol Biol Evol 1987; 44:351
7	Hras-1	Harvey rat sarcoma virus oncogene	clone BS9		*Ellis	J Virol 1980; 36:408
			"		*Sakaguchi	PNAS 1984; 81:525
7	Int-2	mammary tumor integration site-2	Int-2a	+	1-Silver	ref: Moore, J Virol 1986; 60:1156
			int-2c, int-2f		*Casey	Mol Cell Biol 1986; 6:502
			"		*Dickson	Cell 1984; 37:529
			int-2f, int-2g		*Wilkinson	EMBO J 1988; 7:691
		Int-2 flank	probe39A		*Pathak	JNCI 1987; 78:327
7	Kal	kallikrein gene family	mGK-1		*Mason	Nature 1983; 303:300
			mGK-5, mGK-6		*Drinkwater	Nuc Acid Res 1987; 15:10052
			pMF-1, pMF-2		*Fahnestock	Nuc Acid Res 1986; 14:4823
			pMK-1		*Rickards	J Biol Chem 1982; 257:2758
		kallikrein (EGF-BP)	MB1-73, MB2-20A		*Blaber	Biochemistry 1987; 26:6742
		kallikrein (EGF-BP type C)	mGK-9		*Drinkwater	Biochemistry 1987; 26:6750
		kallekrein (EGF-BP type A)	mGK-22		*Drinkwater	Biochemistry 1987; 26:6750
		kallekrein (EGF-BP type B)	mGK-13		*Drinkwater	Biochemistry 1987; 26:6750
7	Ldh-1	lactate dehydrogenase-A chain	λm15		*Li	Eur J Biochem 1985; 149:215

Chr	Symbol&	Gene Name	Probe@	RFLV#	Holder‡	Reference∞
7	Lhb	luteinizing hormone β	unnamed		*fr:Tepper	ref: Kourides, PNAS 1984; 81:517
			[H] DGa		*Naylor	Somatic Cell Mol Genet 1983; 8:757
7	"L1Md-9"	L1 repeat (in Hbb)	unnamed		*Burton	Nuc Acid Res 1985; 13:5071
			"		*Shyman	Nuc Acid Res 1985; 13:5085
7	Mag	Myelin-associated glycoprotein	[R] unnamed	+	*D'Eustachio	J Neurochem 1988; 50:589
			[R] pMAG1.2		*Barton	Genomics 1987; 1:107
			"		*Arquint	PNAS 1987; 84:600
7	Mtv-1	mammary tumor virus locus-1	[V] MMTV(C3H) probe	+	*MacInnes	Virology 1984; 132:12
			"	+	*Traina	J Virol 1981; 41:735
7	Mylpf	myosin light chain, fast skeletal muscle, phosphorylatable	[R] p103		*Katcoff	PNAS 1980 77:960
			"		*Czosnek	EMBO J 1982; 1:1299
7	Ngfa	nerve growth factor α	λMSP-19, λMSP-26		*Evans	EMBO J 1985; 4:133
7	Ngfa,Ngfg	nerve growth factor α, γ	λMSP-22, λMSP-28		*Evans	EMBO J 1985; 4:133
			λMSP-51, λMSP-53		*Evans	EMBO J 1985; 4:133
7	Ngfg	nerve growth factor γ	pSM676	+	1-Gross	ref: Howles, Nuc Acid Res 1984; 12:2791
7	Odc-6	ornithine decarboxylase related seq-6	pMK3048	+	1-Elliott	ref: Richards-Smith, MNL 1984; 71:46
7	Pth	parathyroid hormone	[R] rPTHs-1		*Heinrich	J Biol Chem 1984; 259:3320
			"		*Lalley	Cytogenet Cell Genet 1987; 44:92
7	"Rip"	Repression of LP-450 16α	pf3, pf46	+	*Noshiro	Biochemistry 1988; 27:6434
7	Rps16	ribosomal protein S16	S16		*Wagner	Mol Cell Biol 1985; 5:2560
			S16-I2		*Wiedemann	Somatic Cell Mol Genet 1987;13:777
7	Rrm1	ribonucleotide reductase M1	probe 1		*Thelander	Ann Rev Biochem 1979; 48:133
			"		*Brissenden	Exp Cell Res 1988; 174:302
			probe2		*Lammers	Struct Bonding (Berl) 1983; 54:27
			"		*Brissenden	Exp Cell Res 1988; 174:302
			p247, p201,Mλ-1		*Caras	J Biol Chem 1985; 260:7015
		ribonucleotide reductase M1 mutant	unnamed cDNA		*Caras	Mol Cell Biol 1988; 8:2698
7	Saa-1	serum amyloid A-1	pRS48	+	*Taylor	Mol Gen Genet 1984; 195:491
			"		*Morrow	PNAS 1981; 78:4718
			pmSAP3		*Ishikawa	Nuc Acid Res 1987; 15:7186
7	Saa-1, -2, -3, -ps	serum amyloid A-1,A-2,A-3, pseudo	unnamed		*Lowell	J Biol Chem 1986; 261:8442
7	Tgfb	transforming growth factor β	unnamed		*Fujii	Somatic Cell Genet 1986; 12:281
			"		*Derynck	J Biol Chem 1986; 261:4377
7	"Tg74"	transgene (human insulin)	[H] 11 kb frag insulin	+	*Michalova	Hum Genet 1988; 80:247
7	Th	tyrosine hydroxylase	[R] pHR3.0	+	fr:Chikaraishi	ref: Brilliant, J Neurogenet 1987; 4:259
7	Tyr (c)	tyrosinase	PTY-1		*Kwon	PNAS 1987; 84:7473
			[H] Pmel34,Pmel40,Pmel16		*Kwon	PNAS 1987; 84:7473
			Tyrs-32, Tyrs-33		*Yamamoto	Jpn J Genet 1987; 62:271
			Tyrs-J2, Tyrs-J19		*Yamamoto	Jpn J Genet 1987; 62:271
			pmcTyr1	+	*Ruppert	EMBO J 1988; 7:2715
			pmcTyr54, pmcTyr63		*Ruppert	EMBO J 1988; 7:2715
			pmcTyr10, pmcTyr40		*Ruppert	EMBO J 1988; 7:2715
			pmcTyr20, pmcTyr24		*Ruppert	EMBO J 1988; 7:2715
			pmcTyr3, pmcTyr18		*Ruppert	EMBO J 1988; 7:2715
			pmcTyr2, pmcTyr4		*Ruppert	EMBO J 1988; 7:2715
8	Aprt	adenine phosphoribosyl transferase	pSAM-1		*Sikela	Gene 1983; 22:219
			"		*Farber	Cytogenet Cell Genet 1986;42:198
			pSAM3.1		*Sikela	Gene 1983; 22:219
			"		*Dush	PNAS 1985; 82:2731
8	Aprt-ps	adenine phosphoribosyl transferase, pseudo	p15		*Dush	J Cell Biol 1983; 97:134a
			"		*Farber	Cytogenet Cell Genet 1986;42:198
			"	+	*Dush	Mol Cell Biol 1986; 6:4161
8	Bv-1	C57BL/10 endogenous ecotropic virus	[V] pEnv	+	1-Jenkins	J Virol 1981; 43:26
8	Fabph-2	fatty acid binding protein, heart-2	[R] H-FABP cDNA		*Heuckeroth	J [illegible] Chem 19[illegible]; 262:9709
8	Mt-1	metallothionein-1	mMT-1		*fr:Palmiter	ref: Cox, Hum Genet 1984; 64:61
			unnamed		*Durnam	PNAS 1980; 77:6611
			LmtI, M135		*Brzezinski	Cytobios 1987; 52:33
			pBMTH-1		*fr: Mallon	ref: Glanville, Nature 1981; 292:267
8	Mt-2	metallothionein-2	unnamed		*Searle	Mol Cell Biol 1984; 4:1221
8	Mtv-21	mammary tumor virus locus-21	Mtv21-3'C, Mtv21-5'C		*Peters	J Virol 1986; 59:535
8	Plat	plasminogen activator, tissue	pUC9-A33, pSP65-MT_1		*Huarte	Genes & Devel 1987; 1:1201
			pSP64-MT_2, pSP64-MT_3		*Huarte	Genes & Devel 1987; 1:1201
			pEE2		*fr:Rickles	ref: Rajput, Somatic Cell Mol Genet 1987; 13:581
8	"Spleen-ps"	pseudo-gene, spleen	MScR1.0		1-Weis	J Immunol 1987; 138:3488
8	Tat	tyrosine aminotransferase	pmTAT-SH3.7	+	5-Schutz	ref: Muller, J Mol Biol 1985; 184:367
			λmTAT-1, pmTAT-PE0.6		*Muller	J Mol Biol 1985; 184:367
			pmTAT-EA0.65		"	J Mol Biol 1985; 184:367
			[R] pcTAT-3		*Scherer	PNAS 1982; 79:7205
			[R] prTAT-EE1.05		*Hashimoto	PNAS 1984; 81:6637
8	Ucp	mitochondrial uncoupling protien	pCIN-1	+	1-Jacobson	J Biol Chem 1985; 260:16250

Chr	Symbol&	Gene Name	Probe@	RFLV#	Holder‡	Reference∞
8	Um	uvomorulin	clone F5		*Ringwald	EMBO J 1987; 6:3647
			F5H3	+	5-Kemler	ref: Eistetter, PNAS 1988; 85:3489
			F20		*Kemler	PNAS 83 1986; 83:1364
8	Zfp-4	zinc finger protein-4	pmFP2	+	2-Willison	1st Inter Worksh on Mouse Gene Map 1987
			E7		*Willison	Mouse News Letter 1987; 79:84
9	Apoa-1	apolipoprotein A-I	pMI4		*Birkenmeier	PNAS 1986; 83;2516
			AI clone 1804		*Miller	PNAS 1983; 80:1511
			"		*Meisler	Genetics 1986; 113:713
9	Apoa-4	apolipoprotein A-IV	pBAH-1		2-Kinniburgh	Mol Cell Biol 1986; 6:3807
			pApoA-IV FL-3		2-Kinniburgh	Mol Cell Biol 1986; 6:3807
			"		"	Nuc Acid Res 1985; 13:1953
			unnamed		*Williams	Mol Cell Biol 1986; 6:3807
9	Cdg-3	unnamed	pB10.AT3γ-1		*Krissansen	Immunogenetics 1987; 26:258
			"		*Krissansen	J Immunol 1987; 138:3513
9	Cp	ceroplasmin	[R] Rcp-1-3, [R] pRCo02		*Baranov	Chromosoma 1987; 96:60
9	Crbp,Crbp-2	cellular retinol binding protein-1 & -2	[R] pCRP,PCRPII9	+	*Demmer	J Biol Chem 1987; 262:2458
9	d	dilute	p0.3		*Copeland	Cell 1983; 33:379
			"	+	*Rinchik	Genetics 1986; 112:321
			"	+	*Moore	Genetics 1988; 119:933
			p.Emv-3	+	*Copeland	J Virol 1984; 49:437
			[V] pEnv	+	*Jenkins	J Virol 1981; 43:26
9	d	dilute	λDSE-RI(D)		*Rinchik	Genetics 1986; 112:321
			pR1.3, p0.5, p0.7		*Rinchik	Genetics 1986; 112:321
			p94.1	+	*Rinchik	Genetics 1986; 112:321
9	Ets-1	E26 avian leuemia oncogene-1	[H] pRD700		*Watson	Anticanc Res 1986; 6:631
			[V] E1.28		*Watson	Anticanc Res 1986; 6:631
9	Gia	Gi protein, α chain	unnamed (several)		*Sullivan	PNAS 1986; 83:6687
			pGM1.3		*Ashley	J Biol Chem 1987; 262:15299
9	Gsta	glutathione-S-transferase Ya subunit	[R] plasmid A24		*Daniel	Arch Biochem Biophys 1983; 227:261
			"		*Czosnek	Nuc Acid Res 1984; 12:4825
			λmYa1		*Daniel	DNA 1987; 6:317
			λmL1, λmL3,λmL7,λmL8		*Czosnek	Nuc Acid Res 1984; 12:4825
9	Ltf	lactotransferrin	pT267		*Teng	Biochem J 1986; 240:413
			"		*Teng	Somatic Cell Mol Genet 1987; 13:689
9	Mylc	myosin light chain, alkali, cardiac ventricle	pA29	+	2-Buckingham	ref: Robert, Nature 1985; 314:181
			pGLC450	+	1-Robert	Nature 1985; 314:181
9	Ncam	neural cell adhesion molecule	pEC501	+	2-D'Eustachio	PNAS 1985; 82:7631
			λmN1,pEC502		*D'Eustachio	PNAS 1985; 82:7631
			A14		*Barthels	Nuc Acid Res 1988; 16:4217
			pM1.3		*Goridis	EMBO J 1985; 4:631
			clone NM, clone HB4		*Santoni	Nuc Acid Res 1987; 15:8621
			clones: 4.1, 3.1, B22		*Santoni	Nuc Acid Res 1987; 15:8621
			clones: Bg3, Bg2		*Santoni	Nuc Acid Res 1987; 15:8621
			DW1, DW2, DW42,		*Santoni	Nuc Acid Res 1987; 15:8621
			DW60, DW61		*Santoni	Nuc Acid Res 1987; 15:8621
		Ncam-120	DW3,M9,5.7c		*Barthels	EMBO J 1987; 6:907
9	P450-1	cytochrome P1-450, polycyclic hydrocarbon inducible	pP1-450FL	+	1-Simmons	J Biol Chem 1983; 258:9585
			unnamed	+	*Hildebrand	BBRC 1985; 130:396
			pP1450-57		*Tukey	PNAS 1984; 81:3163
			unnamed		*Kimura	EMBO J 1987; 6:1929
9	P450-1 (mutant)		unnamed(c1)		*Kimura	EMBO J 1987; 6:1929
			unnamed(c37)		*Kimura	EMBO J 1987; 6:1929
9	P450-3	cytochrome P3-450 polycyclic hydrocarbon inducible	pP3-450FL		1-Simmons	J Biol Chem 1983; 258:9585
			"	+	*Prochazka	Science 1987; 237:286
			"	+	1-Nebert	ref: Hildebrand, BBRC 1985; 130:396
			"		*Kimura	Nuc Acid Res 1984; 12:2917
			pP3450-21		*Tukey	PNAS 1984; 81:3163
9	Thy-1	thymus cell antigen	TM-8		*Davis	Nature 1983; 301:80
			"		*Hiraki	J Immunol 1986; 136:4291
			[R] pT64, pT86		*Seki	Science 1985; 227:649
			unnamed (2 cDNA, 2 genomic)		*Seki	Science 1985; 227:649
			pcD-29-5A		*Hiraki	J Immunol 1986; 136:4291
9	Thy-1.2	thymus cell antigen (allele 1.2)	pcT108	+	*Chang	PNAS 1985; 82:3819
			pBST1		*Giguere	EMBO J 1985; 4:2017
			pThy-2.2		*Chen	Cell 1987; 51:7
			"		*Evans	PNAS 1984; 81:5532
			"		*Ingraham	Mol Cell Biol 1986; 6:2923
9	Trf	transferrin	[R] pRtf17		*Baranov	Chromosoma 1987; 96:60
9	T3d	T3-δ chain	pPEM-T3ϕ, pPEM-T3δ		*VandenElsen	PNAS 1985; 82:2920
9	T3e	T3ε- chain	[H] pDJ1		*Gold	PNAS 1987; 84:1664
			"		*Gold	Nature 1986; 321:431
10	Ifg	interferon γ	unnamed IFN-γ cDNA		*Gray	PNAS 1983; 80:5842
			"		*Naylor	Somatic Cell Mol Genet 1984; 10:531

Chr	Symbol&	Gene Name	Probe@	RFLV#	Holder‡	Reference∞
10	Mdm-1	transformed mouse 3T3 cell double minute-1	λc101,λc102		*Cahilly-Snyder	Somatic Cell Mol Genet 1987; 13:235
			λc103,λc104,λc105,p12		*Snyder	J Biol Chem 1988; 263: 17150
10	Mdm-2	transformed mouse 3T3 cell double minute-2	λc201,λc202		*Snyder	J Biol Chem 1988; 263: 17150
10	Myb	meyeloblastosis oncogene	pG4M2b	+	1-Mock, Huppi	ref: Mock, Nuc Acid Res 1987; 15:4700
			λMM1, λgt34-1		*Lavu	Nuc Acid Res 1986; 14:5309
			unnamed		*Shen-Ong	Science 1984; 226:1077
			"		*Mucenski	J Virol 1988; 62:839
			[V] pVM2		*Klempnauer	Cell 1982; 31:453
			"		*Sakaguchi	PNAS 1984; 81:525
			[V] KX1		*Lavu	Nuc Acid Res 1986; 14:5309
10	Pah	phenylalanine hydroxylase	[R] PAH		4-Dahl, Mercer	J Biol Chem 1986; 261:4148
			"	+	*Bode	Genetics 1988; 118:299
			moPAH8		*Ledley	Cytogenet Cell Genet 1988; 47:125
10	Tra-1	tumor rejection antigen (gp96)	pC519, pMA2,		*Srivastava	PNAS 1987; 84:3807
			λMA53, λMA72		*Srivastava	PNAS 1987; 84:3807
			unnamed		* Srivastava	Immunol Today 1988; 9:78
10	Zpf-5	zinc finger protein-5	AC16, pEX2.8		*Chavrier	EMBO J 1988; 7:29
11	Acrb	actylcholine receptor β	bAchR	+	5-Heidmann	Science 1986; 234:866
			β7,β22		*Buonanno	J Biol Chem 1986; 261:16451
11	Cnp-1	cyclic nucleotide phosphodiesterase-1	[R] pCNP-2	+	*Bernier	J Neurosci 1987; 7:2703
			"	+	*Bernier	J Neurosci Res 1988; 20:497
11	Hba	hemoglobin α chain complex	unnamed		*Nishioka	Cell 1979; 18:875
			pCR1aMG	+	*fr:Rougeon	ref: Robert, Gene 1977; 1:229
			α-ad1/2		*Leder	Nature 1981; 293:196
11	Hba-x	hemoglobin X	α-emX		*Leder	Nature 1981; 293:196
11	Hox-2.1	homeo box-2.1	λMo-4,pMo4-1		*Hart	Cell 1985; 43:9
			λMu1		*Hauser	Cell 1985; 43:19
			H24.1		*Jackson	Nature 1985; 317:745
			"		*Munke	Cytogenet Cell Genet 1986; 42:236
11	Hox-2.1	homeo box-2.1	pRRF5		*Joyner	Nature 1985; 314:173
			"		*Hauser	Cell 1985; 43:19
			"	+	*Mock	Genetics 1987; 116:607
			"	+	*Mock	Nuc Acid Res 1987; 15:2397
			RRF12	+	*Lonai	DNA 1987; 6:409
			probe 1, probe 2		*Krumlauf	Development 1987; 99:603
11	Hox-2.2	homeo box 2.2	pMo4-6	+	*Hart	Genetics 1988; 118:319
			"		*Hart	Cell 1985; 43:9
			p6	+	*Lonai	DNA 1987; 6:409
			c5		*Schughart	PNAS 1988; 85:5582
11	Hox-2.2/2.3	homeo box 2.2, 2.3	Ch22, p4	+	*Lonai	DNA 1987; 6:409
11	Hox-2.3	homeo box-2.3	λMo-1		*Hart	Cell 1985; 43:9
			L13,L2,L23,pR0.8,pR1.2		*Meijlink	Nuc Acid Res 1987; 15:6773
11	Hox-2.4	homeo box-2.4	λMo-3		*Hart	Cell 1985; 43:9
11	Hox-2.6	homeo box-2.6	Ch19	+	*Lonai	DNA 1987; 6:409
11	Iap	Intercisternal A particle, Il-3 associated	IAP-IL3		*Ymer	Nuc Acid Res 1986; 14:5091
			λclones: 3.2, 3.4		*Ymer	Nature 1985; 317:255
			λclones: 3.5, 4.9		*Ymer	Nature 1985; 317:255
11	Il-3	interleukin-3	pIL3	+	*Ihle	J Immunol 1987; 138:3051
			pMu2A1		*Barkiw	EMBO J 1987; 6:617
			pILM-3		*Fung	Nature 1984; 307:233
			fpGV-IL12	+	*fr:Fung	ref: Buchberg, Oncogene Res 1988; 2:149
			pILM11		*Campbell	Eur J Biochem 1985; 150:297
			λclones: 4.1, 4.10		*Ymer	Nature 1985; 317:255
			λclones: 4.8, 3.3		*Ymer	Nature 1985; 317:255
			unnamed		*Ymer	Nuc Acid Res 1986; 14:5901
			λMGM3-2, λMGM5-6,		*Miyatake	PNAS 1985; 82:316
			λMGM9-7, λMGM12-4		*Miyatake	PNAS 1985; 82:316
11	Il-3? (TCGF)		clones: 5G, B4, B5		*Yokota	PNAS 1984; 81:1070
			clones: B5, B6, B8, B9		*Yokota	PNAS 1984; 81:1070
11	Cola-1	procollagen type I, α 1	Mov-13		*Harbers	PNAS 1984; 81:1504
			"		*Jaenish	Cell 1983; 32:209
			[H] Hf677		*Chu	Nuc Acid Res 1982; 10:5925
			"		*Schnieke	Nature 1983; 304:315
			[H] pg1H-1		*fr:D.Rowe	ref: Schnieke, Nature 1983; 304:315
11	Csfg	colony stimulating factor, granulocyte	λMGC-3, -5, -10, -11		*Tschiya	Eur J Biochem 1987; 165:7
			pMGCBG		*Tschiya	Eur J Biochem 1987; 165:7
			pMG2		*Tsuchiya	PNAS 1986; 83:7633
			"	+	*Buchberg	Oncogene Res 1988; 2:149

Chr	Symbol&	Gene Name	Probe@	RFLV#	Holder‡	Reference∞
1	Ly-5	lymphocyte antigen-5	clones: 42,45,47,49,70		*Saga	Mol Cell Biol 1988; 8:4889
1	Ly-17	lymphocyte antigen-17 (FcR)	β-2, pFcR13,		*Lewis	Nature 1986; 324:372
			pFcRAb, pFcR11		*Lewis	Nature 1986; 324:372
			FcRα, J774		*Ravetch	Science 1986; 234:718
			"	+	*Seldin	J Exp Med 1988; 167:688
			"	+	*Seldin	Genomics 1988; 2:48
1	Mov-7	Moloney leukemia virus-7	pMov-7		*Chumakov	J Virol 1982; 42:1088
1	Mov-7(f)	Mov-7 flank	p7-1-1		*Munke	Cytogenet Cell Genet 1986; 43:140
1	Mtv-7	mammary tumor virus locus-7	[V]MMTV(C3H) probe	+	*MacInnes	Virology 1984; 132:12
			"	+	*Traina	J Virol 1981; 40:735
1	Mylf	myosin light chain, alkali, fast skeletal	pGLC450	+	1-Robert	Nature 1985; 314:181
1	Odc-1	ornithine decarboxylase-related seq-1	pMK934	+	1-Elliott	Mouse News Letter 1985; 71:46
			pODC934		1-Berger	J Biol Chem 1984; 259:7941
1	Ren-1	renin-1, structural	M5, M9, M14, M16		*Chirgwin	Somatic Cell Mol Genet 1984; 10:633
			r49probe		*Chirgwin	Somatic Cell Mol Genet 1984; 10:633
1	Ren-1,-2	renin-1,-2	pSM479	+	1-Gross	ref: Piccini, Cell 1982;30:205
			"	+	*Dickinson	Genetics 1984; 108:651
			pRn1-4	+	1-Panthier	EMBO J 1982; 1:1417
			pRn1-3	+	1-Rougeon	ref: Panthier, Nature 1982; 298:90
			pRn5.3,pRn4.7		*Panthier	Nature 1982; 298:90
			Id-2		*Field	Hypertension 1984; 6:597
			"	+	*Seldin	J Exp Med 1988; 167:688
			"	+	*Seldin	Genomics 1988; 2:48
1	Ren-2	renin-2	pDD-1D2	+	1-Gross	PNAS 1984; 81:5489
			"		*Dickinson	Genetics 1984; 108:651
1	Rnu1-1	U1 small nuclear RNA-1	pSP64/-U1		*Lund	Science 1985; 229:1271
1	Rnu1-ps1	U1 small nuclear RNA-1 pseudogene	unnamed	+	*Blatt	Somatic Cell Mol Genet 1988; 14:133
1	Sap	serum amyloid P-component	pmSAP3	+	1-Maruyama	ref: Ishikawa, Nuc Acid Res 1987; 15:7186
1	Spna-1	α-spectrin-1	pMαSp1		*Huebner	PNAS 1985; 82:3790
			"		*Curtis	Gene 1985; 36:357
			"	+	*fr:Curtis	ref: Huppi, 1988, J Immunol 141:2807
			"	+	*fr: Huebner	ref: Seldin, Genomics 1988; 2:48
			pB129		*Cioe	PNAS 1985; 82:1367
			"	+	*Birkenmeier	PNAS 1988; 85:8121
			"	+	*Seldin	J Exp Med 1988; 167:688
			"	+	*Seldin	Cytogenet Cell Genet 1987; 45:52
1	Vil	villin	[H] unnamed Vil cDNA		*Pringault	EMBO J 1986; 5:3119
			"	+	*Rousseau-Merck	Hum Genet 1988; 78:130
2	a	agouti (see Emv-15)	pRI	+	*Lovett	Genetics 1987; 115:747
			"	+	*Carlson	Mol Cell Biol 1989
2	Abl	Abelson murine leukemia oncogene	v-abl		*Srinivasan	PNAS 1981 78:2077
			"		*Salinas	Eur J Biochem, 1986; 160:149
			pAbl clone3		*Goff	Cell 1980 23:777
			"		*Frankel	PNAS 1985; 82:6600
			TS c-abl (testis specific)		*Meijer	EMBO J 1987; 6:4041
			λabl1, λabl2, λabl3		*Wang	Cell 1984; 36:349
			λabl4, λabl5		*Wang	Cell 1984; 36:349
2	Acra	acetylcholine receptor α	aAChR	+	5-Heidmann	Science 1986; 234:866
			αprobe		*Isenberg	Nuc Acid Res 1986; 14:5111
2	Actc-1	actin, cardiac α	pAF81		*Alonso	J Mol Evol 1986; 23:11
			"		*Minty	Cell 1983; 30:185
			"	+	1-Buckingham	J Mol Biol 1983; 167:77
			λgA8,λIG10	+	*Garner	EMBO J 1986; 5:2559
			pmC1		*Leader	Biosci Rpt 1986; 6:741
			[R] pC18		*Czosnek	EMBO J 1983; 2:1977
			"	+	*Crosby	Genomics 1989; 5: xx (in press)
2	Ada	adenosine deaminase	cGAM4.5, cGAM2.4		*Ingolia	Mol Cell Biol 1986; 6:4458
			pADA5-29		1-Kellems	J Biol Chem 1985;260:10299
			pADA5-43, pADA5-39		*Yeung	J Biol Chem 1985;260:10299
			pADA5-23, pADA5-12		*Yeung	J Biol Chem 1985;260:10299
2	Akv-3	AKR leukemia virus inducer-3	[V] pEnv	+	1-Jenkins	J Virol 1981; 43:26
			p.Emv-13		*Copeland	J Virol 1984; 49:437
			"		*Copeland	Cell 1983; 33:379
2	B2m	β-2 microglobulin	pβ2-m2		1-Daniel,Kourilsky	EMBO J 1983; 2:1061
			probe 929, Ch4β2-C57		*Parnes	Cell 1982; 29:661
			frag B2m clone	+	*fr:Parnes	ref: Carlson, Mol Cell Biol 1988, 8:5528
		β2m(flank,exonI)	probe A		*Parnes	EMBO J 1986; 5:103
		β2m(exon2,intron)	probe B		*Parnes	EMBO J 1986; 5:103
		β2m(intronI)	probe C		*Parnes	EMBO J 1986; 5:103
		β2m(mutant clone)	λB2bp3', λB2bp5'		*Parnes	EMBO J 1986; 5:103
2	Cas-1	catalase-1	pmCAT-34		1-Shaffer	J Biol Chem 1987; 262:12908
2	Emv-15	endogenous ecotropic MuLV-15	λ129-AY	+	*Siracusa	Genetics 1987; 117:85
			pP0.5		*Lovett	PNAS 1987; 84:2853
			"	+	*Lovett	Genetics 1987; 115:747

Chr	Symbol&	Gene Name	Probe@	RFLV#	Holder‡	Reference∞
11	Csfgm	colony stimulating factor, granulocyte macrophage specific	pGM'ϕ17		*Barlow	EMBO J 1987; 6:617
			unnamed		*fr:Arai(DNAX)	ref: Gough, Nature 1984; 309:763
			"		*Ihle	J Immunol 1987; 138:3051
			pE1-11		*Miyatake	EMBO J 1985; 4:2561
			"	+	*Buchberg	Oncogene Res 1988; 2:149
			pGM37, pGM38		*Gough	Nature 1984; 309:763
			pGH5'19		*Gough	Nature 1984; 309:763
			"	+	*Sola	J Virol 1988; 62:3973
11	Csfm	colony stimulating factor, macrophage	unnamed		*Rajavashisth	PNAS 1987; 84:1157
			unnamed		*DeLamarter	Nuc Acid Res 1987; 15:2389
			unnamed		*Baumbach	J Virol 1988; 62:3151
			pUS-1		*Baumbach	Mol Cell Biol 1987; 7:664
			"		*Buchberg	Oncogene Res 1988; 2:149
			cCSF-10		*fr:Kawasaki	ref: Baumbach, J Virol 1988; 62:3151
			[H] cCSF-17		*Kawasaki	Science 1985; 230:291
			[H] 3ACSF-69		*Wong	Science 1987; 235:1504
11	Erba	avain erythroblastosis oncogene A	unnamed (Oncor)	+	*Buchberg	Oncogene Res 1988; 2:149
11	Erbb	avain erythroblastosis oncogene B	unnamed (Oncor)	+	*Buchberg	Oncogene Res 1988; 2:149
			"	+	*Silver	Mol Cell Biol 1985; 5:1784
			[H] HER64-3		*Ullrich	Nature 1984; 309:418
			"		*Munke	J Mol Evol 1987; 25:134
11	Erbb-2	avain erythroblastosis oncogene B-2	neuc(t)/sp6400	+	*fr:Barbacid	ref: Buchberg, Oncogene Res 1988; 2:149
11	Evi-2	ecotropic viral integration site-2	p597.1	+	1-Copeland	ref: Buchberg, Oncogene Res 1988; 2:149
11	Gfap	glial fibrillary acidic protein	pG18, pG13		*Bulcarek	Nuc Acid Res 1985; 13:5527
			probe G1		*Lewis	PNAS 1984; 81:2743
			"	+	*Bernier	J Neurosci Res 1988; 20:497
11	Gh	growth hormone	mGH		*Jackson-Grusby	Endocrinology 1988; 122:2462
11	Il-4	interleukin-4	unnamed		*Yokota	PNAS 1986; 83:5894
			clone2AE3		*Lee	PNAS 1986; 83:2061
			λMIL9-3, λMIL16-1		*Otsuka	Nuc Acid Res 1987; 15:333
			pCB1		*Brown	Cell 1987; 50:809
			pCB1 derivative	+	*D'Eustachio	J Immunol 1988; 141:3067
			clone 2A-E3		*Lee	PNAS 1986; 83:2061
			pSP6KmIL4-293		*Noma	Nature 1986; 319:640
			pSP6KmIL4-374		*Noma	Nature 1986; 319:640
11	Krt-1	keratin gene complex-1	pK3.68(b2)	+	*Meruelo	Immunogenet 1987; 25:361
			p3-68, p4-2		5-Compton	unpublished, Jackson Lab
			λEB1, pEB1		*Ichinose	Gene 1988; 70:85
			clones 2,3,10,14		*Ichinose	Gene 1988; 70:85
			Endo B cDNA		*Singer	J Biol Chem 19896; 261:538
		keratin, Endo B-β1	SK3112		*Oshima	Genes & Devel 1988; 2:505
		keratin, Endo B-β2	EB31,EB25,EBp11		*Oshima	Genes & Devel 1988; 2:505
		keratin, Endo B-β3	EB1		*Oshima	Genes & Devel 1988; 2:505
		keratin, Endo B-β5	EBp3		*Oshima	Genes & Devel 1988; 2:505
11	"Krt-1ps"	keratin pseudogene	clone 4		*Oshima	Genes & Devel 1988; 2:505
11	Mov-9	Moloney leukemia virus-9	p9-1-2		*Munke	Cytogenet Cell Genet 1986; 43:140
			pMov-9		*Chumakov	J Virol 1982; 42:1088
11	Mov-13	Moloney leukemia virus-13	p13		*Munke	Cytogenet Cell Genet 1986; 43:140
			unnamed		*Harbers	PNAS 1984 81;1504
			"		*Jaenish	Cell 1983; 32:209
11	Myla	myosin light chain, alkali, cardiac atrial	pGLC260	+	*Lonai	DNA 1987; 6:409
			MLC1A	+	2-Barton	J Muscle Res Cell Motl 1985; 6:461
			"	+	*Robert	Nature 1985; 314:181
11	Myhs-e	myosin heavy chain, skeletal muscle, embryonic	pMHC2.2		*Weydert	PNAS 1985; 82:7183
			"		*Weydert	Cell 1987; 49:121
11	Myhs-p	myosin heavychain, skeletal muscle, perinatal	pMHC16.2A		*Weydert	PNAS 1985; 82:7183
			"		*Weydert	Cell 1987; 49:121
11	Myhs-f	myosin heavy chain, skeletal muscle, adult fast	pMHC32		*Weydert	J Biol Chem 1983; 258:13867
			"		*Weydert	Cell 1987; 49:121
			[R] MHC15		*Nudel	Nuc Acid Res 1980; 8:2133
			"		*Czosnek	EMBO J 1982; 1:1299
11	Myhs	myosin heavy chain, skeletal	[R] p82		*Nudel	Nuc Acid Res 1980; 8:2133
			"		*Czosnek	EMBO J 1982; 1:1299
11	Rel	reticuloendotheliosis oncogene	unnamed	+	*fr:Chen	ref: Buchberg, Oncogene Res 1988; 2:149
			[H] pPHHS rel-1		*Brownell	Mol Cell Biol 1985; 5:2826
11	Rnu1a-1	U1 small nuclear RNA-1a	pUE236	+	5-Michael	Somatic Cell Mol Genet 1986; 12:215
			pU1a-236, pU1a-261		*Howard	Nuc Acid Res 1986; 14:9811
			pU1a-214		*Howard	Nuc Acid Res 1986; 14:9811
11	Rnu1a-ps	U1 small nuclear RNA-1a psuedo	pUP325		*Michael	Somatic Cell Mol Genet 1986; 12:215
			pU1-ps325		*Howard	Nuc Acid Res 1986; 14:9811

Chr	Symbol&	Gene Name	Probe@	RFLV#	Holder‡	Reference∞
11	Rpo2-1	RNA polyerase II-1	unnamed		*Pravtcheva	Somatic Cell Mol Genet 1986; 12:523
			BE2.9, pH19-4, pH22-1		*Bartolomei	Mol Cell Biol 1987; 7:586
			pE26-7, pE26-4, λGA1		*Bartolomei	Mol Cell Biol 1987; 7:586
			pEXRPII3'		*Bartolomei	Mol Cell Biol 1987; 7:586
			pHRPII, pEE14.3		*Bartolomei	Mol Cell Biol 1987; 7:586
11	Sparc	secreted acidic cysteine rich glycoprotein	pPE220	+	1-Mason	EMBO J 1986; 5:1831
			"	+	*Mason	EMBO J 1986; 5:1465
			pF9.33, pF9.52		*Mason	EMBO J 1986; 5:1465
			pF9.54, pPE.30		*Mason	EMBO J 1986; 5:1465
11	Tk-1	thymidine kinase-1	pMtk4,pMtk9		*Lin	Mol Cell Biol 1985; 5:3149
			pMtk116, pMtk322		*Lin	Mol Cell Biol 1985; 5:3149
			pMtk535, pMtk536		*Lin	Mol Cell Biol 1985; 5:3149
11	Trp53	transformation-related protein 53	Trp53-1	+	*fr:Benchimol	ref: Buchberg, Oncogene Res 1988; 2:149
			pp53-203,-208		*Oren	PNAS 1983; 80:56
			pp53-422,-271		*Oren	EMBO J 1983; 2:1633
			pp53-176,pCH53-16		*Zabut-Houri	Nature 1983; 306:594
			clone 27.1a		*Jenkins	Nuc Acid Res 1984; 12:5609
			Ch53-7		*Bienz	EMBO J 1984; 3:2179
11	Trp53 (rearranged)		λ53-22		*Rovinski	Mol Cell Biol 1987; 7:847
11	Zfp-1	zinc finger protein-1	pmFP1, E10	+	2-Willison	ref:Ashworth Mouse News Lett 1987;79:84
			"	+	5-Willison	1st Inter Worksh on Mouse Gene Map 1987
12	Aat	α-1-antitrypsin	pC1.2	+	1-D'Eustachio	J Exp Med 1984; 160:827
			λA,λB,λC,λD		*Krauter	DNA 1986; 5:29
			λH,λM,λN		*Krauter	DNA 1986; 5:29
			pliv3,pliv3b		*Krauter	DNA 1986; 5:29
			p1796		*Hill	Mol Cell Biol 1985; 5:2114
			"	+	*Birkenmeier	PNAS 1988; 85:8121
			pG3.5	+	*Blank	Genetics 1988; 120:1073
12	Apob	apolipoprotein B	[R] 2.9 kb cDNA	+	*Lusis	J Biol Chem 1987; 262:7594
12	Crip	cysteine -rich intestinal protein	pMI3	+	5-Birkenmeier	PNAS 1986; 83:2516
			[R] pR13	+	*Birkenmeier	PNAS 1988; 85:8121
12	"DAbl-12"	DNA, Abelson MuLV integration flank	DAbl-12	+	*Bauer	Nuc Acid Res 1988; 16:8200
12	D12Nyu1	DNA segment, NYU-1	M13p7-97	+	1-*D'Eustachio	J Exp Med 1984; 160:827
12	D12Nyu2	DNA segment, NYU-2	M13p19-25	+	1-*D'Eustachio	J Exp Med 1984; 160:827
12	D12Nyu3	DNA segment, NYU-3	M13p20-1	+	1-*D'Eustachio	J Exp Med 1984; 160:827
12	D12Nyu4	DNA segment, NYU-4	M13p2615	+	1-*D'Eustachio	J Exp Med 1984; 160:827
12	D12Nyu5	DNA segment, NYU-5	M13p30-3	+	1-*D'Eustachio	J Exp Med 1984; 160:827
12	D12Nyu6	DNA segment, NYU-6	pBR0 49-9	+	*Blank	Genetics 1988; 120:1073
12	D12Nyu7	DNA segment, NYU-7	pBR0 39L5	+	*Blank	Genetics 1988; 120:1073
12	D12Nyu8	DNA segment, NYU-8	pUC0 38H10	+	*Blank	Genetics 1988; 120:1073
12	D12Nyu9	DNA segment, NYU-9	pUC0 57UXI7	+	*Blank	Genetics 1988; 120:1073
12	D12Nyu10	DNA segment, NYU-10	pUC0 43G14	+	*Blank	Genetics 1988; 120:1073
12	D12Mcg1	DNA segment, Med. Col. of Georgia-1	pRC2.3	+	1-Whitney	ref: Cobb, Biochem Genet 1987; 25:401
12	D12Rp54	DNA segment, Roswell Park-54	pLV54	+	*Hill	Mouse News Letter 1982; 67:36
12	Fcr	Fc receptor γ	pFc24		*Hibbs	PNAS 1986; 83:6980
12	Fos	FBJ osteosarcoma oncogene	[V] pfos1	+	5-Curran	J Virol 1982; 44:674
			"	+	*Birkenmeier	PNAS 1988; 85:8121
			"	+	1-*D'Eustachio	J Exp Med 1984; 160:827
			[V] pFBJ-2		*VanBeveren	Cell 1983; 32:1241
12	Fv-4	Friend virus susceptibility-4	RR2, RR7, RR24		*Dandekar	J Virol 1987; 61:308
			pEc-env (MuLV env)	+	*Kozak	PNAS 1984; 81:834
12	Hox-4	homeo box-4	Ch23		*Lonai	DNA 1987; 6:409
12	Igh	immunoglobulin heavy chain complex	CH603α6, CH603α125		*Davis	Nature 1980; 283:733
			pSγ3		*Lang	Nuc Acid Res 1982; 10:611
			"		*Brown	Mol Cell Biol 1987; 7:450
			pμ3741		*Schibler	Cell 1978; 15:1495
			"		*Brown	Mol Cell Biol 1987; 7:450
			MEP200		*Wood	PNAS 1983 80:3030
			γ2b(11) 12		*fr:Marcu	ref: Schibler, Cell 1978;15:1495
			"		*Brown	Mol Cell Biol 1987; 7:450
			pγSa-3		*fr:Eckhart	ref: Shimizu, Cell 1982;28:499
			"		*Brown	Mol Cell Biol 1987; 7:450
12	Igh-J	immunoglobulin heavy chain, j region	JH1-JH4		*Alt	EMBO J 1984; 3:1209
			"		*Mucenski	Mol Cell Biol 1986; 6:4236
			JH800 (J4)		*fr:Alt	ref: Brown, Mol Cell Biol 1987; 7:450
			VκB5,VκB6 (JH4)		*Trepicchio	J Immunol 1985; 134:2734
12	Igh-J	immunoglobulin heavy chain, j region	J-H probe		*Maki	PNAS 1980; 77:2138
			"		*Blackwell	Cell 1984; 37:105
			pJ11 (J3,J4)		*fr:C.Berger	ref: Marcu, Cell 1980; 22:187
			"		*Brown	Mol Cell Biol 1987; 7:450
			"		*Vasmel	Leukemia 1987; 1:155
			CosJ-17		*Wood	PNAS 1983 80:3030
			ME184-8 (embryonic J)		*Sakano	Nature 1980; 286:676

Chr	Symbol&	Gene Name	Probe@	RFLV#	Holder‡	Reference∞
12	Igh(D-J)	immunoglobulin heavy chain, D-J region	DSP2.3,p6-1		*Alt	PNAS 1982; 79:4118
			"		*Brown	Mol Cell Biol 1987; 7:450
			pDhFL16-1		*fr:Riblett	ref: Silver, JNCI 1986; 77:793
			"	+	*fr:Reth	ref: Lehle, Eur J Imunol 1988; 18:1275
			D probe, 5'Dprobe		*Blackwell	Cell 1984; 37:105
			$D_{SP2.2}$, $D_{FL16.1}$		*Kurosawa	Nature 1981; 290:565
			"		*Kurosawa	J Exp Med 1983; 155:201
12	Igh-1(γ2a)	immunoglobulin heavy chain, (C γ2a)	MEP768		*Roeder	PNAS 1981; 78:475
12	Igh-2(α)	immunoglobulin heavy chain, (C α)	CHSpa29		*Davis	Nature 1980; 283:733
			[Mp]clone 1A		*Osborne	Genetics 1988; 119:925
12	Igh-2(α)(f)	flank of Igh-2 (constant Cα)	p3'α		*Stavnezer	ref: Brown, Mol Cell Biol 1987; 7:450
			"	+	*Blank	Genetics 1988; 120:1073
12	Igh-3(γ2b)	immunoglobulin heavy chain, (Cγ2b)	M141-p21		*Sakano	Nature 1980; 286:676
			MEP3		*Sakano	Nature 1980; 286:676
			MEP5, MEP12, MEP593		*Roeder	PNAS 1981; 78:475
12	Igh-3(γ2b)(f)	5' flank of Igh-3(γ2b)	pBR1.4		*fr:Marcu	ref: Lang, Nuc Acid Res 1982; 10:611
			"		*Brown	Mol Cell Biol 1987; 7:450
12	Igh-4(γ1)	immunoglobulin heavy chain, (C γ)	MEP15, MEP10		*Roeder	PNAS 1981; 78:475
12	Igh-6	immunoglobulin heavy chain, (Cϕ)	RIS73		*Kurosawa	J Exp Med 1983; 155:201
12	Igh-6(μ)	immunoglobulin heavy chain, (Cμ)	B1-3		*Kurosawa	J Exp Med 1983; 155:201
			CHSpm27		*Davis	Nature 1980; 283:733
12	Igh-7(ε)	immunoglobulin heavy chain-7 (IgE)	pIgε-7		*Ishida	EMBO J 1982; 9:1117
			"	+	*Shinkai	Immunogenetics 1988; 27:288
			pMEb-1		*Shinkai	Immunogenetics 1988; 27:288
12	Igh-7(ε)	immunoglobulin heavy chain, (Cε)	unnamed		*Liu	PNAS 1982 79:7852
			"		*Gritzmacher	J Immunol 1987 139:603
12	Igh-7(ε)	immunoglobulin heavy chain, (Sε)	unnamed		*fr:Orida	ref: Liu, J Immunol 124:2728
			"		*Gritzmacher	J Immunol 1987 139:603
12	Igh-7(ε)	immunoglobulin heavy chain, (Cε,Sε)	unnamed		*Gritzmacher	J Immunol 1987 138:324
			"		*Gritzmacher	J Immunol 1987 139:603
12	Igh-8(γ3)	immunoglobulin heavy chain, (C γ3)	MEP311, MEP535		*Roeder	PNAS 1981; 78:475
12	Igh-8(γ3)	immunoglobulin heavy chain, (C γ3)	ME736		*Roeder	PNAS 1981; 78:475
12	Igh-C(μ,γ,ε,α)	immunoglobulin heavy chain, constant μ,γ,ε,α	unnamed		1-Rougeon	Cell 1983; 32:515
			"		1-Rougeon	Gene 1981; 13:365
			"		1-Rougeon	Gene 1980; 12:77
12	Igh-V	immunoglobulin heavy chain, variable	37.1		*Hartman	EMBO J 1984; 3:3023
			"		*Manheimer-Lory	PNAS 1986; 83:8293
			CHSpV_H3		*Manheimer-Lory	PNAS 1986; 83:8293
			unnamed		1-Rougeon	Nuc Acid Res 1983; 11:7887
			"		1-Rougeon	Nuc Acid Res 1981; 9:4099
			VhMOPC-21	+	*Craig	ref: Brodeur, Eur J Immunol 1984; 14:922
			VGAM3-8		*Winter	EMBO J 1985; 4:2861
			"	+	*Blankenstein	Eur J Immunol 1[illegible]; 17:1351
			"	+	*fr:Blankenstein	ref: Lehle, Eur J Immunol 1988; 18:1275
			VhMOPC-141	+	*Kelly-Siebenlist	ref: Brodeur, Eur J Immunol 1984; 14:922
			M67-19		*fr:Marcu	ref: Lang, Nuc Acid Res 1982; 10:611
			"		*Brown	Mol Cell Biol 1987; 7:450
12	Igh-V	Igh- variable (BALB)	unnamed		*Liu	Nuc Acid Res 1987; 15:6296
12	Igh-V	Igh- variable (V_H10)	Vh10 probe	+	*Kofler	J Immunol 1988; 140:4031
12	Igh-V	Igh- variable (Q52 family)	300.19	+	*Yancopoulos	Nature 1984; 311:727
			"	+	*Blankenstein	Eur J Immunol 1987; 17:1351
			"	+	*fr:Blankenstein	ref: Lehle, Eur J Immunol 1988; 18:1275
			clone Q52J		*Sakano	Nature 1981; 290:562
			MEP203		*Sakano	Nature 1981; 290:562
			VhQUPC-52	+	*Shapiro	ref: Brodeur, Eur J Immunol 1984; 14:922
			pVhQ52N	+	*fr:Brodeur	ref: Tutter, Immunogenetics 1988; 28:125
			MEP103		*Sakano	Nature 1980; 286:676
			VhOXAZOLONE	+	*Brodeur	Eur J Immunol 1984; 14:922
12	Igh-V	Igh- variable (S107 family)	S107		*Perlmutter	Adv Immunol 1984; 35:1
			pVhS107	+	*Brodeur	Eur J Immunol 1984; 14:922
			"		*Marcu	Cell 1980 22:187
			"	+	*Blankenstein	Eur J Immunol 1987; 17:1351
			"	+	*Tutter	Immunogenetics 1988; 28:125
			"	+	*fr:Blankenstein	ref: Lehle, Eur J Immunol 1988; 18:1275
			p107V1		*Early	Cell 1980; 19:992
			"	+	*fr:Hood	ref: Hilbert, J Immunol 1988; 140:4364
12	Igh-V	Igh- variable (S107 family)	Vh38C	+	*Nelson	Mol Cell Biol 1983; 3:1317
			Vh6G6	+	*Clarke	J Exp Med 1984; 159:773
12	Igh-V	Igh-variable (24X family)	VhX24, pL7		*Hartman	EMBO J 1984; 3:3023
			"	+	*Brodeur	Eur J Immunol 1984; 14:922
			V_H441		*Ollo	Nuc Acid Res 1981; 9:4099
			"		*Hartman	EMBO J 1984; 3:3023
			V_H441	+	*Blankenstein	Eur J Immunol 1987; 17:1351

Chr	Symbol&	Gene Name	Probe@	RFLV#	Holder‡	Reference∞
12	Igh-V	Igh-variable (V31 family)	V31	+	*Winter	EMBO J 1985; 4:2861
			"	+	*Blankenstein	Eur J Immunol 1987, 17:1351
			"	+	*fr:Blankenstein	ref: Lehle, Eur J Immunol 1988; 18:1275
12	Igh-V	Igh-variable (V31 and J558 families)	λ1.8.7, λ1.8.18	+	*Blankenstein	Eur J Immunol 1987; 17:1351
12	Igh-V	Igh- variable (J558 family)	pY19-10		*Manheimer-Lory	PNAS 1986; 83:8293
			clone A1		*Yancopoulous	Cell 1985; 40:271
			pV.A1 Pst (clone A1)		*Schuler	EMBO J 1988; 7:2019
			93G7		*Sims	Science 1982; 216:309
			CDR-1, CDR-2	+	*Rathbun	J Mol Biol 1988; 202:383
			V_H 21, V_H122T, V_H28,	+	*Rathbun	J Mol Biol 1988; 202:383
			V_H186, V_H43Y	+	*Rathbun	J Mol Biol 1988; 202:383
			V186		*Bothwell	Cell 1981; 24:625
			"	+	*Blankenstein	Immunogenetics 1987; 26:237
			V_H83, V_H9143		*Rathbun	J Mol Biol 1988; 202:383
			V_H1.8, V_H43X, V_H15		*Rathbun	J Mol Biol 1988; 202:383
			J558V_H		*Livant	Cell 1986; 47:461
			pVhJ558	+	*Tutter	Immunogenet 1988; 28:125
			pα(J558)[13]		*Marcu	Cell 1980; 22:187
			"		*Brown	Mol Cell Biol 1987; 7:450
			VhJ558	+	*Brodeur	Eur J Immunol 1984; 14:922
			VhDx11	+	*Early	ref: Brodeur, Eur J Immunol 1984; 14:922
			VhT1035	+	*Brodeur	Eur J Immunol 1984; 14:922
			VhNP^b	+	*Brodeur	Eur J Immunol 1984; 14:922
			VhHEL-3F4	+	*Brodeur	Eur J Immunol 1984; 14:922
			VhS117N	+	*Brodeur	Eur J Immunol 1984; 14:922
			VhARS	+	*Sims	Science 1982; 216:309
			VhMPC-11		*Lang	Nuc Acid Res 1982, 10:611
			Vh70Z	+	*Nelson	Mol Cell Biol 1983; 3:1317
12	Igh-Vps	Igh- variable (J558 pseudo)	VAR104, VAR34	+	*Blankenstein	Immunogenetics 1987; 26:237
			VAR100		*Blankenstein	Immunogenetics 1987; 26:237
			$V_H\Psi$1.3		*Rathbun	J Mol Biol 1988; 202:383
			$V_H\Psi$122B	+	*Rathbun	J Mol Biol 1988; 202:383
12	Igh-V	Igh- variable (J606 family)	VhJ606	+	*Crews	ref: Brodeur, Eur J Immunol 1984; 14:922
			"	+	*Blankenstein	Eur J Immunol 1987; 17:1351
			"	+	*fr:Blankenstein	ref: Lehle, Eur J Immunol 1988; 18:1275
			pBV14,Vh39		*Hartman	EMBO J 1984; 3:3023
			pBV14RB7		*Wang	Cell 1985; 43:659
			"		*Tutter	Immunogenetics 1988; 28:125
12	Igh-V	Igh- variable(3609P family)	pVh3635	+	*fr:Brodeur	ref: Tutter, Immunogenetics 1988; 28:125
			V_H3609P		*Riblet	In: Genetics and Molecular Immunology 1986 (ed. Herzenberg et al)
12	Igh-V	Igh- variable (3660 family)	M460		*Dzierzak	J Exp Med 1985; 162:1494
			VGAM3.2	+	*Winter	EMBO J 1985; 4:2861
			"	+	*Near	PNAS 1984; 81:2167
			"	+	*Blankenstein	Eur J Immunol 1987; 17:1351
			"	+	*fr:Blankenstein	ref: Lehle, Eur J Immunol 1988; 18:1275
			pRN5.15	+	*Near	PNAS 1984; 81:2167
			"		*Tutter	Immunogenetics 1988; 28:125
12	Igh-V	Igh- variable (7183 family)	p129-48		*Manheimer-Lory	PNAS 1986; 83:8293
			Vh7183	+	*Coleclough	ref: Brodeur, Eur J Immunol 1984; 14:922
			VhSAPC-15	+	*Brodeur	Eur J Immunol 1984; 14:922
			"	+	*Tutter	Immunogenetics 1988; 28:125
			81X		*Yancopoulous	Nature 1984; 311:727
			"		*Manheimer-Lory	PNAS 1986; 83:8293
			"	+	*Blankenstein	Eur J Immunol 1987; 17:1351
			Vh37-77	+	*Clarke	ref: Brodeur, Eur J Immunol 1984; 14:922
			Vh37-65N	+	*Clarke	ref: Brodeur, Eur J Immunol 1984; 14:922
			VDJ_3-1	+	*Yancopoulos	ref: Brodeur, Eur J Immunol 1984; 14:922
12	"Igh-en"	Igh enhancer	Eμ		*Chen	Cell 1987; 51:7
			"		*Gillies	Cell 1983; 33:717
12	Lamb-1	laminin, B1 subunit	pPE386, pPE49		*Elliot	In Vitro Cell Dev Bio 1985; 21:477
			p24, λB1-11		*Sasaki	PNAS 1987; 84:935
12	Mtv-9	mammary tumor virus locus-9	Mtv9-5'C		*Peters	J Virol 1986; 59:535
			"	+	*Blank	Genetics 1988; 120:1073
			Mtv9-3'C		*Peters	J Virol 1986; 59:535
12	Nmyc	neuroblastoma myc-related oncogene	pLM1, clone91		1-Taya	EMBO J 1986; 5:1215
			pN7.7		*Yancopoulos	PNAS 1985; 82:5455
			"		*DePinho	PNAS 1986; 83:1827
			pM.2		*Zimmerman	Nature 1986; 319:780
			unnamed		*Katoh	Nuc Acid Res 1988; 16:3589

Chr	Symbol&	Gene Name	Probe@	RFLV#	Holder‡	Reference∞
12	Odc-8	ornithine decarboxylase related-8	pODC48	+	*Colombo	Nuc Acid Res 1988; 16:9075
12	Pol-7	viral polymerase gene-7	B65.5PV2 derivative	+	*Rossomando	Immunogenetics 1986; 23:233
12	Pomc-1	pro-opiomelanocortin-α	λa9, pMKSU16		*Uhler, Herbert	J Biol Chem 1983; 258:9444
12	Rnu1a-2	U1a2 small nuclear RNA	mU1a2	+	*Lund	Somatic Cell Mol Genet 1988; 14:143
12	Spi-1	serine protease inhibitor -1	pLv1796	+	1-Hill	Mol Cell Biol 1985; 5:2114
12	Spi-2	serine protease inhibitor-2	pLv54	+	1-Hill	Mol Cell Biol 1985; 5:2114
12	Spnb-1	β-spectrin-1	pβ58		*Laurila	Somatic Cell Mol Genet 1987;13:93
			"		*Cioe	Blood 1987; 70:915
			"	+	*Birkenmeier	PNAS 1988; 85:8121
12	"Tg-GH"	human growth hormone transgene	λHUGH/4-1		*Covarrabias	Mol Cell Biol 1987; 7:2243
13	Ctla-2	cytotoxic T lymphocyte associated protein-2	M41G12		*Brunet	Nature 1986; 322:268
13	Ctla-3	cytotoxic T lymphocyte associated protein-3	M11E6		*Brunet	Nature 1986; 322:268
			"	+	*Sola	J Virol 1988; 62: 3973
13	"D13?"	flank of inserted human TTR seq	p59		*Wakasugi	Devel Genet 1988; 9:203
13	"D13?"	DNA segment, testies specific	PL7		*Leroy	Devel 1987; 101 (supp):177
13	D13Pas1	DNA segment, testis specific transcript	PL5		*Leroy	Devel 1987; 101 (supp): 177
			"	+	*Sola	J Virol 1988; 62:3973
13	D13Pas2	transgene flank, integration site hepatitis B	4/12		*Hadchouel	Nature 1987; 329:454
13	Dhfr	dihydrofolate reductase	pdhfr, pSVMdhfr		*Killary	Somatic Cell Mol Genet 1986;12:641
			pSVMdhfr		*Lee	Nature 1981; 294:228
13	Fim-1	Friend MuLV integration site-1	167-PP15		1-Gisselbrecht	ref: Sola, J Virol 1988; 62:3973
			"		*Sola	J Virol 1986; 60:718
			probe51-EH09	+	*Sola	J Virol 1986; 60:718
13	Hist1 (H2b,H2a,H3)	histone gene complex 1	MM291		*Graves	J Mol Biol 1985; 183:179
13	Hist1 (H3.2,H2b, H3.1,H2a)	histone gene complex 1	MM221		*Graves	J Mol Biol 1985; 183:179
13	Hist1 (H3)	histone gene complex 1	H3.1-291		*Taylor	J Mol Evol 1986; 23:242
			H3.1-221, H3.2 221		*Taylor	J Mol Evol 1986; 23:242
			"		*Sittman	Nuc Acid Res 1983; 11:6679
13	Hist1 (H3)	histone gene complex 1	MM221, MM291		*Graves	J Mol Biol 1985; 183:179
13	Hist1-2bps	histone gene complex -1 H2b pseudogene	MM291A		*Liu	Nuc Acid Res 1987; 15:3023
13	Plf	proliferin	PLF-1, PLF-2		*Wilder	Mol Cell Biol 1986; 6:3283
			"		*Linzer	PNAS 1985; 82:4356
			PLF42, PLF149		*Linzer	EMBO J 1987; 6:2281
13	Pl-1	placental lactogen-1	mPRL,mPL-I,		*Jackson-Grusby	Endocrinology 1988; 122:2462
			mPLF,mPRP		*Jackson-Grusby	Endocrinology 1988; 122:2462
13	Pl-2	placental lactogen -2	mPL-II		*Jackson-Grusby	Endocrinology 1988; 122:2462
			clone B, clone G		*Jackson	PNAS 1986; 83:8496
13	Tcrg	T cell receptor γ chain	pHD54	+	*Owen	J Immunol 1986; 137:1044
			"		*Murre	Nature 1985; 316:1
			clone γ7.1	+	*Rathbun	Immunogenetics 1988; 27:121
			VTg5	+	*Kranz	Science 1985; 227:941
			"	+	*Sola	J Virol 1988; 62:3973
			pB10.AT3γ-1		*Krissansen	J Immunol 1987; 138:3513
			pTZCγ4		*fr:Iwamoto	ref: Woolf, Nuc Acid Res 1988; 16:3863
			pB5Cγ		*fr:McElligott	ref: Hayday, Cell 1985; 40:259
			"		*Woolf	Nuc Acid Res 1988; 16:3863
			pSP11γV,pSP17γV		*fr:Raulet	ref: Garman, Cell 1986; 45:733
			"		*Woolf	Nuc Acid Res 1988; 16:3863
			pUCVγ1.1		*Woolf	Nuc Acid Res 1988; 16:3863
13	Tcrg	T cell receptor γ, constant Cγ1, Cγ2, Cγ3	pCγ2		*fr:Iwamoto	ref: Schuler, EMBO J 1988; 5:2139
13	"Tg174"	transgene (human insulin)	[H] 11 kb frag insulin	+	*Michalova	Hum Genet 1988; 80:247
14	Ctla-1	cytotoxic T lymphocyte associated protein-1	M41D12		*Brunet	Nature 1986; 322:268
			clones 7.1.2, A.1.1		*Brunet	Nature 1986; 322:268
			C1.1		*Brunet	Nature 1986; 322:268
			"	+	*fr: Brunet	ref: Harper, Immunogenetics 1988; 28:439
14	"Glud1"	glutamate dehydrogenase-like	[H] Gdh-X		*Hanauer	Nuc Acid Res 1987; 15:6038
			"	+	*fr: Hanauer	ref: Harper, Immunogenetics 1988; 28:439
14	hr	hairless	MX40B/BgX	+	1-Stoye	Cell 1988; 54:383
			MX40A, MX40B	+	1-Stoye	Cell 1988; 54:383
14	Ms6-5	minisatellite 6-5	[H] probes 33.6, 33.15		*Jeffreys	Nuc Acid Res 1987; 15:2823
14	Ms15-7	minisatellite 15-7	[H] probes 33.6, 33.15		*Jeffreys	Nuc Acid Res 1987; 15:2823
14	Myhc-a	myosin heavy chain, cardiac muscle, adult	pMHC101, pMHC141	+	2-Buckingham	ref: Weydert, PNAS 1985; 82:7183
14	Nras-ps	neuroblastoma ras oncogene pseudogene	probes: N8.5, N2.7		*Ryan	Nuc Acid Res 1984; 12:6063
			unnamed		*Chang	Oncogene Res 1987; 1:129
14	Plau	plasimogen activator, urokinase	pDB15		*Belin	Eur J Biochem 1985; 148:225
			"		*Collart	J Immunol 1987; 139:949
14	Rib-1	ribonuclease-1, pancreatic	pMPR-1	+	1-Elliott	Cytogenet Cell Genet 1986; 42:110
14	Tcra	T cell antigen receptor α chain	T1.2	+	5-Dembic	Nature 1985; 314:271
			ppαDO	+	*Epstein	J Exp Med 1986; 163:759
			"		*Yague	Cell 1985; 42:81
			pHDS58		*Saito	Nature 1984; 312:36
			"	+	*Berman	Immunogenetics 1986; 24:328
			"		*Kranz	Science 1985; 227:941
			Va7.2		*Kotzin	J Exp Med 1987; 165:1237

Chr	Symbol&	Gene Name	Probe@	RFLV#	Holder‡	Reference∞
14	Tcra	T-cell receptor antigen (constant region)	Tcr_{α}		*Winoto	Nature 1985; 316:832
			"	+	*fr:Winoto	ref: Harper, Immunogenetics 1988; 28:439
14	Trp53-ps	transformation related p53, pseudogene	pCH53-11		*Zabut-Houri	Nature 1983; 306:594
15	Gdc-1	glycerolphosphate dehydrogenase-1	pH8,pC8	+	1-Kozak	PNAS 1983; 80:3020
			gdc-1	+	1-Kozak	Genetics 1985; 110:123
			"	+	*fr:Kozak	ref: Duncan, Mol Cell Biol 1988; 8:2705
			"	+	*fr:Kozak	ref: Huppi, Immunogenetics 1988; 27:215
15	Hba-3ps	hemogloblin α-3, pseudogene	α-Ψ3		*Leder	Nature 1981; 293:196
			"		*fr:Leder	ref: Adolph, Cytogenet Cell Genet 1988; 47:189
15	Hox-3	homeo box-3	λMo-EA, pMo-EA		*Awgulewitsch	Nature 1986; 320:328
15	Hox-3.1	homeo box-3.1	λm1,λm2		*Breier	EMBO J 1986; 5:2209
			pm31-1, m31C		*Breier	EMBO J 1986; 5:2209
15	Hox-3.2	homeo box-3.2	m32-9		*Breier	EMBO J 1988; 7:1329
15	Ins-3	insulin I or insulin pseudogene	pI19	+	*Meruelo	Immunogenetics 1987; 25:361
15	Int-1	mammary tumor integration site-1	pMT25		1-Nusse	ref: Ooyen, Cell 1984; 39:233
			probe C & D		*Nusse	Nature 1984; 307:131
			"		*fr: Nusse	ref: Aldoph, Cytogen Cell Gene 1987; 44:65
			probes: 2, 26, 17A, 104A		*Fung	Mol Cell Biol 1985; 5:3337
			pint-1E, pint-1E5		*Wilkinson	Cell 1987; 50:79
15	Krt-2	keratin gene complex-2	pK4.1(b1)	+	*Meruelo	Immunogenetics 1987; 25:361
			pK435(a1)	+	"	"
			pKG59(4.0Pst)		2-Compton	J Biol Chem 1985; 260:5867
			pKG59(6.0Bam)		2-Compton	J Biol Chem 1985; 260:5867
			p6-94		5-Compton	unpublished, Jackson Lab
			pE8		*Morita	Gene 1988; 68:109
		cytokeratin Endo A	pKA56		*Ouellet	Gene 1988; 70:75
			RecXVI		*Vasseur	PNAS 1985; 82:1155
			clone α1		*Vasseur	PNAS 1985; 82:1155
15	Krt-2ps	cytokeratin Endo A, pseudogene	clone α2		*Vasseur	PNAS 1985; 82:1155
15	Ly-6	lymphocyte antigen-6	pKLy6.1-2R	+	*LeClair	EMBO J 1986; 5:3227
			"	+	*LeClair	PNAS 1987; 84:1638
			"	+	*fr:LeClair	ref: Duncan, Mol Cell Biol 1988; 8:2705
			"	+	*fr:LeClair	ref: Huppi, Immunogenetics 1988; 27:215
			pKLy6.1-2D		*LeClair	EMBO J 1986; 5:3227
			pKLy6.1-5W		*LeClair	EMBO J 1986; 5:3227
15	Ly-6.2	lymphocyte antigen-6.2 allele	B18,B28		*Palfree	Immunogenetics 1987; 26:389
15	Ly-6C	lymphocyte antigen-6C	B17,B34		*Palfree	J Immunol 1988; 140:305
			clone 24		*Bothwell	J Immunol 1988; 140:2815
15	Mlvi-1	Moloney -MuLV integration site-1	pTS25E/P		*Tsichlis	J Virol 1985; 56:258
			"		*Mucenski	J Virol 1988; 62:839
			pTS26		*Kozak	Mol Cell Biol 1985; 5:894
			"		*Adolph	Cytogenet Cell Genet 1988; 47:189
15	Mlvi-2	Moloney-MuLV integration site-2	pTS10		*Tsichlis	Mol Cell Biol 1984; 4:997
			"		*Mucenski	J Virol 1988; 62:839
			"		*Adolph	Cytogenet Cell Genet 1988; 47:189
15	Myc	myelocytomatosis oncogene	λE1, λST4-2,		*Cory	EMBO J 1985; 4:675
			λST4-1, λST4-3		*Cory	EMBO J 1985; 4:675
			pM104BH		*Shen-Ong	Cell 1982; 31:443
			p104E.5		*fr:M. Cole	ref: Mucenski, J Virol 1988; 62:839
			pMcmyc54		*fr:Marcu	ref: Stanton, Nature 1983; 303:401
			"		*Zimmerman	Nature 1986; 319:780
			"	+	*Huppi	Immunogenetics 1988; 27:215
				+	*fr:Marcu	ref: Duncan, Mol Cell Biol 1988; 8:2705
			pSVc-myc-1		*Land	Nature 1983; 304:596
			Ryc7.4		*Marcu	PNAS 1983; 80:519
15	Niard	non-immunoglobulin associated rearranging DNA (near Myc)	pα25BH3.4		*Marcu	PNAS 1983; 80:519
15	P450-2D	cytochrome P450IID	unnamed		*Kimura	Nuc Acid Res 1986; 14:6765
15	Pva	parvalbumin	clone 9f		*Berchtold	PNAS 1985; 82:1414
			"		*Kluxen	Eur J Biochem 1988; 176:153
15	Pvt-1	plasmacytoma variant translocation-1	cW28, cW22, cW18.1		*Cory	EMBO J 1985; 4:675
			cW24.1, cW18.2, cW17		*Cory	EMBO J 1985; 4:675
			λW, λA4-1, λA4-2		*Cory	EMBO J 1985; 4:675
			cA4-5, cA4-3, cA4-12		*Cory	EMBO J 1985; 4:675
			Probe B		*fr: Cory	ref: Adolph, Cytogenet Cell Genet 1988; 47: 189
			Probe E		*Graham	Nature 1985; 314:740
15	Rpl30	ribosomal protein L30	rpL30-1		*Wiedemann	Mol Cell Biol 1984; 4:2518
			L30-I3		*Wiedemann	Somatic Cell Mol Genet 1987;33:77
15	Sis	simian sarcoma oncogene	pSSV11 (Oncor)	+	*Meruelo	Immunogenetics 1987; 25:361
			p3040 (Oncor)	+	*Duncan	Mol Cell Biol 1988; 8:2705
			"	+	*Huppi	Immunogenetics 1988: 27:215

Chr	Symbol&	Gene Name	Probe@	RFLV#	Holder‡	Reference∞
15	Tgn (cog)	thyroglobulin	[R] pRT57	+	*Taylor	PNAS 1987; 84:1986
			[R] pRT27.15, pRT36.1		*Taylor	PNAS 1987; 84:1986
15	Xmmv-55	xenotropic-MCF leukemia virus-55	pXenv	+	*Hogarth	Immunogenetics 1987; 25:21
15	Xmmv-72	xenotropic -MCF leukemia virus-72 (Pol-5)	B65.5PV2	+	*Meruelo	Immunogenetics 1987; 25:361
			"	+	*Rossomando	Immunogenetics 1986; 23:233
16	Akv-2	AKR leukemia virus inducer-2	AKR lD242	+	5-Epstein	Immunogenetics 1984; 19:527
16	App	amyloid β (A4) percursor protein	unnamed		*Philip	Am J Hum Genet 1987; 41:A233
			[H] clone B2.3		*Robakis	PNAS 1987; 84:4190
			"	+	*Reeves	Mol Brain Res 1987; 2:215
			pMAZ11,pMAZ61		*Yamada	BBRC 1987; 149:665
			[H] FB68L, [H] HL124	+	*Cheng	PNAS 1988; 85:6032
16	Cola-2	procollagen type II	pAZ1003, pAZ1009		*Liau	J Biol Chem 1986; 261:11362
			"		*Schmidt	Nature 1985; 314:286
			pAZ1001		*Liau	J Biol Chem 1986; 261:11362
			"		*Schmidt	Nature 1985; 314:286
			λ-colm212		*Schmidt	J Biol Chem 1984; 259:7411
			[H] Hf32		*Myers	PNAS 1981; 78:3516
			"		*Schnieke	Nature 1983; 304:315
16	D21S16h	homolog human DNA segment D21S16	[H] pGSE9	+	*Cheng	PNAS 1988; 85:6032
16	D21S52h	homolog human DNA segment D21S52	[H] 511-1H/511-2P	+	*Cheng	PNAS 1988; 85:6032
16	D21S58h	homolog human DNA segment D21S58	[H] 524-5P	+	*Cheng	PNAS 1988; 85:6032
16	"D16?"	DNA segment, nonstringent hybridization	pYMT2/B		*Bishop	Nuc Acid Res 1987; 15:2959
16	Ets-2	E26 avian leukemia oncogene-1, 5' domain	P1.27		2-Sherman	PNAS 1983; 5465
			"	+	*Reeves	Cytogenet Cell Genet 1987; 44:76
			pA3		*Watson	PNAS 1988; 85:7862
16	Ets-2	E26 avian leukemia oncogene-1, 5' domain	unnamed (H33 homol)	+	*Watson	PNAS 1986; 83:1792
			[H] H33		*Watson	Anticancer Res 1986; 6:631
			"	+	*Cheng	PNAS 1988; 85:6032
16	Igl-1	immunoglobulin λ chain	pABλ -1		*Bothwell	Nature 1981; 290:65
			"		*Miller	PNAS 1981; 78:3829
			pCλ14E	+	5-Epstein	Immunogenetics 1986; 23:78
			MOPC104E	+	*Scott	Nature 1982; 300:757
			"	+	*Reeves	Cytogenet Cell Genet 1987; 44:76
16	Igl-1	immunoglobulin λ chain	$p\lambda_{II}$-1		*Bothwell	Nature 1981; 290:65
			$p\lambda_{I}$-13		*Bothwell	Nature 1981; 290:65
16	Igl-C	immunoglobulin λ chain, constant	pAB1-1		*Kindt	Eur J Immunol 1986; 15:535
			KA9,KA11,KH5		*Miller	PNAS 1981; 78:3829
			pEJ3-C_{λ}		*Miller	PNAS 1981; 78:3829
			"		*Bothwell	Nature 1981; 290:65
16	Igl-C1	immunoglobulin λ chain, constant-1	unnamed		*fr:Bothwell	Nature 1981; 290:65
			"		*Mami	J Immunol 1987; 138:3980
16	Igl-C2	immunoglobulin λ chain, constant-2	unnamed		*fr:Hozumi	ref. Alonso, J Immunol 1985; 135:614
			"		*Mami	J Immunol 1987; 138:3980
16	Igl-C2-3	immunoglobulin λ chain, constant-2,3	MOPC315 plasmid$_{II}^{-1}$	+	*Hilbert	J Immunol 1988; 140:4364
16	Igl-C4	immunoglobulin λ chain, constant-4	clone 8		*Mami	J Immunol 1987; 138:3980
16	Igl-C5	immunoglobulin λ chain, constant-5	clone 71-2	+	*Mami	EMBO J 1988; 7:117
			pCL5s	+	*Mami	EMBO J 1988; 7:117
16	Igl-J2,J4	immunoglobulin λ chain, J region	unnamed		*Blomberg	PNAS 1982; 79:530
			"		*Mami	J Immunol 1987; 138:3980
16	Igl-V (V1)	immunoglobulin λ chain, variable-1	Vλ1		*fr:Simon	ref: Picard, PNAS 1983; 80:417
			"		*Dildrop	Eur J Immunol 1987; 17:731
16	Igl-V (V1-2)	immunoglobulin λ chain, variable-1,2	MOPC104E cDNA		*Scott	Nature 1982; 300:757
			"	+	*fr:Scott	ref: Hilbert, J Immunol 1988; 140:4364
16	Igl-V (V2)	immunoglobulin λ chain, variable-2	Vλ2		*fr:Simon	ref: Bothwell, Nature 1981; 290:65
			"		*Dildrop	Eur J Immunol 1987; 17:731
			pEJ1-V_{λ}		*Miller	PNAS 1981; 78:3829
			"		*Bothwell	Nature 1981; 290:65
			clone 9A8		*Scott	Nature 1982; 300:757
			"	+	*fr:Scott	ref: Hilbert, J Immunol 1988; 140:4364
16	Igl-Vx	immunoglobulin λx variable chain	Y31		*Sanchez	PNAS 1987; 84:9185
16	"Igll"	immunoglobulin λ chain like	7pB5, 7pB12,		*Kudo	EMBO J 1987; 6:103
			7pB18, 7pB28, pZ183-1a		*Kudo	EMBO J 1987; 6:103
16	Mtv-6	mammary tumor virus locus-6	unnamed		*Callahan	J Virol 1984; 49:1005
			MMTVLTR	+	5-Reeves	Cytogenet Cell Genet 1987; 44:76
16	Mx-1	myxovirus (influenza virus) resistance	pMx34, pMx41		1-Haller	ref: Staeheli, Cell 1986; 44:147
			pMx34	+	*Staeheli	Mol Cell Biol 1988; 8:4518
			pMx34 frag	+	*Reeves	J Virol 1988; 62:4372
			pMx41	+	*Staeheli	J Virol 1986; 58:967
16	Pgk-1ps1	phosphoglycerate kinase-1 pseudogene 1	clone B24		*Adra	Somatic Cell Mol Genet 1988; 14:69

Chr	Symbol&	Gene Name	Probe@	RFLV#	Holder‡	Reference∞
16	Prm-1	protamine-1	unnamed		*Hecht	Somatic Cell Mol Genet 1986; 12:203
			mP1		*Kleene	Biochemistry 1985; 24:719
			pMP1-1	+	*Reeves	Cytogenet Cell Genet 1987; 44:76
16	Prm-2	protamine-2	mP2-1,mP2-2		*Yelick	Mol Cell Biol 1987; 7:2173
16	Smst	somatostatin	p1-16-13		2-Goodman	J Biol Chem 1982; 257:1156
			"	+	*Reeves	Cytogenet Cell Genet 1987; 44:76
			[H] pghS7-2.7		*Shen, Rutter	Science 1984; 224:168
					*Lalley	Cytogenet Cell Genet 1987; 44:92
16	Sod-1	superoxide dismutase-1, soluble	SOD4.1	+	5-Reeves	Cytogenet Cell Genet 1987; 44:76
			[H] pSG1-10	+	*Cheng	PNAS 1988; 85:6032
16	Vpreb-1	pre-B lymphocyte gene-1	pZ121		*Kudo	EMBO J 1987; 6:2267
			"		*Bauer	EMBO J 1988; 7:111
16	Vpreb-2	pre-B lymphocyte gene-2	7pB60, 7pB70		*Kudo	EMBO J 1987; 6:2267
17	"anonymous"		cosmid 3-4-1	+	*Kasahara	PNAS 1987; 84:3325
17	"anonymous"		cosmid 1-2-4		*Kasahara	PNAS 1987; 84:3325
17	"anonymous"(testis)		cosmid 1-1-1		*Kasahara	PNAS 1987; 84:3325
17	Bf	complement component factor B	pBmB2		*Sackstein	J Biol Chem 1985; 258:14693
			cosmids: D-7, D-12,		*Chaplin	PNAS 1983; 80:6947
			cosmids: D-9, D-16,		*Chaplin	PNAS 1983; 80:6947
			cosmids: D-24, D-6,		*Chaplin	PNAS 1983; 80:6947
			cosmids: D-14		*Chaplin	PNAS 1983; 80:6947
17	C3	complement component-3	pMLC3-7, pMLC-4		1-Fey	PNAS 1982 79:7619
			pMLC3-1		1-Fey	PNAS 1982; 79:7619
			"		*Lyon	Immunogenetics 1988; 27:375
17	C4	complement component-4	pC412I, pC427A		*Sepich	PNAS 1985; 82:5895
			pFC4/10		*Nonaka	PNAS 1984; 81:6822
			pMC4/21	+	5-Tosi	Phi Trans R Soc L 1984; 306:389
			"	+	*Levi-Strauss	PNAS 1985; 82:1746
			pMC4/7	+	5-Tosi	Phi Trans R Soc L 1984; 306:389
			"	+	*Levi-Strauss	PNAS 1985; 82:1746
			"		*Golubric	Immunogenetics 1984; 21:247
			cosmid 16.2		*Stavenhagen	Mol Cell Biol 1987; 7:1716
			cosmids: E-7, E-64,		*Chaplin	PNAS 1983; 80:6947
			cosmids: E-58, E-69		*Chaplin	PNAS 1983; 80:6947
			cosmids: F-4, F-13		*Chaplin	PNAS 1983; 80:6947
			cosmids: F-5, F-7		*Chaplin	PNAS 1983; 80:6947
			cosmids: E-42, E-15,		*Chaplin	PNAS 1983; 80:6947
17	C4	complement component-4	cosmids: E-26		*Chaplin	PNAS 1983; 80:6947
17	$C4^{w7}$	complement component-4 (w7 allele)	cosmids: cl-41, cl-21,		*Nakayama	J Immunol 1987; 138:620
			cosmids: cl-40, cl-4,		*Nakayama	J Immunol 1987; 138:620
			cosmids: cl-20, cl-32		*Nakayama	J Immunol 1987; 138:620
17	$C4^{w7}$	complement component-4 (w7 allele)	cosmids: cl-33, cl-38,		*Nakayama	J Immunol 1987; 138:620
			cosmids: cl-39, cl-15		*Nakayama	J Immunol 1987; 138:620
17	Cbs	cystathionine β-synthase	[R]p610		*Kraus	PNAS 1986; 83:2047
			"		*Munke	Am J Hum Genet 1988; 42:550
17	Crya-1	α crystallin	pMαACr1		*King	Science 1982; 215:985
			"		*Quinlan	Genes & Devel 1987; 1:637
			pMαACR2, pMαA	+	1-Skow	Genetics 1985; 110:723
			"	+	1-Skow	Genetics?? 1987
			unnamed		*fr:Piatigorski	ref: King, Sci 1982; 215:985
					*Bucan	Genes & Devel 1987; 1:376
17	Cryb	β crystallin	pMβCr1		*Inana	J Biol Chem 1982; 257:9064
			"		*Quinlan	Genes & Devel 1987; 1:637
17	D17Leh48	DNA segment, Lehrach-48 (*t*)	Tu48	+	1-Lehrach	ref: Fox, Cell 1985; 40:63
			"		*Rohme	Cell 1984; 36:783
17	D17Leh54M	DNA segment, Lehrach-54M	p5411-R2	+	2-Bucan	Genes & Devel 1987; 1:376
17	D17Leh66E	DNA segment, Lehrach-66E	p66M-RT	+	1-Hermann	Cell 1986; 44:469
17	D17Leh66D	DNA segment, Lehrach-66D	p66M-RT	+	1-Hermann	Cell 1986; 44:469
			"	+	*Schimenti	J Mol Biol 1987; 194:583
17	D17Leh66	DNA segment, Lehrach-66 (*t*)	Tu66	+	1-Fox	Cell 1985; 40:63
			"		*Rohme	Cell 1984; 36:783
			"	+	5-Schimenti	J Mol Biol 1987; 194:583
			Ca-52,Ca-45,	+	5-Schimenti	J Mol Biol 1987; 194:583
			Bb-53,Bb-59	+	5-Schimenti	J Mol Biol 1987; 194:583
			Cg3-38,Bb-40	+	5-Schimenti	J Mol Biol 1987; 194:583
17	D17Leh80	DNA segment, Lehrach-80 (*t*)	Tu80	+	1-Lehrach	unpublished
17	D17Leh89	DNA segment, Lehrach-89 (*t*)	Tu89	+	1-Lehrach	Cell 1984; 35:783
			"	+	*Bucan	Genes & Devel 1987; 1:376
17	D17Leh94	DNA segment, Lehrach-94	Tu94	+	1-Lehrach	ref: Rohme, Cell 1984; 36:783
			"		*Bucan	Genes & Devel 1987; 1:376
17	D17Leh108	DNA segment, Lehrach-108 (*t*)	Tu108	+	2-Fox	Cell 1985; 40:63
			"		*Rohme	Cell 1984; 36:783

Chr	Symbol&	Gene Name	Probe@	RFLV#	Holder‡	Reference∞
17	D17Leh111	DNA segment, Lehrach-111	Tu111	+	1-Lehrach	unpublished
17	D17Leh116	DNA segment, Lehrach-116	Tu116	+	2-Lehrach	unpublished
17	D17Leh119	DNA segment, Lehrach-119 (*t*)	Tu119	+	*Lehrach	ref: Hermann, Cell 1986; 44:469
			"	+	*MacMurray	Genetics 1988; 120:545
			probeII		*Hermann	Cell 1986; 44:469
			C6.90,C6.95		*Herrmann	Cell 1987; 48:813
17	D17Leh119I,II	DNA segment, Lehrach-119I, 119II	C4.74, C4.38,		*Herrmann	Cell 1987; 48:813
			C4.47, C4.1, C4.37,		*Herrmann	Cell 1987; 48:813
			C2.119, C4.41, C4.43		*Herrmann	Cell 1987; 48:813
17	D17Leh119I	DNA segment, Lehrach 119I	p119A-R	+	1-Hermann	Cell 1986; 44:469
17	D17Leh122	DNA segment, Lehrach 122(*t*)	Tu122	+	1-Fox	Cell 1985; 40:63
					*Rohme	Cell 1984; 36:783
17	D17Leh173	DNA segment, Lehrach 173	Tu173	+	3-Lehrach	unpublished
17	D17Leh180	DNA segment, Lehrach 180	Tu180	+	*Rohme	Cell 1984; 36:783
			"		*Bucan	Genes & Devel 1987; 1:376
17	D17Leh443	DNA segment, Lehrach 443	Tu443	+	1-Lehrach	ref: Rohme, Cell 1984; 36:783
			"	+	*Bucan	Genes & Devel 1987; 1:376
17	D17Leh467	DNA segment, Lehrach 467	Tu467	+	*Rohme	Cell 1984; 36:783
			"		*Bucan	Genes & Devel 1987; 1:376
17	D17Leh508	DNA segment, Lehrach 508	Tu508	+	2-Lehrach	unpublished
17	D17Leh525	DNA segment, Lehrach 525	Tu525	+	2-Bucan	Genes & Devel 1987; 1:376
17	D17Leh550	DNA segment, Lehrach 550	Tu550	+	2-Lehrach	unpublished
17	D17Rp1	DNA segment, Roswell Park-1	pMK1109	+	1-Elliott,Mann	Mouse News Letter 1985; 71:48
17	D17Rp17	DNA segment, Roswell Park-17	pMK174	+	1-Mann	Genetics 1986; 114:993
			"	+	*Sarventnick	Genetics 1986; 113:723
17	D17Tu1	DNA segment, Tubingen-1	clone 3.4.1 derivative	+	*Kasahara	PNAS 1987; 84:3325
17	D17Tu2	DNA segment, Tubingen-2	clone 9.4.3 derivative	+	*Figueroa	Genetics 1987; 117:101
17	D21S56h	homolog human DNA segment D21S56	[H] 520-10R	+	*Cheng	PNAS 1988; 85:6032
17	H-2ps	major histocompatibility, pseudo class I	clone 27.1	+	*Steinmetz	Cell 1981; 25:683
17	H-2	major histocompatibility, class I	pH-2IIa, pH-2II	+	*Steinmetz	Cell 1981; 24:125
			pH-2I	+	*Steinmetz	Cell 1981; 24:125
			pH2d37C	+	5-Lalanne	Cell 1985; 41:469
			C1432	+	*Flaherty	PNAS 1985; 82:1503
		major histocompatibility, class I (5')	407Bam		*Weiss	Nature 1984; 310:650
17	H-2D	major histocompatibility-D	pH-2^{d}-1		*Bregegere	Nature 1981; 292:78
			"		*Kvist	PNAS 1981; 78:2772
			pH-2^{d}-3		*Bregegere	Nature 1981; 292:78
17	H-2D^{k}	major histocompatibility-D^{k}	pH-2III	+	*Steinmetz	Cell 1981; 24:125
17	H-2K	major histocompatibility-K	mp8.17.1		*Winoto	PNAS 1983; 80:3425
			"		*Mains	Immunogenetics 1986; 23:357
			probe 2(or D)		*Steinmetz	Cell 1986; 44:895
			probe 4(or C)		*Steinmetz	Cell 1986; 44:895
			probe 5(or B)		*Steinmetz	Cell 1986; 44:895
			cosmids: H6.13, H6.3		*Steinmetz	Cell 1986; 44:895
		major histocompatibiltiy-K, unique seq	probe 8D		*Kress	J Biol Chem 1983; 258:13929
			"	+	*Uehara	EMBO J 1987; 6:83
17	H-2K^{k}	major histocompatibility-K^{k}	H2Kf8		*Minamide	Immunogenetics 1988; 27:148
17	H-2K^{b}	flank, major histocompatibility K^{b}	LS1/1, Bm1-18,		*Weiss	Nature 1984; 310:650
			Bm1R3/1		*Uehara	EMBO J 1987; 6:83
17	H-2K^{d}	major histocompatibility-K^{d}	p1954	+	*Forejt	Genet Res 1988; 51:111
17	H-2L^{d}	major histocomatibility-L^{d}	pMHC-1		1-Seidman	ref: Evans, PNAS 1982; 79:1994
			"		*Margulies	Nature 1982; 295:168
			probes: 1, 2, 3, 3A,	+	*Rubocki	PNAS 1986; 83:9606
			probes: 4, 5, 6, 7		*Sun	J Exp Med 1985; 162:1588
17	Hba-4ps	hemoglobulin α-4, pseudogene	apsi4		1-Leder	Nature 1981; 293:196
			"	+	*D'Eustachio	J Exp Med 1984;159:958
			α-Ψ4	+	1-Silver	ref: Fox, Immunogenetics 1984; 19:125
17	Hprt-ps	Hprt, pseudo	λHPT30	+	*Isamat	Somatic Cell Mol Genet 1988; 14:359
17	H-2I	I region of H-2	cosmid H3.31		*Muller	EMBO J 1987; 6:369
			cosmids: H1.1, H2.1		*Steinmetz	EMBO J 1984; 3:2995
			cosmids: H2.19, H6.19		*Steinmetz	EMBO J 1984; 3:2995
			cosmid H9.19		*Steinmetz	EMBO J 1984; 3:2995

Chr	Symbol&	Gene Name	Probe@	RFLV#	Holder‡	Reference∞
17	H-2IA	Ia region of H-2	probe14(or 1)		*Steinmetz	Cell 1986; 44:895
			cosmids: II10.6, II5.9		*Steinmetz	Cell 1986; 44:895
			I-Agen	+	5-McConnell	J Immunol 1986; 136:3076
			I-Aspec	+	5-Gallahan	J Virol 1986; 61:218
			pIaβ-1	+	*Seidman	ref: Epstein, J Exp Med 1985; 161:1219
			probes 6, 7, 12.2	+	*Steinmetz	Cell 1986; 44:895
			"	+	*Uematsu	Immunogenetics 1988; 27:96
			probes 8,10,13		*Steinmetz	Cell 1986; 44:895
			"		*Uematsu	Immunogenetics 1988; 27:96
			probe A, probe C	+	*Uematsu	EMBO J 1986; 5:2123
			"	+	*Uematsu	Immunogenetics 1988; 27:96
17	H-2IAα	Ia α region of H-2	pAAC6		*Benoist	PNAS 1983; 80:534
17	H-2IAβ	Ia β region of H-2	p2894	+	1-Wu	ref: Robinson, J Immunol 1983; 131:2025
			unnamed (B6.C-H-2^{bm12})		*McIntyre	Nature 1984; 308:551
		I-$A\beta^{w30-1}$ allele (Tuw7)	I-$A\beta^{w30-1}$ genomic		*Golubic	Genet Res 1987; 50:137
		I-$A\beta^{w36-1}$ allele (Tuw8)	I-$A\beta^{w36-1}$ genomic		*Golubic	Genet Res 1987; 50:137
17	H-2IAβ	I-$A\beta^{w31-2}$ allele (Tuw10)	I-$A\beta^{w31-2}$ genomic		*Golubic	Genet Res 1987; 50:137
17	H-2IE	Ie region of H-2	probe16(or10)		*Steinmetz	Cell 1986; 44:895
			probe15(or10)		*Steinmetz	Cell 1986; 44:895
			cosmids: II4.14, II12.6		*Steinmetz	Cell 1986; 44:895
17	H-2IEα	Ie α region of H-2	clone 32.1		*fr:Hood	ref: Steinmetz, Nature 1982; 300:35
			"		*Artzt	Devel Genet 1987; 8:1
			probe 7		*fr:L. Hood	ref: McNicholas, Science 1982; 218:1229
			"	+	*Lafuse	Immunogenetics 1986; 24:352
17	H-2IEβ	Ie β region of H-2	B1,Ba		*Steinmetz	Nature 1982; 300:35
			"	+	*Lafuse	Immunogenetics 1986; 24:352
			probe 4		*fr:L. Hood	ref: Steinmetz, Nature 1982; 300:35
			"		*Lafuse	Immunogenetics 1986; 24:352
		I-$E\beta^{w31-1}$ allele (Tuw8)	I-$E\beta^{w31-1}$ genomic		*Golubic	Genet Res 1987; 50:137
		I-$E\beta^{w31}$ allele (Tuw10)	I-$E\beta^{w31}$ genomic		*Golubic	Genet Res 1987; 50:137
17	H-2IEβ2	Ie β2 region of H-2	probe 1, 2, 3, 4, 5	+	*Braunstein	EMBO J 1986; 5:2469
17	Int-3	mammary tumor integration site -3	int-3	+	5-Gallahan	J Virol 1986; 51:218
			pA1, pCz-4, pCz-5		*Gallahan	J Virol 1987; 61:66
			probe A		*Gallahan	J Virol 1987; 61:218
17	Mb1	class I MHC subfamily gene (telomeric Qa)	Mb1	+	*Singer	Immunogenetics 1988; 28:13
17	Ms15-3	minisatellite 15-3	[H] probes 33.6, 33.15		*Jeffreys	Nuc Acid Res 1987; 15:2823
17	Oh21-1, -2	steroid 21-hydroxylase-1, active gene	pM21OH	+	1-Meo	ref: Amor, PNAS 1985; 82:4453
17	Oh21	steroid 21-hydroxylase	[H] pC21/3c	+	*Gillet	Immunogenetics 1988; 27:133
			"		*White	PNAS 1986; 83:5111
			clone 3.3, clone 9.2	+	*Amor	PNAS 1985; 82:4453
			E-7, E-26		*Chaplin	PNAS 1986; 83:9601
			[B] pcBP-450c21-9		*Yoshioka	J Biol Chem 1986; 258:4106
			"		*Gotoh	Endocrinology 1988; 123:1923
17	Pim-1	proviral integration , MCF	unnamed		*Nagarajan	PNAS 1986; 83:2556
			probe A	+	*Nadeau	Genetics 1987; 117:533
			C5C clone		*Cuypers	Cell 1984; 37:141
			PrA (C5C subclone)		*Vijaya	J Virol 1987; 61:1164
17	Pgk-1ps2	phosphoglycerate kinase-1 pseudogene 2	?		*Adra	Somatic Cell Mol Genet 1988; 14:69
17	Pgk-2	phosphoglycerate kinase-2	Probe G (pMPGK-2)		*Boer	Mol Cell Biol 1987; 7:3107
			"		*Adra	Somatic Cell Mol Genet 1988; 14:69
			clone B12		*Adra	Somatic Cell Mol Genet 1988; 14:69
17	Q1-Q3	Q lymphocyte antigen-1,-2, -3	B1.24, H26		*Weiss	Nature 1985; 310:650
			"		*Robinson	Immunogenetics 1988; 27:79
			c50.2.1		*Winoto	PNAS 1983; 80:3425
			"	+	*O'Neill	Immunogenetics 1986; 24:368
17	Qa-2,3	Qa lymphocyte antigen-2, -3	p46.11		*Winoto	PNAS 1983; 80:3425
			"		*Lyon	Immunogenetics 1988; 27:375
17	Q4,Q6,Q8	Q lymphocyte antigen-4,-6,-8	c1432	+	*Flaherty	PNAS 1985; 82:1503
			"	+	*O'Neill	Immunogenetics 1986; 24:368
17	Q4, Q5	Q lymphocyte antigen-4,-5	Bm1-2, B2.5		*Weiss	Nature 1985; 310:650
			"		*Robinson	Immunogenetics 1988; 27:79
17	Q7/Q9	Q lymphocyte antigen-7,-9 (intron)	c5.23	+	**O'Neill	Immunogenetics 1986; 24:368
17	Q10	Q lymphocytpe antigen-10 region gene	10Q10 probe		*Cosman	PNAS 1982; 79:4947
17	Slp	sex- limited protein	pSlp/1		*Nonaka	PNAS 1984; 81:6822
			pSlp12J		*Sepich	PNAS 1985; 82:5895
			cosmids: 15.3, 38.3		*Stavenhagen	Mol Cell Biol 1987; 7:1716
			cosmids: 11.1, 13.4		*Stavenhagen	Mol Cell Biol 1987; 7:1716

Chr	Symbol&	Gene Name	Probe@	RFLV#	Holder‡	Reference∞
17	Slp^{w7d}	sex-limited protein (w7d allele)	cosmids: cl-16, cl-25,		*Nakayama	J Immunol 1987; 138:620
			cosmids: cl-35, cl-42		*Nakayama	J Immunol 1987; 138:620
			cosmids: cl-18, cl-27		*Nakayama	J Immunol 1987; 138:620
			cosmids: cl-17, cl-34		*Nakayama	J Immunol 1987; 138:620
17	Slp^{w7c}	sex-limited protein (w7c allele)	cosmids: cl-1, cl-36		*Nakayama	J Immunol 1987; 138:620
17	Slp^{w7b}	sex-limited protein (w7b allele)	cosmids: cl-29, cl-18		*Nakayama	J Immunol 1987; 138:620
17	Slp^{w7a}	sex-limited protein (w7a allele)	cosmids: cl-30, cl-15		*Nakayama	J Immunol 1987; 138:620
			cosmids: cl-22, cl-6,		*Nakayama	J Immunol 1987; 138:620
			cosmid: cl-23		*Nakayama	J Immunol 1987; 138:620
17	Sod-2	superoxide dismutase-2, mitochondrial	unnamed		*Hallewell	Nuc Acid Res 1986; 14:9539
17	Tcp-1	t-complex protein-1	p30	+	5-Willison	Cell 1986; 44:727
17	Tcp-1^{b}	t-complex protein-1 (b allele)	pB1.4	+	*Willison	JEEM 1986; 97:151
			"		*Lyon	Immunogenetics 1988; 27:375
17	Tcr-1	transmission distortion responder (t)	pTcr7, pTcr9, pTcr12		*Schimenti	Cell 1988; 55:71
			pTcr16, pTcr10		*Schimenti	Cell 1988; 55:71
		Tcr? (t-locus,testes)	Cg3-79, Cg3-100		*Schimenti	J Mol Biol 1987; 194:583
17	Tla	thymus leukemia antigen	p37E3		*Kourilsky	Cell 1985; 41:469
		Tla (genes: T14^{c}, T15^{c},T16^{c},T17^{c},gene 37)	c37-16		*Transy	J Exp Med 1987; 166:341
			"		*Sun	J Exp Med 1985; 162:1588
		Tla gene 37, T17^{c}	λ:16		*Transy	J Exp Med 1987; 166:341
		thymus leukemia antigen, gene 37	pH-2^{d}-37		*Lalanne	Cell 1985; 41:469
			"		*Transy	J Exp Med 1987; 166:341
17	Tnf	tumor necrosis factor	pmTα2		1-Jongeneel	Nuc Acid Res 1986;14:7713
			λ6		*Pennica	PNAS 1985; 82:6060
17	Tnfa	tumor necrosis factor, α	p-mTNF		*Fransen	Nuc Acid Res 1985; 13:4417
17	Tnfa/Tnfb	tumor necrosis factor α, β	clone603-38	+	*Gardner	J Immunol 1987; 139:476
			clone603-1	+	*Gardner	J Immunol 1987; 139:476
			clone603-18	+	*Gardner	J Immunol 1987; 139:476
17	Tnfb	tumor necrosis factor, β	unnamed		*Gray	Nuc Acid Res 1987; 15:3937
			pMuLT		*Li	J Immunol 1987; 138:4496
17	Tpx-1	testis specific gene	1-1-1H4, B2-1	+	*Kasahara	Immunogenetics 1989; 29:61
18	Fim-2	Friend MuLV integration site-2	341-PB04	+	1-Gisselbrecht	ref: Sola, J Virol 1988; 62:3973
			clone 17(1)		*Sola	J Virol 1986; 60:718
			341-HS18		*Sola	J Virol 1986; 60:718
18	Fms	feline sarcoma oncogene, McDonough	[H] pSM7C		*Hoggan	J Virol 1988; 62:1055
18	Grl-1	glucocorticoidreductase-1	pW28,clone21		*Danielson	EMBO J 1986; 5:2513
			pN10		*Northrop	J Biol Chem 1986; 261:11064
18	Ii	Ia associated invariant chain	pIi5	+	5-Richards	Immunogenetics 1985; 22:193
			"		*Singer	EMBO J 1984; 3:873
			"		*Koch	EMBO J 1987; 6:1677
			cos10.7		*Yamamoto	J Immunol 1985; 134:3461
			"		*Koch	EMBO J 1987; 6:1677
18	Mbp	myelin basic protein	PP535		*Roth	J Neurosci Res 1986; 16:227
			clones: M44, M72, M78		*Newman	PNAS 1987; 84:886
			λ471, B7,		*Molineaux	PNAS 1986; 83:7542
			pdF191,λL2		*Molineaux	PNAS 1986; 83:7542
			pNZ111, pdF191		*deFerra	Cell 1985; 43:721
			pMBP-1		*Roach	Cell 1983; 34:799
			pMBP-2		*Kimura	J Neurochem 1985; 44:692
			Cos138		*Readhead	Cell 1987; 48:703
18	Mbp	myelin basic protein (deleted)	λ25	+	*Molineaux	PNAS 1986; 83:7542
18	Mbp	myelin basic protein (shiverer mutant)	Cosm 352, Cosm 214,		*Popko	Cell 1987; 48:713
			Cosm 211, Cosm 191,		*Popko	Cell 1987; 48:713
			Cosm 231, Cosm 201,		*Popko	Cell 1987; 48:713
			Cosm 31, Cosm 41c,		*Popko	Cell 1987; 48:713
			Cosm 14, Cosm 91,		*Popko	Cell 1987; 48:713
			Cosm 101, Cosm 111,		*Popko	Cell 1987; 48:713
			Cosm 132, Cosm 302,		*Popko	Cell 1987; 48:713
			Cosm 12c		*Popko	Cell 1987; 48:713
18	Mtv-2	mammary tumor virus locus-2	MMTV (GR-40)		*Michalides	Virology 1985; 142:278
			"		*Hynes	PNAS 1982; 78:2038
			MTV-LTR		*Michalides	Virology 1985; 142:278
18	"Tg171"	transgene (human insulin)	[H] 11kb frag insulin	+	*Michalova	Hum Genet 1988; 80:247
18	Zpf-6	zinc protein-6	DF2, DF6		*Lemaire	PNAS 1988; 85:4691
			OC3.1		*Sukhatme	Cell 1988; 53:37
			mgEgr-1.1, p2.4, p6.6		*Tsai-Morris	Nuc Acid Res 1988; 16:8835

Chr	Symbol&	Gene Name	Probe@	RFLV#	Holder‡	Reference∞
19	D19Rp19	DNA segment, Roswell Park-29	pMK1711	+	1-Elliott	Mouse News Letter 1988; 80:180
19	"Ly-44"	lymphocyte antigen-44	pmB1-1, pmB1-3		*Tedder	J Immunol 1988; 141:4388
19	P450-2C	cytochome P450IIC	[R] pTF-1	+	*Meehan	PNAS 1988; 85:2662
			"		*Friedberg	Biochemistry 1986; 25:7975
			pM8-1	+	*Meehan	PNAS 1988; 85:2662
			pPB5-21, pPB3-15		*Meehan	PNAS 1988; 85:2662
19	Pomc-2	pro-opiomelanocortin-β	λβ7, pMKSU16		*Uhler, Herbert	J Biol Chem 1983; 258:9444
X	Ags	a-galactosidase	[H] λAG18	+	*fr: Bishop	ref: Mullins, Genomics 1988; 3:187
X	Araf	raf-related oncogene	mA-raf		*Huebner	PNAS 1986; 83:3934
			p19-1		*Huleihel, Rapp	Mol Cell Biol 1986; 6:2655
			"	+	*Avner	Somatic Cell Mol Genet 1987; 13:267
X	Cybb	cytochrome b-245, β polypeptide	[H] hs7	+	5-Avner	Trends in Genetics 1988; 4:18
			"		*Nathans	Science 1986; 232:193
X	Cf-8	coagulation factor VIII	[H] p61-51	+	*fr: Knopf	ref: Mullins, Genomics 1988; 3:187
			[H] p114.12	+	5-Lawn RM	ref: Avner, Trends in Genetics 1988; 4:18
			[H] pKPN		*Toole	Nature 1984; 312:342
			"	+	*Cavanna	Genomics 1988; 3:337
X	Cf-9	coagulation factor IX	[H] F9	+	5-Avner	PNAS 1987; 84:1629
			"		*Cameurio	Nature 1983; 306:701
			[H] pKT218	+	*fr: McGraw	ref: Mullins, Genomics 1988; 3:187
X	Dmd	Duchenne muscular dystrophy sequences	MC2-6	+	*Hoffman	Science 1987; 238:347
			"		*Chamberlain	Somatic Cell Mol Genet 1987; 13:671
			[H] FSM-5		*Monaco	Trends in Genetics 1987; 3:33
X	Dmd	Duchenne muscular dystrophy sequences	[H] pCA1	+	*Brockdorff	Nature 1987; 328:166
			[H] pCa1b		*Cross	EMBO J 1987; 6:3277
			"	+	*Cavanna	Genomics 1988; 3:337
X	"DX?"	DNA segment, testis specific probe	PL10		*Leroy	Devel 1987; 101 (supp): 177
X	"DX?"	DNA segment, nonstringent hybridization	pYMT2/B		*Bishop	Nuc Acid Res 1987; 15:2959
X	DXPas1	DNA segment, Pasteur Institute-1	probe 45 (or MSDX1)	+	2-Avner	EMBO J 1985; 4:3695
X	DXPas2	DNA segment, Pasteur Institute-2	probe 52 (or MDSX2)	+	2-Avner	EMBO J 1985; 4:3695
X	DXPas3	DNA segment, Pasteur Institute-3	probe 66 (or MSDX3)	+	2-Avner	EMBO J 1985; 4:3695
X	DXPas4	DNA segment, Pasteur Institute-4	probe 87 (or MSDX4)	+	2-Avner	EMBO J 1985; 4:3695
X	DXPas5	DNA segment, Pasteur Institute-5	probe 100 (or MSDX5)	+	2-Avner	EMBO J 1985; 4:3695
X	DXPas6	DNA segment, Pasteur Institute-6	[H] probe C11	+	2-Avner	PNAS 1987; 84:5330
			"		*Mandel	Cold Spr Hbr Symposium 1986; 51:195
X	DXPas7	DNA segment, Pasteur Institute-7	[H] probe M2C	+	2-Mandel	Cold Spr Hbr Symposium 1986; 51:195
X	DXPas8	DNA segment, Pasteur Institute-8	[H] probe St14		2-Mandel,Oberle	PNAS 1985; 82:2824
			"	+	*Avner	PNAS 1987; 84:1629
X	DXPas9	DNA segment, Pasteur Institute-9	probe 80 (or MDFXY)	+	2-Bishop	ref: Avner, Trends in Genetics 1988; 4:18
X	DXPas10	DNA segment, Pasteur Institute-10	"	+	2-Bishop	ref: Avner, Trends in Genetics 1988; 4:18
X	DXPas11	DNA segment, Pasteur Institute-11	"		2-Bishop	ref: Avner, Trends in Genetics 1988; 4:18
X	DXPas12	DNA segment, Pasteur Institute-12	"		2-Bishop	ref: Avner, Trends in Genetics 1988; 4:18
X	DXPas13	DNA segment, Pasteur Institute-13	[H] CCR1	+	5-Avner	Heilig: Nature 1987; 328:168
			"		*Worton	Science 1986; 226:1447
X	DXPas14	DNA segment, Pasteur Institute-14	[H] E87-25	+	5-Mandel	ref: Heilig, Nature 1987; 328:168
X	DXPas15	DNA segment, Pasteur Institute-15	[H] p212	+	5-Pearson	ref: Avner, Trends in Genetics 1988; 4:18
X	DXPas17	DNA segment, Pasteur Institute-17	CP8.6	+	4-Avner	Trends in Genetics 1988; 4:18
X	DXPas18	DNA segment, Pasteur Institute-18	L9	+	2-Avner	Trends in Genetics 1988; 4:18
X	DXPas19	DNA segment, Pasteur Institute-19	L11	+	4-Avner	Trends in Genetics 1988; 4:18
X	DXPas20	DNA segment, Pasteur Institute-20	[H] CX52.5	+	5-Pearson	ref: Avner, Trends in Genetics 1988; 4:18
X	DXPas22	DNA segment, Pasteur Institute-22	CR5	+	4-Avner	ref: Avner, Trends in Genetics 1988; 4:18
X	DXSmh10	DNA segment, St. Mary's Hospital-10	DXSmh10 (MDSX10)		1-Brown	ref: Fisher, PNAS 1985; 82:5846
			"	+	*Brockdorff	EMBO J 1987; 6:3291
X	DXSmh14	DNA segment, St. Mary's Hospital-14	DXSmh14	+	1-Brown	ref: Brockdorff, EMBO J 1987; 6:3291
X	DXSmh23	DNA segment, St. Mary's Hospital-23	DXSmh23	+	1-Brown	ref: Brockdorff, EMBO J 1987; 6:3291
X	DXSmh36	DNA segment, St. Mary's Hospital-36	DXSmh36 (MDRX36)		1-Brown	ref: Fisher, PNAS 1985; 82:5846
			"	+	*Brockdorff	EMBO J 1987; 6:3291
X	DXSmh43	DNA segment, St. Mary's Hospital-43	DXSmh43 (MDSX43)		1-Brown	ref: Fisher, PNAS 1985; 82:5846
			"	+	*Brockdorff	EMBO J 1987; 6:3291
X	DXSmh44	DNA segment, St. Mary's Hospital-44	DXSmh44 (MDSX44)	+	1-Brown	ref: Fisher, PNAS 1985; 82:5846
			"		*Brockdorff	EMBO J 1987; 6:3291
X	DXSmh59	DNA segment, St. Mary's Hospital-59	DXSmh59	+	1-Brown	ref: Brockdorff, EMBO J 1987; 6:3291
X	DXSmh64	DNA segment, St. Mary's Hospital-64	DXSmh64	+	1-Brown	ref: Brockdorff, EMBO J 1987; 6:3291
X	DXSmh66	DNA segment, St. Mary's Hospital-66	DXSmh66	+	1-Brown	ref: Brockdorff, EMBO J 1987; 6:3291
X	DXSmh67	DNA segment, St. Mary's Hospital-67	DXSmh67	+	1-Brown	ref: Brockdorff, EMBO J 1987; 6:3291
X	DXSmh91	DNA segment, St. Mary's Hospital-91	DXSmh91	+	1-Brown	ref: Brockdorff, EMBO J 1987; 6:3291
X	DXSmh93	DNA segment, St. Mary's Hospital-93	DXSmh93	+	1-Brown	ref: Brockdorff, EMBO J 1987; 6:3291
X	DXSmh117	DNA segment, St. Mary's Hospital-117	DXSmh117	+	1-Brown	ref: Brockdorff, EMBO J 1987; 6:3291
X	DXSmh120	DNA segment, St. Mary's Hospital-120	DXSmh120 (MDSX120)	+	1-Brown	ref: Fisher, PNAS 1985; 82:5846
			"		*Brockdorff	EMBO J 1987; 6:3291
X	DXSmh129	DNA segment, St. Mary's Hospital-129	DXsmh129		*Brockdorff	EMBO J 1987; 6:3291
			"	+	*Cavanna	Genomics 1988; 3:337

Chr	Symbol&	Gene Name	Probe@	RFLV#	Holder‡	Reference∞
X	DXSmh141	DNA segment, St. Mary's Hospital-141	DXSmh141 (MDRX141)	+	1-Brown	ref: Fisher, PNAS 1985; 82:5846
			"		*Brockdorff	EMBO J 1987; 6:3291
X	DXSmh172	DNA segment, St. Mary's Hospital-172	DXSmh172 (MDSX172)		1-Brown	ref: Fisher, PNAS 1985; 82:5846
			"	+	*Brockdorff	EMBO J 1987; 6:3291
X	DXSmh191	DNA segment, St. Mary's Hospital-191	DXSmh191	+	1-Brown	ref: Brockdorff, EMBO J 1987; 6:3291
X	DXSmh219	DNA segment, St. Mary's Hospital-219	DXSmh219 (MDSX219)	+	1-Brown	ref: Fisher, PNAS 1985; 82:5846
			"		*Brockdorff	EMBO J 1987; 6:3291
X	DXSmh222	DNA segment, St. Mary's Hospital-222	DXSmh222 (MDRX222)		1-Brown	ref: Fisher, PNAS 1985; 82:5846
			"	+	*Brockdorff	EMBO J 1987; 6:3291
X	DXSmh225	DNA segment, St. Mary's Hospital-225	DXSmh225 (MDRX225)		1-Brown	ref: Fisher, PNAS 1985; 82:5846
			"	+	*Brockdorff	EMBO J 1987; 6:3291
X	DXSmh255	DNA segment, St. Mary's Hospital-255	DXSmh255	+	1-Brown	ref: Brockdorff, EMBO J 1987; 6:3291
X	DXWas17	DNA segment, Univ of Washington-17	probe 17A	+	2-Disteche	Cytometry 1982; 2:282
X	DXWas31	DNA segment, Univ of Washington-31	probe 31A	+	2-Disteche	Cytometry 1982; 2:282
X	DXWas68	DNA segment, Univ of Washington-68	probe 68-36	+	2-Disteche	Cytogenet Cell Genet 1985; 39:262
X	DXWas70	DNA segment, Univ of Washington-70	probe 70-38	+	2-Disteche	Cytogenet Cell Genet 1985; 39:262
			"		*Disteche	Nuc Acid Res 1987; 15:4393
X	Gdx	Anonymous gene near G6pd	G28 + derivatives	+	1-Avner,Amar	PNAS 1987; 84:1629
X			"	+	*Cavenna	Genomics 1988; 3:337
X	G6pd	glucose-6-phosphate dehydrogenase	[H] pGDP3		*Avner	PNAS 1987; 84:1629
			"	+	*fr: Luzzatto	ref: Mullins, Genomics 1988; 3:187
X	Hprt	hypoxanthine guanine phoshoribosyl transferase	pHPT5		1-Caskey	ref:Konecki, Nuc Acid Res 1982; 10:6763
			"		*Melton	PNAS 1984 81:2147
			"	+	*Avner	PNAS 1987; 84:1629
X	Hprt	hypoxanthine guanine phoshoribosyl transferase	pMEV	+	*fr:Johnson	ref: Mullins, Genomics 1988; 3:187
X	Hprt(3')		pHPT2		*Brennard	PNAS 1982; 79:1950
X	Hprt(intron1)		pHPRλ13-in1		*Lock	Mol Cell Biol 1986; 6:914
			"		*Jones	Somatic Cell Mol Genet 1987;13:325
X	Odc-13	ornithine decarboxylase related seq-13	pMK934	+	1-Elliott	Mouse News Lett 1984; 71:46
			? probably pMK934	+	*Stephenson	Nuc Acid Res 1988; 16:1642
			pODC934		1-Berger	J Biol Chem 1984; 259:7941
X	Otc	sparse fur (wildtype)	unnamed		*Veres	Science 1987; 237:415
			"		*Veres	J Biol Chem 1986; 261:7588
			pOTC		*fr:Caskey	ref: Mullins, Mol Cell Biol 1987; 7:3916
			"	+	*fr:Caskey	ref: Mullins, Genomics 1988; 3:187
X	Otc	sparse fur	unnamed		*Veres	Science 1987; 237:415
			[R] pOTC-1		*Takiguchi	PNAS 1984; 81:7412
			"		*Ohtake	BBRC 1987; 146:1064
X	Pgk-1	phosphoglycerate kinase-1	λPGKM63		*Mori	Gene 1986; 45:275
			"	+	*Cavanna	Genomics 1988; 3:337
			[H] probe A (pHPGK7-e)		5-Michelson	PNAS 1983; 80:472
			"		"	ref: Adra, Som Cell Mol Genet 1988; 14:69
			probe E (pMPGK5-b)		5-Michelson	ref: Adra, Som Cell Mol Genet 1988; 14:69
			clone B12.5, B17		*Adra	Somatic Cell Mol Genet 1988; 14:69
X	Plp	myelin proteolipid protein	λPLP 1, λPLP 2		*Hudson	PNAS 1987; 84:1454
			λ23,λ28		*Moriguchi	Gene 1987; 55:333
			[R] p27		*Milner	Cell 1985; 42:931
X	Plp	myelin proteolipid protein (jp mutant)	p-23	+	2-Dautigny	Nature 1986; 321:867
			pJ-31, pJ-81		*Nave	PNAS 1986; 83:9264
		myelin proteolipid protein (wildtype)	pC-4		*Nave	PNAS 1986; 83:9264
X	Rsvp	red sensitive visual pigment	[H] hs7	+	*fr: Nathans	ref: Mullins, Genomics 1988; 3:187
X	Syn-1	synapsin I	[R] p5E2	+	5-deGennaro	ref: Avner, Trends in Genetics 1988; 4:18
			"		*Yang-Feng	PNAS 1986; 83:8679
X	Timp	tissue inhibitor of metalloproteinase`	unnamed	+	*Jackson	Nuc Acid Res 1987; 15:4357
			p3/10		*Gewert	EMBO J 1987; 6:651
			"	+	*fr:Williams	ref: Mullins, Genomics 1988; 3:187
X	Xlr	X-linked family of B cell surface antigens	pX310, pM1	+	2-Cohen	Nature 1985; 314:369
X&Y	"?"	unnamed	pY80/B, pY371/B,		*Avner,Bishop	Mouse News Letter 1986; 74:98
			pY302/B		*Avner,Bishop	Mouse News Letter 1986; 74:98
X&Y		Mov-15	pMov-15/1	+	*Harbers	Nature 1986; 324:682
Y	"DY?"	Y chr repeat	AC11	+	*Nishioka	Genetics 1986; 113:417
Y	"DY?"	male sequence	ACC2, ACC3		*Nishioka	Genome 1987; 29:380
Y	"DY?"	Y chr sequence	pYB10		5-Eicher	ref: Propst, Oncogene Res 1988; 2:227
Y	"DY?"	Y chr sequence	pY1, pY2, pY3	+	1-Lamar	Cell 1984; 37:171
Y	"DY?"	Y chr sequence	pY353/B	+	*Bishop	Nature 1985; 315:70
Y	"DY?"	Y chr sequence	pYMT2/B		*Bishop	Nuc Acid Res 1987; 15:2959
Y	"DY?"	Y repetitive sequence	M34	+	*Singh	Mol Gen Genet 1988; 212:440
Y	"DY?"	Y sex determining region	Σ[8]	+	*Singh	Mol Gen Genet 1988; 212:440

Probes where gene location is unknown or sequences are on multiple chromosomes

Chr	Symbol&	Gene Name	Probe@	RFLV#	Holder‡	Reference∞
?	"Actg"	actin, gamma	GA63		*Tokunaga	Mol Cell Biol 1988; 8:3929
?	"Actg-ps"	γ-actin pseudo	λmA81,λmA36,		*Man	Nuc Acid Res 1987; 15:3291
?			λmA118,λmA119		*Man	Nuc Acid Res 1987; 15:3291
?	"Agly-1"	a1 acid glycoprotein-1	pMAGP2, pMAGP3		*Cooper	J Biol Chem 1986; 261:1849
?	"Agly-2"	a1 acid glycoprotein-2	pMAGP4		*Cooper	J Biol Chem 1986; 261:1849
?	AKR-MuLV+flank		clone 623, clone 614		*Lowy	PNAS 1980; 77:614
?	AKR-MuLV+flank		clone 621		*Lowy	PNAS 1980; 77:614
?	AKV env probe		pAKV-4		*Bautch	J Virol 1986 60;693
			"		*Herr	J Virol 1983 46:70
			pAKV-3, -5		*Herr	J Virol 1983 46:70
			"		*Mucenski	J Virol 1987; 61:2929
			pEC0		*Chattopadhyay	PNAS 1980; 77:5774
?	"Alas"	ALA synthase	pML5-1, pMS20		*Schoehaut	Gene 1986; 48:55
			pMS6, pMS12		*Schoehaut	Gene 1986; 48:55
?	Aldo-1	aldolase-1	unnamed		*Mestek	Nuc Acid Res 1987; 15:10595
?	"Ang"	angiotensin	λMAI,λgMA2,cMA2		*Clouston	Genomics 1988; 2:240
			cMA20,cMA18		*Clouston	Genomics 1988; 2:240
?	"Arg-1"	arginase (liver)	[R]pARGr-2		*Ohtake	BBRC 1987; 146:1064
M	"Bam5"	BAM5 repeat	pMBA-14		*Kashara	PNAS 1987; 84:3325
			"		*Fanning	Nuc Acid Res 1982; 10:5003
M	"B1"	B1 repeat	pMR225		*Flores	Plasmid 1987; 17:257
			pMR225, pMR164,		*Bennett	Mol Cell Biol 1984; 4:1561
			pMR118, pMR50, pMR131		*Bennett	Mol Cell Biol 1984; 4:1561
			p61-41, B1-9, B1-15		*Moshier	Gene 1987; 58:19
			Mm31		*fr:Georgiev	ref: Krayev, Nuc Acid Res 1980; 8:1201
					*Jubier-Maurin	Nuc Acid Res 1987; 15:7395
M	"B2"	B2 repeat	pMR142		*fr: Lueders	Fukumoto, Som Cell Mol Gen 1986; 12:611
			"		*Flores	Plasmid 1987; 17:257
			pMR142, pMR293		*Bennett	Mol Cell Biol 1984; 4:1561
			p49C8,p7E4		*fr:Edwards	Mol Cell Biol 1985; 5:3280
			"		*Gallant	Mol Biol Rpt 1987; 12:49
?	Bmk	B cell myeloid kinase (= hck)	hck cDNA		*Holtzman	PNAS 1987; 8325
			MK21.1		*Ziegler	Mol Cell Biol 1987;7:22
			clone 13.5		*Klemsz	Nuc Acid Res 1987; 15:9600
?	"B-myc"	B-myc	[R] pRM44		*Ingvarsson	Mol Cell Biol 1988; 8:3168
?	"Bgp"	bone gla	unnamed		*Celeste	EMBO J 1986; 5:1885
?	"Campa"	cAMP, α subunit	MCG-1, MCG-3, MCG-5		*Chrivia	J Biol Chem 1988; 263:5739
?	"Campb"	cAMP, β subunit	unnamed		*Chrivia	J Biol Chem 1988; 263:5739
?	Calm	calmodulin	pCaM16		*Putkey	J Biol Chem 1983; 258:11864
			"		*Kluxen	Eur J Biochem 1988; 176:153
?	"Cathb"	preprocathepsin B	λmCB14,		*Chan	PNAS 1986; 83:7721
			λmCB54, λmCB58		*Chan	PNAS 1986; 83:7721
?	"Cathl"	cathepsin L precursor	pcosMMEP, pMMEP-14		*Troen	J Biol Chem 1988; 263:254
?	"Clq"	C1q (B chain complement)	pCl-q-B-C65,		*Wood	Immunol Lett 1988; 17:59
			pClq-B-C61, pClq-B-C301,		*Wood	Immunol Lett 1988; 17:59
			pClq-B-C56, pCl1-B-C78		*Wood	Immunol Lett 1988; 17:59
?	"Clgi"	collagenase inhibitor	16C8		*Edwards	Nuc Acid Res 1986; 14:8863
?	Col3a-	procollagen type III α	pMC51		1-deCrombugghe	J Biol Chem 1985; 260:531
			pCIII-1-C119,pCIII-1-C534		*Wood	Gene 1987; 61:225
			pCIII-1-C572		*Wood	Gene 1987; 61:225
?	Col4a-	procollagen type IV, α	pPE1180		*Kurkinen	Nuc Acid Res 1983; 11:6199
?	Col4a-1	procollagen type IV, α-1	pPE41, pPE131,		*Kurkinen	J Biol Chem 1987; 262:8496
			pPE90, pPE132, pPE123		*Kurkinen	J Biol Chem 1987; 262:8496
			MIV-5,MIV-6, MIV-7,MIV-8		*Killen	J Mol Biol 1988; 263:8706
			p21, pF-5		*Killen	J Mol Biol 1988; 263:8706
			pCIV1-C92, pCIV-1-C177		*Wood	Febs Lett 1988; 227:1
			pCIV-1C308, pCIV-1-C87		*Wood	Febs Lett 1988; 227:1
			pCIV-1-PE12, pCIV-1-PE16		*Wood	Febs Lett 1988; 227:1
?	Col4a-2	procollagen type IV, α-2	pPE10A		*Kurkinen	J Biol Chem 1987; 262:8496
			HA3		*Kurkinen	Nature 1985; 317:177
?	"Ctl-x"	unknown,expresses in CTLs	B10, C11		*Lobe	PNAS 1986; 83:1448
?	"Dpolb"	DNA polymerase β	λMGB01		*Yamaguchi	Mol Cell Biol 1987; 7:2012
?	"Ecl"	EC repeat	pMR161, pMR272		*Bennett	Mol Cell Biol 1984; 4:1561
?	Eftu	elongation factor Tu	λmTu-1, pWR1		*Roth	Mol Cell Biol 1987; 7:3929
?	"Ela-2"	elastase,typeII	pMPe7, λME4, λME1		*Stevenson	Nuc Acid Res 1986; 14:8307
?	"Ercc-1"	excision repair gene	λcDME		*vanDuin	Nuc Acid Res 1988; 16:5305
			clones 1B, 5A, 4A, 3.1		*vanDuin	Nuc Acid Res 1988; 16:5305
M	"Ecol"	1.3kb EcoRI repeat	pMD006, pMD007		*Dubnick	J Mol Evol 1983; 19:115
			pMD335, pMD302,		*Dubnick	J Mol Evol 1983; 19:115
			pMD350 ,pMD416,		*Dubnick	J Mol Evol 1983; 19:115
			pMD437, pMD106,		*Dubnick	J Mol Evol 1983; 19:115
			pMD131, pMD135		*Dubnick	J Mol Evol 1983; 19:115

Chr	Symbol&	Gene Name	Probe@	RFLV# Holder‡	Reference∞
M	Emv-?	Emv loci	unnamed	*Jenkins	J Virol 1982; 43:26
			"	*Chattopadhyay	PNAS 1980; 77:5774
M	Env-?	env gene (xenotropic)	pXenv	*Bedigian	Immunogenetics 1986; 23:156
				*fr:Bedigian	ref: Colombo, Oncogene 1988; 2:395
			"	*Buckler	J Virol 1982; 41:228
			MCF247, NZB-IV-6	*Hoggan	J Virol 1986; 60:980
			"	*O'Neill	J Virol 1986; 58:359
?	"Enoa"	α-enolase	pBEα1, pBEα2, pBEα3,	*Lazar	BBRC 1986; 141:271
		(non-neuron)	pBEα4, pBEα5, pBEα6,	*Lazar	BBRC 1986; 141:271
			pBEα7, pBEα8	*Lazar	BBRC 1986; 141:271
?	"Enob"	β-enolase(neuron)	pBEg1, pBEg2, pBEg3	*Lazar	BBRC 1986; 141:271
10 or 15	Fabph-3	fatty acid binding protein, heart-1	[R] H-FABP cDNA	*Heuckeroth	J Biol Chem 1987; 262:9709
?	"Fas"	fatty acid synthase	pFAS-1	*Paulauskis	J Biol Chem 1988; 263:7049
?	"Fil"	Filaggrin	pFM3-2, pFM6-1A2	*Rothnagel	J Biol Chem 1987; 262:15643
?	"Fth"	ferritin, heavy chain	unnamed cDNAs	*Torti	J Biol Chem 1988; 263:12638
			"	*Yoshitaka	Nuc Acid Res 1988; 16:10373
?	"Gltb"	β-1,4-galactosyltransferase	MGT-1, MGT-2	*Shaper	J Biol Chem 1988; 263:10420
			MGT-P1 through-P7	*Shaper	J Biol Chem 1988; 263:10420
?	"Glus"	glutamine synthetase	SE1.9, Glus-genomic	*Bhandari	PNAS 1988; 85:5789
			[CH] pGSRK-1	*Burns	BBRC 1986; 134:146
?	"Gnrhp"	GnRH assoc. protein	λmGP-1, λmGP-2	*Mason	Science 1986; 234:1366
?	"Hist1-ps"	H3.3-like pseudo	MH611,MH321,MH921	*Wellman	Gene 1987; 59:29
?	"Hf"	HF gene	unnamed	*Gershenfeld	Science 1986; 232:854
?	Hsp68	heat shock protein 68	pXS3.0	*Aujame	Can J Genet Cytol 1986;28:1064
			pmHS214	*Aujame	Mol Cell Bio 1985; 5:1780
?	Hsp70	heat shock protein 70	pM1.8	*fr:Morimoto	ref: Krawczyk, Mol Bio Rpt 1987; 12:27
			pM9.5	*Hunt	ref: Zakeri, Mol Cell Biol 1988; 8:2925
			unnamed	*Giebel	Devel Biol 1988; 125:200
?	"Hsp70.2"	heat shock protein 70-related	λ621, pM3.8	*Zakeri	Mol Cell Biol 1988; 8:2925
?	"Hsp70-ps"	heat shock protein 70, pseudo	λ4, pMHS213	*Lowe	PNAS 1984; 81:2317
			"	*Zakeri	Mol Cell Biol 1988; 8:2925
?	"Hsp89"	heat shock protein 89	pC6-23	*Levine	Mol Cell Biol 1984; 4:2142
M	IAP	intracisternal A particles	pMIA1	*Lueders	PNAS 1980; 77:3571
			"	*Kuff	Chromosoma 1986; 93:213
			MIA14,MIA48,	*Kuff	PNAS 1986; 83:6583
			pMIA34	*Lueders	PNAS 1980; 77:3571
			"	*Kuff	Mol Cell Biol 1981; 1:216
M	IAP	intracisternal A particles	clones 8.3, 9.5, 10.2	*Grossman	Nuc Acid Res 1987; 15:3823
			"	*Moore	J Immunol 1986; 136:4283
			"	*Martens	PNAS 1985; 82:2460
			clones B12,D8,D20	*Grossman	Nuc Acid Res 1987; 15:3823
			L31, p3A67	*Aota	Gene 1987; 56:1
			A-81	*Cole	J Virol 1981; 38:680
1 or 3	Il-2	interleukin -2	MMT-1, MT-20	*Yokota	PNAS 1985; 82:68
			MT-28, MT-18	*Yokota	PNAS 1985; 82:68
			pMIL2-45, pMIL2-20	*Fuse	Nuc Acid Res 1984; 12:9323
			pMIL2-20	*fr:Taniguchi	ref: Kashima, Nature 1985; 313:402
			"	*fr: Taniguchi	ref: D'Eustachio, J Immunol 1988; 141:3067
			MIL-2G70	*fr: Taniguchi	ref: D'Eustachio, J Immunol 1988; 141:3067
?	"Il2r"	interleukin-2 receptor	pmIL-2R-1, pmIL2R-2,	*Shimuzu	Nuc Acid Res 1985; 13:1505
			pmIL-2R-3	*Shimuzu	Nuc Acid Res 1985; 13:1505
?	"Il-5"	interleukin-5	clone 4G	*Yokota	PNAS 1987; 84:7388
			pSP6K-mTRF23	*Kinashi	Nature 1986; 324:70
?	"Il-6"	interleukin-6	pHP1A4, pHP1B5	*VanSnick	Eur J Immunol 1988; 18:193
?	"Int-41"	MMTV integration site-41	pBλ41, pAS311	*Garcia	EBMO J 1986; 5:127
?	"Preb-1"	pre-B cell marker	pZ183	*Sakaguchi	EMBO J 1986; 5:2139
?	"Igf-a"	preproIGF-1A	migf1-1, migf1-2	*Bell	Nuc Acid Res 1986; 14:7873
			migf1-3, migf1-9	*Bell	Nuc Acid Res 1986; 14:7873
?	"Igf-b"	preproIGF-1B	migf1-4, migf1-13	*Bell	Nuc Acid Res 1986; 14:7873
			migf1-10, migf1-13	*Bell	Nuc Acid Res 1986; 14:7873
?	"Kcb-1"	mouse brain K+ channel-1	MBK1.6, MBK1.18,	*Tempel	Nature 1988; 332:837
			MBK1.9, MBK1.14	*Tempel	Nature 1988; 332:837
M	L1Md	L1 repeat	1.35,X	*Papaconstantinou	Nuc Acid Res 1984; 12:6575
M	L1Md-A2	L1 repeat	CE102	*Loeb	Mol Cell Biol 1986; 6:168
M	L1Md-A2	L1 repeat (M. domesticus)	M2-25, M1-27, M2-27,	*Edgell	Developmental Control of Globin Gene
			M2-22, M2-23C, M2-20,	"	Expression, Alan R Liss, Inc, p. 107, 1987
			M1-25, M2-19	"	"
M	L1Md-A13	L1 repeat	CE113	*Shehee	J Mol Biol 1987; 196:757
M	LINE1	L1 repeat	pbS18	*Schmeckpeper	Nuc Acid Res 1981; 9:1853
			pMR134	*Voliva	Nuc Acid Res 1983; 11:8847
			clones 1G-1 through 1G-10	*Jubier-Maurin	Nuc Acid Res 1987; 15:7395
			clones 4AG 1 through 4AG-10	*Jubier-Maurin	Nuc Acid Res 1987; 15:7395
			pMR290	*fr: Hastie	ref: Bennett, EMBO J 1984; 3:467
			"	*Jubier-Maurin	Nuc Acid Res 1987; 15:7395

Chr	Symbol&	Gene Name	Probe@	RFLV#	Holder‡	Reference∞
M	LINE1	L1 repeat	pBfL-5		*fr:Fanning	ref: Fanning, Nuc Acid Res 1983; 11:5073
			"		*Jubier-Maurin	Nuc Acid Res 1987; 15:7395
			L5.8		*fr:Mottez	ref: Mottez, Nuc Acid Res 1986; 14:3119
			"		*Jubier-Maurin	Nuc Acid Res 1987; 15:7395
?	"Ldh-1ps"	LDH-A, pseudo	λM10		*Li	Mol Biol Evol 1986; 3:330
?	"Ldh-3"	LHD-C	mC31,mC50		*Sakai	Biochem J 1987; 242:619
?	Lif	leukemia inhibitory factor	pLIFNK3		*Gearing	Nuc Acid Res 1988; 16:9857
?	"Lig-1"	lipogenic induced gene	p7.2		*Paulauskis	J Biol Chem 1988; 263:7049
?	"Lig-2"	lipogenic induced gene-2	p5.1		*Paulauskis	J Biol Chem 1988; 263:7049
M	"Lrep"	LLRep3 repeat	λ8B1-2, λ8-B1-1		*Heller	Mol Cell Biol 1988; 8:2797
?	"Lmyc"	L-myc	unnamed		*Zimmerman	Nature 1986; 319:780
?	"Lpl"	lipoprotein lipase	mL5, mL31, mL11		*Kirchgessner	J Biol Chem 1987; 262:8463
?	Lzm	M lysozyme	unnamed		*Cross	PNAS 1988; 85:6232
?	"Mac-1a"	macrophage antigen α chain (integrin)	λMM-1		*Sastre	PNAS 1986; 83:5644
?			clones: 106, 118, 217		*Pytela	EMBO J 1988; 7:1371
			clones:202, 204, 308		*Pytela	EMBO J 1988; 7:1371
?	"Mdrs"	multidrug resistance	λDR11, pcDR1.3,		*Gros	Nature 1986; 323:728
			"		*Croop	Canc Res 1987; 47:5982
			pcDR1.4, pcDR1.5		*Gros	Nature 1986; 323:728
			[CH] pDR7.8, pDR1.6		*Gros	PNAS 1986; 83:337
?	"Mdrs-2"	multidrug resistance-2	DR29, DR27		*Gros	Mol Cell Biol 1988; 8:2770
?	"Mes-1"	Mes-1	pB1		*Williams	Mol Cell Biol 1986; 6:4558
?	"Mga-ps"	MgA-Y1pseudo	λmA19		*Leader	Gene 1985; 36:369
?	"Mgp-1"	membrane glycoprotein-1 (CD3 complex?)	pZ176, m-mb-1-W-8		*Sakaguchi	EMBO J 1988; 7:3457
?	"Micap2"	mouse microtubule associated protein 2	unnamed (cDNAs)		*Wang	Nuc Acid Res 1988; 11369
?	"Mifr"	MIF repeat	pMR134, pMR257		*Bennett	Mol Cell Biol 1984; 4:1561
			pMR134		*Flores	Plasmid 1987; 17:257
			pMR257		*Bennett	Mol Cell Biol 1984; 4:1561
M	"Ms"	minisatellite	pSP64.2.5EI		*Georges	Nuc Acid Res 1987; 15:7193
M	"Ms"	minisatellite	mo-1	+	*Kominami	Nuc Acid Res 1988; 16:1197
M	"Mins"	minor satellite	PMR150		*Voliva	Nuc Acid Res 1983; 11:8847
?	"Mip"	Mip	MuMIP clone 52		*Davatellis	J Exp Med 1988; 167:1939
			MuMIP clone 32		*Davatellis	J Exp Med 1988; 167:1939
?	Mov-2	Moloney leukemia virus-2*	pMov-2		*Harbers	Nuc Acid Res 1982; 10:2521
?	"Mx-2"	myxovirus (influenza virus) resistance 2	unnamed cDNA	+	*Staeheli	Mol Cell Biol 1988; 8:4524
M	"Mtr"	MT repeats	λM68910-1, λM68910-2		*Bastien	Gene 1987; 57:81
			CebA-847, MT 1, MT-36		*Heinlein	Nuc Acid Res 1986; 14:6403
M	"?"	F-MuLV (flank)	E57BS, 341-PB04		*Sola	J Virol 1986; 60:718
M	"?"	F-MuLV (env)	p0.8		*Silver	J Virol 1986; 57:526
M	"?"	Mo-MuLV	p8.2		*fr. Goff	ref: Schwartzberger,J Virol 1983; 46:538
			"		*fr: Goff	ref:Colombo, Oncogene 1988; 2:395
M	"?"	MuLV envelope	pEC_{env}		*Chan	PNAS 1979; 77:5779
M	"?"	MuLV specific	pC6		*Panthier	Virol 1987; 138:409
M	"?"	pro-MuLV (RFM)	pRFM#6		*Nikbakht	J Gen Virol 1987; 68:683
?	"RFM-R"	solitary MuLV LTR from RFM/Un	pRFM2		*Kuemmerle	Virology 1987; 160:379
			pRFM7,pRFM8		*Kuemmerle	Virology 1987; 160:379
?	"RFM-Rf"	RFM-R flank from pRFM7	probe 7A, 7B	+	*Kuemmerle	Virology 1987; 160:379
?	"RFM-Rf-2"	RFM-R flank from pRFM8	probe 8A, 8B	+	*Kuemmerle	Virology 1987; 160:379
M	"?"	MMTV (GR)	λMMTV-GR		*Buetti	Cell 1981; 23:335
			MMTV Rep		*Callahan	PNAS 1982; 79:4113
M	"?"	MMTV (C3H)	pMMTV(C3H)		*Callahan	PNAS 1982; 79:4113
			p201		*fr: Callahan	ref: Major, Nature 1981, 289:253
			"		*fr: Callahan	ref: Colombo, Oncogene 1988; 2:395
			p4.2A		*fr:Majors	ref:Traina-Dodge, Genetics 1985; 111:597
M	"?"	MURRS (murine retrovirus rel seq)	MuRRS-1, MuRRs-2	+	*Schmidt	Nuc Acid Res 1985; 13:3461
			MuRRs-5, MuRRs-6	+	*Schmidt	Nuc Acid Res 1985; 13:3461
?	"nf-68"	neurofilament NF68	NF68		*fr:Lewis	ref: Somerville, Genome 1988; 30:499
?	"nfl"	neurofilament, small subunit	NF-L cDNA		*Julien	Mol Brain Res 1986; 1:243
?	"nfl"&"nfm"	neurofilament, small & midsize subunits	cosIC3		*Julien	Mol Brain Res 1986; 1:243
?	"nfm"	neurofilament, midsize subunit	NF-M cDNA, λNF-M		*Julien	Mol Brain Res 1986; 1:243
?	"nfh"	neurofilament, large subunit	cos3A1		*Julien	Mol Brain Res 1986; 1:243
			"		*Julien	Gene 1988; 68:307
			cos2A1, cos1A1		*Julien	Mol Brain Res 1986; 1:243
?	"Nhcp-17"	non-histone chromosomal protein (HMG-17)	pM17c		*Landsman	Nuc Acid Res 1988; 16:10386
?	"Nhcp-I"	non-histone chormosomal protein (HGM-I)	HGMI cDNA		*Johnson	J Biol Chem 1988; 263:18338
?	Odc-?	ornithine decarboxylase-?	pODC16		*Janne	Ann NY Acad Sci 1984;438:72
?	Odc-?	ornithine decarboxylase-?	pOD12.7		*Coffino	Nuc Acid Res 1988; 16:2731
?	"P450-rel"	C-P-$450_{16\alpha}$ related (5 genes)	clones 5,21,52,54,58		*Wong	Biochemistry 1987; 26:8683
?	"P450-16a"	C-P-45016α (male-specfic)	p16α-1		*Harada	PNAS 1985; 82:2024
			"		*Wong	Biochemistry 1987; 26:8683
			p16-α2,p16α-16		*Wong	Biochemistry 1987; 26:8683
?	"P450-t"	P-450, testis	unnamed cDNA		*Virbasius	J Biol Chem 1988; 263:6791
?	"Per-1"	Per-related (Drosophilia)	pSP64.2.5EI		*Georges	Nuc Acid Res 1987; 15:7193
			clone 115		*Ishida	Nuc Acid Res 1988; 16:3581

Chr	Symbol&	Gene Name	Probe@	RFLV#	Holder‡	Reference∞
?	"Pif4a"	protein synthesis initiation factor 4A	FA1, FA2		*Reddy	Gene 1988; 70:231
			unnamed genomic clones		*Nielsen	EMBO J 1988; 7:2097
?	"Pif-4aps	Pif-4a pseudogenes	clones A,B,C,E,F,G,H,I,J		*Nielsen	EMBO J 1988; 7:2097
?	"Pif-4ar"	Pif-4a retroposon	clone D		*Nielsen	EMBO J 1988; 7:2097
?	"Pif-4a-2"	protein synthesis initiation factor 4A-II	unnamed genomic clones		*Nielsen	EMBO J 1988; 7:2097
M	Pgk-1ps?	Pgk-1 related sequences	clones B6, B8, B13		*Adra	Somatic Cell Mol Genet 1988; 14:69
?	"Phkg"	Phk γ subunit	pγ-Phk3,pγ-Phk9,pγ-Phk2		*Chamberlain	PNAS 1987; 84:2886
?	"Rdsp"	rDNA spacer	pMrA		*Kuhn	EMBO J 1987; 6:3487
?	"Rr"	R repeat	pMR290		* Bennett	Mol Cell Biol 1984; 4:1561
			"		*Flores	Plasmid 1987; 17:257
?	"Rpo3"	RNA polymerase III	p71		*Gallant	Mol Bio Rpt 1987; 12:49
?	Rnr-?	ribosomal RNA	SALE, SALB, PP		*Bourbon	DNA 1988; 7:181
			Mr974, pMR974		*Grummt	Nuc Acid Res 1979; 6:1351
?	"Rpr"	45S pre-rRNA promotor	pMr6000		*Kuhn	EMBO J 1987; 6:3487
?	Rnu?	U1 small nuclear RNA, brain specific	λAB3118		*Kobayashi	Nuc Acid Res 1988; 16:360
?	Rnu?	U1 small nuclear RNA, brain specific	λAB1290		*Anzai	J Neurochem 1986; 47:63
?	Rnu1.1, Rnu1.2	U1 small nuclear RNA	unnamed		*Moussa	Nuc Acid Res 1987; 15:3622
?	Rnu1-ps2	U1 small nuclear RNA-1 pseudogene-2	unnamed	+	*Blatt	Somatic Cell Mol Genet 1988; 14:133
?	Rnu1b2	U1 small nuclear RNA-1b2	pU1b-136		*Howard	Nuc Acid Res 1986; 14:9811
?	Rnu2	U2 small nuclear RNA -2	U2.47, p16-41		*Moshier	Gene 1987; 58:19
?	"Rnu2-B1"	B1 repeats assoc with Rnu2	B1-9, B1-15		*Moshier	Gene 1987; 58:19
?	"Rpl27"	Ribosomal protein L27	cCL3		*Belhumeur	Nuc Acid Res 1987; 15:1019
?	"RpS6"	Ribosomal protein S6	unnamed		*Lalanne	Nuc Acid Res 1987; 15:4990
M	"Sat"	satellite DNA	p16		*fr:Fittler	ref: Horz, Nuc Acid Res 1981; 9:683
M	"Sat"	satellite DNA	pSAT1, pSAT2		*Radic	Cell 1987; 50:1101
M	"Spcd"	small polydispersed circular DNA	pSP (many)		*Sunnerhagen	Nuc Acid Res 1986; 14:7823
?	Surf-1	surfeit gene-1	T2A, T1B, IDE		*Williams	Mol Cell Biol 1986; 6:4558
?	Surf-2	surfeit gene-2	H1, T3C, J3		*Williams	Mol Cell Biol 1986; 6:4558
?	"Surf-3"	surfeit gene-3	MY3		*Williams	PNAS 1988; 85:3527
?	"Surf-4"	surfeit gene-4	RMC4		*Williams	PNAS 1988; 85:3527
?	"Tau"	Tau protein	pTA1, pTA2		*Drubin	J Cell Biol 1984; 98:1090
			pTA2E, pTA3,		*Lee	Science 1988; 239:285
			pTA3E, pTA2E'		*Lee	Science 1988; 239:285
?	"Thbp"	thyroid hormone binding protein (p55)	unnamed		*Gong	Nuc Acid Res 1988; 16:1203
?	Tp-1	transition protein-1	pCE3,HSAR700		*Kleene	Devel Biol 1983; 98:455
			"		*Kleene	BBA 1988; 950:215
?	Tp-2	transition protein-2	pMTP2-1, pMTP2-2		*Kleene	J Biol Chem 1987; 262:17272
?	"Tyms"	thymidylate synthase	TSB-9, TSB-16,		*Deng	J Biol Chem 1986; 261:16000
			TS-PPH-1, pMTS-3		*Deng	J Biol Chem 1986; 261:16000
			pMTS-1, pMTS-2, pMTS-4		*Geyer	J Biol Chem 1984; 259:7206
?	"Tps-1"	tumor promoter sensitive gene	pHA		*Lerman	Int J Canc Res 1986; 37:293
			pRG-11, P3R, P2R		*Garrity	Gene 1988; 68:63
?	"Try-1"	trypsin T_a	λMT4		*Stevenson	Nuc Acid Res 1986; 14:8307
?	"Try-2"	trypsin T_b	λMT5		*Stevenson	Nuc Acid Res 1986; 14:8307
?	"Try-3"	trypsin T_d	λMT2		*Stevenson	Nuc Acid Res 1986; 14:8307
?	"Try-3,-4"	trypsin T_c,T_d	λMT9		*Stevenson	Nuc Acid Res 1986; 14:8307
?	"Try-5"	trypsin T_e	λMT12		*Stevenson	Nuc Acid Res 1986; 14:8307
?	"Tuba"	tubulin (α1)	[R] unnamed		*Lemischka	J Mol Biol 1981; 151:101
			"		*Schnieke	Nature 1983; 304:315
?	"U3ltr"	U3 ecotropic LTR	U3LTR		*Quint	J Virol 1984; 50:432
			"		*Mucenski	J Virol 1987; 61:2929
			unnamed		*Khan	Nuc Acid Res 1987; 15:7640
?	Wap	whey acidic protein	pSVc-myc-1		*fr:Henninghauser	ref:Schoenenberger, EMBO J 1988; 7:169
?	"ypt-1"	ypt-1	F9-104, F9-12,		*Haubruck	EMBO J 1987; 6:4049
			pcD-YPT82, pcD-YPT71		*Haubruck	EMBO J 1987; 6:4049
?	Zfp-1	zinc finger protein-1	λmkr1		*Chowdhury	Cell 1987; 48:771
?	Zfp-2	zinc finger protein-2	λmkr2		*Chowdhury	Cell 1987; 48:771
			cmKr2A, cmKr2B		*Chowdhury	EMBO J 1988; 7:1345
?	Zfp-17	zinc finger protein-17	phage 4.1,4.2,4.3		*Chavrier	Mol Cell Biol 1988; 8:1319
?	Zfp-19	zinc finger protein-19	phage 6.1		*Chavrier	Mol Cell Biol 1988; 8:1319
?	Zfp-21	zinc finger protein-21	phage 8		*Chavrier	Mol Cell Biol 1988; 8:1319
?	Zfp-22	zinc finger protein-22	phage 9.1,9.3		*Chavrier	Mol Cell Biol 1988; 8:1319
?	Zfp-33	zinc finger protein-33	phage 20		*Chavrier	Mol Cell Biol 1988; 8:1319
			pEX2.8, AC16		*Chavrier	EMBO J 1988; 7:29
?	Zfp-?	zinc finger protein-?	mkr3		*Chowdhury	Nuc Acid Res 1988; 16:9995
?	Zfp-?	zinc finger protein-?	mkr4		*Chowdhury	Nuc Acid Res 1988; 16:9995
?	Zfp-?	zinc finger protein-?	mkr5		*Chowdhury	Nuc Acid Res 1988; 16:9995

Retroviral and Cancer Related Genes of the Mouse (*Mus musculus*)
(2N = 40)

June, 1989

Christine A. Kozak
National Institute of Allergy and Infectious Diseases
National Institutes of Health
Bethesda, MD 20892

Several types of mouse genes are included in this map. Tables 1-3 list retroviral sequences inherited through the germline and identified by their expression as viral proteins or infectious virus, or by hybridization to segments of the viral genome. Such proviral sequences are often restricted to particular breeding lines or strains, and more complete information on strain distribution is available elsewhere (38,39,64,69). Table 4 lists cellular genes which affect virus expression, virus-induced disease, and tumor antigen expression. The remaining tables provide information on the cellular genes associated with the development of neoplasia: the non-germline viral integration sites disrupted in virus-induced tumors (Table 5) and oncogenes, growth factors, and their receptors (Table 6).

TABLE 1. Endogenous Mouse Mammary Tumor Viruses

Locus	Chromosome	Reference
Mtv-1[1]	7	152,154
Mtv-2[1]	18	94
Mtv-3	11	111
Mtv-6	16	18,123
Mtv-7	1	86,94,152
Mtv-8	6	126
Mtv-9	12	6,18,34
Mtv-11	14	18,62,86,119
Mtv-13	4	101,39
Mtv-17	4	94
Mtv-21	8	117

[1]Expressed as infectious virus.

Abbreviations: MuLV, murine leukemia virus; MMTV, mouse mammary tumor virus; MCF MuLV, mink cell focus-forming MuLV; *env*, viral envelope gene; *pol*, viral polymerase gene; RadLV, radiation leukemia virus; NDV, Newcastle disease virus; X-MuLV, xenotropic MuLV.

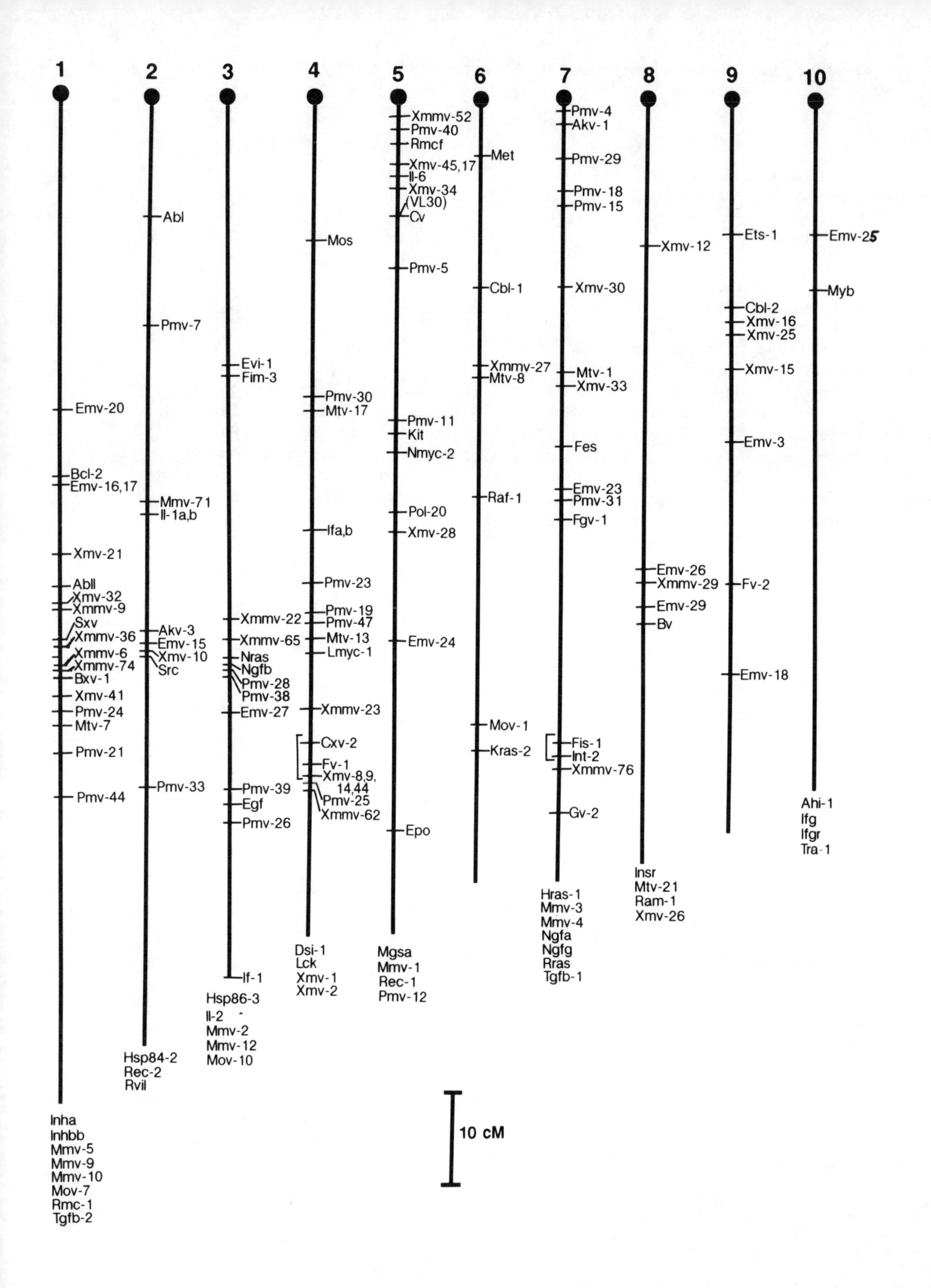
1
2
3
4
5
6
7
8
9
10
Emv-20
Bcl-2
Emv-16,17
Xmv-21
Abll
Xmv-32
Xmmv-9
Sxv
Xmmv-36
Xmmv-6
Xmmv-74
Bxv-1
Xmv-41
Pmv-24
Mtv-7
Pmv-21
Pmv-44
Inha
Inhbb
Mmv-5
Mmv-9
Mmv-10
Mov-7
Rmc-1
Tgfb-2
Abl
Pmv-7
Mmv-71
Il-1a,b
Akv-3
Emv-15
Xmv-10
Src
Pmv-33
Hsp84-2
Rec-2
Rvil
Evi-1
Fim-3
Xmmv-22
Xmmv-65
Nras
Ngfb
Pmv-28
Pmv-38
Emv-27
Pmv-39
Egf
Pmv-26
If-1
Hsp86-3
Il-2
Mmv-2
Mmv-12
Mov-10
Mos
Pmv-30
Mtv-17
Ifa,b
Pmv-23
Pmv-19
Pmv-47
Mtv-13
Lmyc-1
Xmmv-23
Cxv-2
Fv-1
Xmv-8,9,
14,44
Pmv-25
Xmmv-62
Dsi-1
Lck
Xmv-1
Xmv-2
Xmmv-52
Pmv-40
Rmcf
Xmv-45,17
Il-6
Xmv-34
(VL30)
Cv
Pmv-5
Pmv-11
Kit
Nmyc-2
Pol-20
Xmv-28
Emv-24
Epo
Mgsa
Mmv-1
Rec-1
Pmv-12
Met
Cbl-1
Xmmv-27
Mtv-8
Raf-1
Mov-1
Kras-2
Pmv-4
Akv-1
Pmv-29
Pmv-18
Pmv-15
Xmv-30
Mtv-1
Xmv-33
Fes
Emv-23
Pmv-31
Fgv-1
Fis-1
Int-2
Xmmv-76
Gv-2
Hras-1
Mmv-3
Mmv-4
Ngfa
Ngfg
Rras
Tgfb-1
Xmv-12
Emv-26
Xmmv-29
Emv-29
Bv
Insr
Mtv-21
Ram-1
Xmv-26
Ets-1
Cbl-2
Xmv-16
Xmv-25
Xmv-15
Emv-3
Fv-2
Emv-18
Emv-25
Myb
Ahi-1
Ifg
Ifgr
Tra-1
10 cM

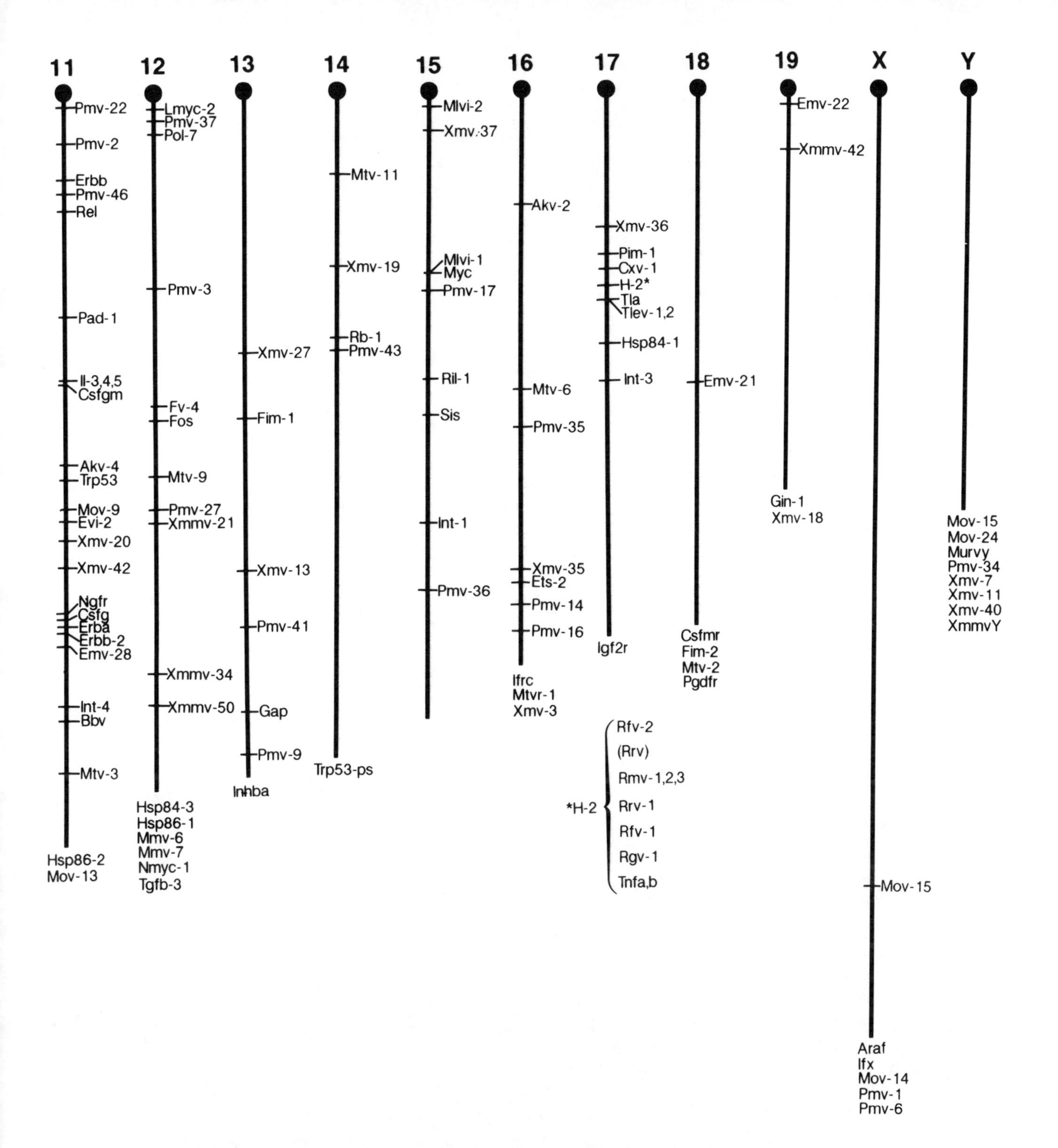

Mouse Linkage Map of Retroviral and Cancer Related Genes

TABLE 2. Endogenous Ecotropic Murine Leukemia Viruses

Type	Locus	Other Designations	Chromosome	Reference
AKV MuLV (N-tropic)	*Cv*[1,2]	*Emv-1*	5	60,70,73
	Emv-2[1]	*Bv*	8	73
	Emv-3[1]	*Dbv, Sev*	9	63,73
	Akv-1[1]	*Emv-11*	7	130
	Akv-2[1]	*Emv-12*	16	72
	Akv-3	*Emv-13*	2	149
	Akv-4[1]	*Emv-14*	11	149
	Emv-15[1]		2	22
	Emv-16[1], *17*[1]		1	15
	Emv-18[1]		9	103
	Emv-20		1	151
	Emv-21		18	151
	Emv-22		19	151
	Emv-23	*C58v-4*	7	124,136,151
	Emv-24		5	151
	Emv-25		10	151
	Emv-26	*C58v-1*	8	73,151
	Emv-27		3	151
	Emv-28		11	151
	Emv-29		8	151
	Fgv-1[1]		7	73
B-tropic	*Bbv*[1]		11	73
Wild mouse	*Fv-4*	*Akvr-1*	12	4,34,113
Moloney	*Mov-1*[1]		6	13
	Mov-7		1	107
	Mov-9[1]		11	107
	Mov-10		3	107
	Mov-13[1]		11	106,107,134
	Mov-14[1]		X	143
	Mov-15[1]		X,Y	48
	Mov-24[1]		Y	141

[1]Expressed as infectious virus.
[2]Integrated into retrovirus-like VL30 sequences (55).

TABLE 3. Endogenous Nonecotropic Murine Leukemia Viral Genes

Type[1]	Locus[2]	Chromosome	Reference
Xenotropic	*Bxv-1*[3]	1	71
	Xmv-1,2	4	53
	Xmv-3	16	53
	Xmv-21,32,41	1	38
	Xmv-10	2	38
	Xmv-8,9,14,44	4	38
	Xmv-45,17,34,28	5	38
	Xmv-30,33	7	38
	Xmv-26,12	8	38
	Xmv-16,25,15	9	7,38
	Xmv-20,42	11	7,38
	Xmv-27,13	13	38
	Xmv-19	14	38
	Xmv-37	15	38
	Xmv-35	16	38
	Xmv-36	17	38
	Xmv-18	19	38
	Xmv-7,11,40	Y	38
MCF (Polytropic)	*Mmv-5,9,10*	1	53
	Mmv-71[4]	2	53,92
	Mmv-2,12	3	53
	Mmv-1	5	53
	Mmv-3,4	7	53
	Mmv-6,7	12	53
	Pmv-24,21,44	1	39
	Pmv-7,33	2	39
	Pmv-28,38,39,26	3	39
	Pmv-30,23,19,47,25	4	7,39,160
	Pmv-40,5,11,12	5	7,39
	Pmv-4,29,18,15,31	7	7,39
	Pmv-2,22,46	11	39
	Pmv-37,3,27	12	39
	Pmv-41,9	13	39
	Pmv-43	14	146
	Pmv-17,36	15	39,92,103,160
	Pmv-35,14,16	16	39
	Pmv-1,6	X	39
	Pmv-34	Y	39
Xenotropic or MCF	*Xmmv-6,9,36*	1	7,160
	Xmmv-74	1	92
	Xmmv-22	3	9
	Xmmv-65	3	39
	Xmmv-23	4	7
	Xmmv-52	5	160
	Xmmv-27	6	7
	Xmmv-76	7	138
	Xmmv-29	8	7
	Xmmv-21,34	12	7
	Xmmv-50	12	160
	Xmmv-42	19	160
	XmmvY	Y	7

TABLE 3. (Cont.)

Type	Locus	Chromosome	Reference
Other	*Pol-20*	5	127
	Pol-7	12	127
	Tlev-1,2	17	114
	Murvy	Y	33

[1]Some of the viral *env* segments used as hybridization probes react specifically with xenotropic or MCF MuLV related sequences, other probes react with both virus types. *Mmv* and *Pmv* loci were identified using probes derived from different regions of the MCF *env*. *Tlev-2* and the *Pol* loci hybridize with probes derived from the retroviral polymerase gene, *Tlev-1* with probes from the *gag*, *pol*, and *env* regions, and *Murvy* with probes related to the wild mouse virus M720.

[2]Based on identical strain distribution patterns in RI lines, and, in some cases, fragment size similarities, the following independently described and mapped proviral loci may be identical: *Pmv-17, Xmmv-55, -72* and *Pol-23; Xmmv-61* and *-9; Xmv-42* and *Xmmv-3; Xmv-16* and *Xmmv-2; Pmv-18* and *Xmmv-35; Pmv-25* and *Xmmv-62; Pmv-5* and *Xmmv-5;* and *Pmv-30* and *Xmmv-8*. For each of these, the locus name listed in the table reflects hybridization with the most specific probe.

[3]Expressed as infectious virus.

[4]Also termed *Xmmv-71*.

Table 4 next page

TABLE 5. Tumor Specific Integration Sites

Locus	Inducing Virus	Chromosome	Reference
Ahi-1	Abelson MuLV	10	118
Dsi-1	Moloney MuLV	4	155
Evi-1	AKV MuLV	3	104
Evi-2	AKV MuLV	11	16,17
Fim-1	Friend MuLV	13	140
Fim-2	Friend MuLV	18	140
Fim-3	Friend MuLV	3	140
Fis-1	Friend MuLV	7	10,137,138
Gin-1	Gross Passage A	19	156
Int-1	MMTV	15	79,93,112,116
Int-2	MMTV	7	10,116,138
Int-3	MMTV	17	41
Int-4	MMTV	11	17
Mlvi-1= Mis-1=Pvt-1	Moloney, MCF MuLVs	15	65,66,76,157
Mlvi-2	Moloney MuLV	15	47,153
Pad-1	MMTV	11	17
Pim-1	Moloney, MCF MuLVs	17	50,108

TABLE 4. Genes Controlling Virus Replication, Viral and Tumor Antigens, and Resistance to Virus-Induced Disease

Locus	Phenotype	Chromosome	Reference
Virus Replication and Disease			
Cxv-1	High levels of X-MuLV	17	161
Fv-1	Resistance to N- or B-tropic MuLVs	4	131
Fv-2	Resistance to Friend SFFV	9	81
Fv-4	Resistance to ecotropic MuLVs	12	6,34,113
If-1	NDV induced interferon	3	96
Ifa	alpha-interferon	4	27,84
Ifb	beta-interferon	4	26,27
Ifg	gamma-interferon	10	109
Ifgr	gamma-interferon receptor	10	87
Ifrc	Interferon receptor	16	23
Ifx	NDV-induced interferon	X	31,115
Mtvr-1	MMTV receptor	16	51
Ram-1	Amphotropic MuLV receptor	8	43
Rec-1	Ecotropic MuLV receptor	5	43
Rec-2	M813 ecotropic MuLV receptor	2	121
Ril-1	Resistance to radiation leukemia	15	52,79,90
Rmc-1	MCF MuLV receptor	1	67
Rmcf	Resistance to MCF MuLVs	5	49
Rfv-1,2	Recovery from Friend MuLV-	17	21
Rgv-1	Resistance to Gross virus	17	82
Rmv-1,2,3	Resistance to Moloney virus	17	30
Rrv-1	Resistance to RadLV	17	83
-	Resistance to RadLV	17	89
Rvil	Resistance to RadLV disease	2	91
Sxv	Susceptibility to X-MuLVs	1	68
Viral antigens			
Cxv-2	High levels of XenCSA	4	102
Gv-2	G_{IX} antigen expression	7	144
Tumor antigens			
Hsp84-1	Heat shock protein sequences	17	99,127
Hsp84-2	related to tumor antigens	2	99
Hsp84-3	"	12	99
Hsp86-4	"	12	100
Hsp86-5	"	11	100
Hsp86-6	"	3	100
Tla	Thymus-leukemia antigen	17	12
Tra-1	Tumor rejection antigen gp96	10	142
Trp53	Transformation-related protein	11	16,25,129
Trp53-ps	Trp53 pseudogene	14	17,25

TABLE 6. Proto-oncogenes, Growth Factors and Their Receptors

Locus	Source/Description	Chromosome	Reference
Abl	Abelson MuLV	2	45
Abl1	Abelson-related gene	1	135
Bcl-2	B cell leukemia	1	97,110
Cbl-1	Murine retrovirus CasNS-1	6	122
Cbl-2	Murine retrovirus CasNS-1	9	122
Csfg	Granulocyte colony stimulating factor	11	17
Csfgm	B-cell macrophage colony stimulating factor	11	3,17
Csfmr=Fms	Colony stimulating factor-1 receptor, McDonough feline sarcoma virus	18	54
Egf	Epidermal growth factor	3	39,162
Epo	Erythropoietin	5	77
Erba(Egfr)	Avian erythroblastosis virus	11	17,163
Erbb	Avian erythroblastosis virus	11	17,139,163
Erbb-2	*Neu* protooncogene	11	3,17
Ets-1	Avian retrovirus E26	9	122,159
Ets-2	Avian retrovirus E26	16	123,159
Fes	Snyder-Theilen feline sarcoma virus	7	8,74,150
Fos	FBJ murine osteosarcoma virus	12	6,34
Gap	GPTase activating protein	13	57
Igf2r	Insulin-like growth factor, type 2	17	78
Il-1a,b	Interleukin-1 alpha and beta	2	2,35
Il-2	Interleukin-2	3	36
Il-3	Interleukin-3	11	3,17,61
Il-4	Interleukin-4	11	36,80,148
Il-5	Interleukin-5	11	80
Il-6	Interleukin-6	5	98
Inha	Inhibin, alpha subunit	1	5
Inhba	Inhibin, beta A subunit	13	5
Inhbb	Inhibin, beta B subunit	1	5
Insr	Insulin receptor	8	158
Kit	H24 feline sarcoma virus	5	20,44
Lck	Lymphocyte specific protein kinase	4	88
Met	Osteosarcoma MNNG-HOS	6	29
Mgsa	Melanoma growth stimulating activity	5	125
Mos	Moloney murine sarcoma virus	4	28,147
Myb	Avian myeloblastosis virus	10	11,59,95,133
Myc	Avian myelocytomalosis virus MC29	15	1,24,34,59,79,93
Lmyc-1	Small cell lung carcinoma-1	4	19
Lmyc-2	Small cell lung carcinoma-2	12	19
Nmyc-1	Neuroblastoma-1	12	19
Nmyc-2	Neuroblastoma-2	5	19
Raf-1	3611 murine sarcoma virus	6	125
Araf	Raf related gene	X	2,58
Hras-1	Harvey murine sarcoma virus	7	74,120,133
Kras-2	Kirsten murine sarcoma virus	6	125,133
Nras	Human neuroblastoma	3	14,46,132
Rras	Ras related gene	7	85
Ngfa	Nerve growth factor, alpha subunit	7	37,56
Ngfb	Nerve growth factor, beta subunit	3	32,162
Ngfg	Nerve growth factor, gamma subunit	7	37,56
Ngfr	Nerve growth factor receptor	11	17

TABLE 6. (Cont.)

Locus	Source/Description	Chromosome	Reference
Pgdfr	Platelet-derived growth factor receptor	18	17
Rb-1	Human retinoblastoma	14	145
Rel	Reticuloendotheliosis virus strain T	11	14
Sis	Simian sarcoma virus	15	1,75,79,93
Src	Rous sarcoma virus	2	8
Tgfb-1	Transforming growth factor, beta-1	7	40
Tgfb-2	Transforming growth factor, beta-2	1	4
Tgfb-3	Transforming growth factor, beta-3	12	4
Tnfa,b	Tumor necrosis factor alpha, beta	17	42,105

REFERENCES

1. Adolph,S., et al. 1987. Cytogenet. Cell Genet. 44:65.
2. Avner, P., et al. 1987. Somat. Cell Mol. Genet. 13:267.
3. Barlow, D. P. et al. 1987. EMBO J. 6:617.
4. Barton, D. E., et al. 1988. Oncogene Res. 3:323.
5. Barton, D.E., et al. 1989. Genomics 5:91
6. Blank, R. D., et al. 1988. Genetics 120:1073.
7. Blatt, C. et al. 1983. Proc. Natl. Acad. Sci. USA. 80:6298.
8. Blatt, C., et al. 1984. Mol. Cell. Biol. 4:978.
9. Blatt, C., et al. 1988. Somat. Cell Mol. Genet. 14:133.
10. Brilliant, M. H., et al. 1987. J. Neurogenet. 4:259.
11. Boydy, J. L. et al. 1988. Genetics 118:229
12. Boyse, E.A. et al. 1964. Nature 201:779.
13. Breindl, M., et al. 1979. Proc. Natl. Acad. Sci. USA 76:1938.
14. Brownell, E., et al. 1986. Am. J. Hum. Genet. 39:194
15. Buchberg, A.M., et al. J. Virol. 60:1175.
16. Buchberg, A. M., et al. 1988. Oncogene Res. 2:149.
17. Buchberg, A. M., et al. 1989. Genetics 122:153.
18. Callahan, R., et al. 1984. J. Virol. 49:1005.
19. Campbell, G. R., et al. 1989. Oncogene Res. 4:47
20. Chabot, B., et al. 1988. Nature 335:88.
21. Chesebro, B., et al. 1978. J. Immunol. 120:1081.
22. Copeland, N. G., et al. 1983. Proc. Natl. Acad. Sci. USA 80:247.
23. Cox, D. R., et al. 1980. Proc. Natl. Acad. Sci. USA 77:2168.
24. Crews, S., et al. 1982. Science 218:1319.
25. Crosnek, H. H. 1984. Mol. Cell. Biol. 4:1638.
26. Dandoy, F., et al. 1985. J. Virol. 56:216.
27. Dandoy, F., et al. 1984. J. Exp. Med. 160:294.
28. Dandoy, F., et al. 1989. Genomics 4: 546.
29. Dean, M., et al. 1987. Genomics 1:167.
30. Debre, P. et al. 1979. J. Immunol. 123:1806.
31. DeMaeyer-Guignard, J., et al. 1983. J. Interferon Res. 3:241.
32. Dracopoli, N. C., et al. 1988. Genomics 3:161.
33. Eicher, E. M., et al. 1989. Genetics 122:181.
34. Erikson, J., et al. 1986. J. Immunol. 136:3137.
35. D'Eustachio, P., et al. 1987. Immunogenet. 26:339.
36. D'Eustachio, P., et al. 1988. J. Immunol. 141:3067.
37. Evans, B. A., et al. 1985. EMBO J. 4:133.
38. Frankel, W. N., et al. 1989. J. Virol. 63:1763
39. Frankel, W. N., et al. J. Virol. In Press.
40. Fujii, D., et al. 1986. Somat. Cell Mol. Genet. 12:281.
41. Gallahan, D., et al. 1986. J. Virol. 61:218.

42. Gardner, S. M., et al. 1987. J. Immunol. 139:476.
43. Gazdar, A. F., et al. 1977. Cell 11:949.
44. Geissler, E. N., et al. 1988. Cell 55:185.
45. Goff, S. P., et al., 1982. Science 218:1317.
46. Guerrero, I., et al. 1984. Science 225:1041.
47. Hameister, H., et al. Cytogenetics and Cell Genet. In Press.
48. Harbers, K., et al. 1986. Nature 324:682.
49. Hartley, J. W., et al. 1983. J. Exp. Med. 158:16.
50. Hilkens, J., et al. 1986. Somat. Cell Mol. Genet. 12:81.
51. Hilkens, J., et al. 1983. J. Virol. 45:140.
52. Hogarth, P. M., et al. 1987. Immunogenet. 25:21.
53. Hoggan, M. D., et al. 1986. J. Virol. 60:980.
54. Hoggan, M. D., et al. 1988. J. Virol. 60:1055
55. Horowitz, J. M., et al. 1987. J. Virol. 61:701.
56. Howles, P. N., 1984. Nucl. Acids Res. 12:2791.
57. Hsieh, C. L. et al. Cytogenet. and Cell Genet. In Press.
58. Huebner, K., et al. 1986. Proc. Natl. Acad. Sci. USA 83:3934.
59. Huppi, K., et al. 1988. Immunogenet. 27:215.
60. Ihle, J. N., et al. 1979. Science 204:71.
61. Ihle, J. N., et al. 1987. J. Immunol. 138:3051.
62. Jeffreys, A. J., et al. 1987. Nucl. Acids Res. 15:2823.
63. Jenkins, N. A., et al. 1981. Nature 293:370.
64. Jenkins, N. A., et al. 1982. J. Virol. 43:26.
65. Jolicoeur, P., et al. 1985. J. Virol. 56:1045.
66. Koehne, C. F., et al. 1989 J. Virol. 63:2366
67. Kozak, C. A. 1983. J. Virol. 48:300.
68. Kozak, C. A. 1985. J. Virol. 55:690.
69. Kozak, C. A., et al. 1987. J. Virol. 16:1651.
70. Kozak, C. A., et al. 1979. Science 204:69.
71. Kozak, C. A., et al. 1980. J. Exp. Med. 152:219.
72. Kozak, C. A., et al. 1980. J. Exp. Med. 152:1419.
73. Kozak, C. A., et al. 1982. J. Exp. Med. 155:524.
74. Kozak, C. A., et al. 1983. J. Virol 47:217.
75. Kozak, C. A., et al. 1983. Science 221:867.
76. Kozak, C. A., et al. 1985. Mol. Cell. Biol. 5:894.
77. Lacombe, C., et al. 1988. Blood 4:1440.
78. Laureys, G., et al. 1988. Genomics 3:224.
79. LeClair, K. P., et al. 1987. Proc. Natl. Acad. Sci. USA 84:1628.
80. Lee, J. S., et al. 1989. Somat. Cell Mol. Genet. 15:143-152.
81. Lilly. F. 1970. J. Natl. Cancer Inst. 45:163.
82. Lilly, F., 1964 et al. Lancet ii:1207.
83. Lonai, P., et al. 1977. J. Exp. Med. 146:1164.
84. Lovett, M., et al. 1984. EMBO J. 3:1643.
85. Lowe, D. g., et al. 1987. Cell 48:137.
86. MacInnes, J. I., et al. 1984. Virol. 132:12.
87. Mariano, T.M., et al. 1987. J. Biol. Chem. 262:5812.
88. Marth, J. D., et al. 1986. Proc. Natl. Acad. Sci. USA 83:7400.
89. Meruelo, D., et al. 1977. J. Exp. Med. 146:1079.
90. Meruelo, D., et al. 1981. J. Exp. Med. 154:1201.
91. Meruelo, D. M., et al. 1983. Proc. Natl. Acad. Sci. USA 80:462.
92. Meruelo, D., et al. 1983. Proc. Natl. Acad. Sci. USA 80:5032.
93. Meruelo, D., et al. 1987. Immunogenet. 25:361.
94. Michalides, R. et al. 1985. Virol. 142:278.
95. Mitchell, M. et al. 1989. Genetics 121:803.
96. Mobraaten, L.E., et al. 1984. J. Hered. 75:233.
97. Mock, B., et al. 1988. Cytogenet. Cell Genet. 47:11.
98. Mock, B. A., et al. 1989. J. Immunol. 142:1372.
99. Moore, S.K., et al. 1987. Gene 56:29.
100. Moore, S. K. et al. 1989. J. Biol. Chen. 264:5343.

101. Morris, V. L., et al. 1979. Virol. 92:46.
102. Morse, H. C., III, et al. 1979. J. Exp. Med. 149:1183.
103. Mowat, M., et al. 1983. J. Virol. 47:471.
104. Mucenski, M. L., et al. 1988. Oncogene Res. 2:219.
105. Muller, U., et al. 1987. Nature 325:265.
106. Munke, M., et al. 1985. Cytogenet. Cell Genet. 40:706.
107. Munke, M., et al. 1986. Cytogenet. Cell Genet. 43:140.
108. Nadeau, H. J., et al. 1987. Genetics 117:533.
109. Naylor, S. L., et al. 1984. Cytogenet. Cell Genet. 37:550.
110. Negrini, M., et al. 1987. Cell 49:455.
111. Nusse, R., et al. 1980. J. Exp. Med. 152:712.
112. Nusse, R., et al. 1984. Nature 307:131.
113. Odaka, T., et al. 1981. J. Natl. Cancer Inst. 67:1123.
114. Pampeno C. L. et al. 1986 J. Virol. 58: 296.
115. Pederson, E. B., et al. 1983. Infect. Immun. 42:740.
116. Peters, G., et al. 1984. Mol. Cell Biol. 4:375.
117. Peters, G., et al. 1986. J. Virol. 59:535.
118. Poirier, Y., et al. 1988. J. Virol. 62:3985.
119. Prakash, O., et al. 1985. J. Virol. 54:285.
120. Pravtcheva, D. D., et al. 1983. Somat. Cell Genet. 9:681.
121. Rapp, U. R., et al. 1980. J. Supramol. Structure 14:343.
122. Regnier, D. C., et al. J. Virol. In Press.
123. Reeves, R. H., et al. 1987. Cytogenet. Cell Genet. 44:76.
124. Rinchik, E. M., et al. 1989. Genomics 4:251.
125. Robbins, J. M., et al. 1986. J. Virol. 57:709.
126. Romano, Y. W., et al. 1989. Immunogenet. 29:142.
127. Rossomando, A., et al. 1986. Immunogenet. 23:233.
128. Rotter, V., et al. 1984. Mol. Cell. Biol. 4:383.
129. Rowe, W. P., et al. 1972. Science 178:860.
130. Rowe, W. P., et al. 1973. Science 180:640.
131. Ryan, J., et al. 1984. Nucl. Acids Res. 12:6063.
132. Sakaguchi, A., et al. 1984. Proc. Natl. Acad. Sci. USA 81:525.
133. Sakaguchi, A. Y., et al. Cytogenet. Cell Genet. In Press.
134. Schnieke, A., et al. 1983. Nature 304:315.
135. Seldin, M. F., et al. 1989. Genomics 4:221.
136. Silver, J. 1985. J. Virol. 55:494.
137. Silver, J., et al. 1986. J. Virol. 57:526.
138. Silver, J. S., et al. 19 J. Virol. 60:1156
139. Silver, J. S., et al. 1985. Mol. Cell. Biol. 5:1784.
140. Sola, B., et al. 1988. J. Virol. 62:3973.
141. Soriano, P., et al. 1987. Genes Dev. 1:366.
142. Srivastava, P. K., et al. 1988. Immunogenet. 28:205.
143. Stewart, C., et al. 1983. Science 221:760.
144. Stockert, E., et al. 1972. Science 178:862.
145. Stone, J.C., et al. 1989. Genomics 5:70.
146. Stoye, J. P., et al. 1988. Cell 54:383.
147. Swan, D., et al. 1982. J. Virol. 44:752.
148. Takahashi, M., et al. 1989. Genomics 4:47.
149. Taylor, B. A., et al. 1985. J. Virol. 56:172.
150. Taylor, B. A., et al. 1989. M.N.L. 83:167.
151. Taylor,B. A., et al. Genomics In Press.
152. Traina, V. L., et al. 1981. J. Virol. 40:735.
153. Tsichlis, P. N., et al. 1984. Mol. Cell. Biol. 4:997.
154. Van Nie, R., et al. 1975. Int. J. Cancer 16:922.
155. Vijaya, S., et al. 1987. J. Virol. 1164.
156. Villermur, R., et al. 1987. Mol. Cell. Biol. 7:512.
157. Villeneuve, L., et al. 1986. Mol. Cell. Biol. 6:1834.
158. Wang, L.-M., et al. 1988. Genomics 3:172.
159. Watson, D. K., et al. 1986. Proc. Natl. Acad. Sci. USA 83:1792.
160. Wejman, J. C., et al. 1984. J. Virol. 50:237.
161. Yetter, R.A., et al. 1983. Proc. Natl. Acad. Sci. USA 80:505.
162. Zabel, B. U., et al. 1985. Proc. Natl. Sci. USA 82:469.
163. Zabel, B. U., et al. 1984. Proc. Natl. Acad. Sci. USA 81:4874.

Gene map of the Rat (*Rattus norvegicus*) 2N=42

August 1989

Göran Levan
Karin Klinga
Department of Genetics
University of Gothenburg
Box 33031
S-400 33 GOTHENBURG
Sweden

Claude Szpirer
Josiane Szpirer
Department of Molecular Biology
Université Libre de Bruxelles
67, Rue de Chevaux
B-1640 RHODE-ST-GENESE
Belgium

Number of genes listed: 150 (Table 1)
Number of genes assigned to specific chromosome: 130 (Table 2)
Number of genes assigned to specific linkage group: 92

Very few rat gene assignments have been confirmed by a second independent group. Therefore, most of the assignments are provisional.

Inconsistent assignments:

RT1. The major histocompatibility locus has been assigned to chromosome 14 (67). We have results that show that RT1, C4 and CBS are syntenic and located on chromosome 20 (52). Since many genes are assigned by their linkage to RT1 it is important that this inconsistency is completely clarified as soon as possible.

IGH. The immunoglobulin heavy chain has been assigned to chromosome 14 (76). A more recent study has presented firm evidence that IGH is on chromosome 6 (71).

ALB. A study using in situ hybridization has suggested that the albumin gene may be on chromosome 2 (82). In contrast, our study in somatic cell hybrids permitted assignment of the albumin gene to chromosome 14 (89). This assignment was corroborated by data from chromosome sorting (10).

PRL. The prolactin gene has been assigned to chromosome 9 (26). A more recent study assigned this gene to chromosome 17 (11).

Linkage groups.

Eleven linkage groups have been identified in the rat (Fig. 1). Five of these may be assigned to specific chromosomes on the basis of the data presented here.

Chromosome 1 - LG I
Chromosome 5 - LG II
Chromosome 6 - LG VIII
Chromosome 9 - LG X
Chromosome 20 - LG IX

Table 1. Alphabetical listing of genes assigned to chromosome (Chr) and/or linkage group (LG) in the rat

Symbol	Chr	LG	Gene description	Reference
A39	13	-	Hepatocyte antigen	73
ABL	3	-	Abelson viral oncogene homologue	93
ACADM	2	-	Medium-chain acyl-CoA dehydrogenase	86
ACO1	5	II	Aconitase 1	1
AFP	14	-	Alpha-fetoprotein	89
AGL	20	IX	Ag-L cell surface alloantigen	53
AK1	3	-	Adenylate kinase 1	50
AK2	5	II	Adenylate kinase 2	106,107
ALB	14	-	Albumin	10,82,89
ALDH	9	X	Aldehyde dehydrogenase	15
ALDOB	5	II	Aldolase B	91
ALP	-	XI	Alkaline phosphatase	1,15,40,41
B1	11	-	Hepatocyte antigen	73
C3	9	X	Complement component 3	91,92
C4	20	IX	Complement component 4	52,97
C6	-	XI	Complement component 6	15,28
CALC	1	I	Calcitonin	50
CBS	20	IX	Cystathionine beta synthase	52
CRYA1	20	IX	Alpha crystallin 1	77
CRYG1	9	X	Gamma crystallin 1	19,20
CRYG2	9	X	Gamma crystallin 2	19,20
CRYG3	9	X	Gamma crystallin 3	19,20
CRYG4	9	X	Gamma crystallin 4	19,20
CRYG5	9	X	Gamma crystallin 5	19,20
CRYG6	9	X	Gamma crystallin 6	19,20
CT	20	IX	Cell surface alloantigen	55
CYP2B1	1	I	Cytochrome P450 b	74
CYP2B2	1	I	Cytochrome P450 e	74
D4RP	5	II	DNA segment from MMU4 (Roswell Park)	88
D5G	5	II	DNA segment from RNO5 (Göteborg)	42
DHFR1	2	-	Dihydrofolate reductase 1 (active)	30
DHFR2	4	-	Dihydrofolate reductase (pseudogene)	30
DSI1	5	II	MoMuLV integration site	88
EAG1	9	X	Endotelial antigen	14
EGFR	14	-	Epidermal growth factor receptor (ERBB1)	91
ENO1	5	II	Enolase 1	105,107
ERBA2	15	-	AEV oncogene homologue A2	91
ERBB2	10	-	AEV homologue B (neu)	85
ES1	-	V	Esterase 1	2
ES2	-	V	Esterase 2	98
ES3	-	V	Esterase 3	62,99,100,104
ES4	-	V	Esterase 4	100,104
ES7	-	V	Esterase 7	57
ES8	-	V	Esterase 8	57
ES9	-	V	Esterase 9	63
ES10	-	V	Esterase 10	64

Table 1. Continued

Symbol	Chr	LG	Gene description	Reference
ES14	-	V	Esterase 14	25,32,103
ES15	-	V	Esterase 15	32
ES16	-	V	Esterase 16	31
F5	13	-	Coagulation factor V	16
FGA	2	-	Fibrinogen A alpha	54
FGB	2	-	Fibrinogen B beta	54,91,92
FGG	2	-	Fibrinogen gamma	54
FH	9	X	Fumarate hydratase	8
FPGS	3	-	Folylpolyglutamate synthetase	50
FUCA	5	II	Alpha-L-fucosidase 1 (tissue)	108
G6PD	X	-	Glucose-6-phosphate dehydrogenase	105
GALK	10	-	Galactokinase	50
GDH	5	II	Glucose dehydrogenase	108
GGTB2	5	II	Glycoprotein 4-beta-galactosyl transferase	88
GH	10	-	Growth hormone	11
GL1	-	VI	Plasma protein	70
GLA	X	-	Alpha-galactosidase	50
GLO1	20	IX	Glyoxalase 1	27,81
GLS	9	X	Glutaminase	61
GLUT1	5	II	Glucose transport	88
GOT1	2	-	Glutamic-oxaloacetic transaminase 1	50
GPI	1	I	Glucose phosphate isomerase	109
GST	1	I	Glutathione-S-transferase, placental enzyme	56
HAO1	20	IX	Hydroxyacid oxidase	15
HBB	1	I	Hemoglobin beta-chain	22,78
HPRT	X	-	Hypoxanthine phosphoribosyl transferase	89,105
HRAS1	1	I	Harvey RSV oncogene homologue 1 (active)	21,90
HRAS2	X	-	Harvey RSV oncogene homologue 1	90
HX	X	-	Cell surface alloantigen	65
HY	Y	-	Cell surface alloantigen	33,65
IFNA	5	II	Interferon alpha	39
IFNB	5	II	Interferon beta	87
IGH1	6	VIII	Immunoglobulin allotype	3
IGH2	6	VIII	Immunoglobulin allotype	4
IGH	6	VIII	Immunoglobulin heavy chain	71,76
IGK	4	-	Immunoglobulin kappa polypeptide	10,72
IGL	11	-	Immunoglobulin lambda polypeptide	85
INS1	1	I	Insulin 1	78
INS2	1	I	Insulin 2	78
IVD	3	-	Isovaleryl-CoA dehydrogenase	86
KRAS2	4	-	Kirsten RSV oncogene homologue 2 (active)	90
LCK1	5	II	Lymphocyte tyrosine kinase 1	87
LCK2	7	-	Lymphocyte tyrosine kinase 2	87
LDHA	1	I	Lactate dehydrogenase A	80,109
MDHL	-	IV	Malate dehydrogenase-like enzyme	59
ME1	8	-	Malic enzyme 1 (soluble)	50
MIS1	7	-	Same as MLVI1 and mouse pvt-1	35

Table 1. Continued

Symbol	Chr	LG	Gene description	Reference
MLVI1	7	-	MoMuLV integration site 1	43,94
MLVI2	2	-	MoMuLV integration site 2	94
MLVI3	15	-	MoMuLV integration site 3	94
MLVI4	7	-	MoMuLV integration site 4	29
MOS	5	II	Moloney sarcoma oncogene	87
MP1	8	-	Mannose phosphate isomerase	105
MUP	5	II	Major urinary protein	49,66,87,96
MYC	7	-	AMV oncogene homologue	83,94
MYCB	3	-	myc-like oncogene	38
MYCL	5	II	myc-like oncogene	36
MYCN	6	VIII	myc-like oncogene	36
NEU	20	IX	Neuraminidase	27,95
ORM	5	II	Orosomucoid (=AGP)	87
PCCB	8	-	Propionyl-CoA carboxylase (beta subunit)	86
PEP3	-	X	Peptidase 3	101
PEPD	1	I	Peptidase D	109
PFK2	X	-	Phosphofructokinase	17
PGD	5	II	Phosphogluconate dehydrogenase	9,44,66,107
PGK	X	-	Phosphoglycerate kinase	50
PGLY1	4	-	P-glycoprotein 170 kDa	29
PGM1	5	II	Phosphoglucomutase 1	106,107
PND	5	II	Pronatriodilatin (=ANF)	87
PRL	17	-	Prolactin	11,26
PTH	1	I	Parathyroid hormone	50
RAF	4	-	3611-MSV viral oncogene homologue	37
RAFAS	13	-	Breakage region for raf-rearrangement	37
RBPC1	5	II	Rat prostatic binding protein	110
RBPC2	5	II	Rat prostatic binding protein	110
RBPC3	5	II	Rat prostatic binding protein	110
RN5S1	12	-	Ribosomal 5s RNA	84
RN5S2	19	-	Ribosomal 5s RNA	84
RNR1	3	-	Ribosomal 18s and 28s RNA gene 1	75,84
RNR2	11	-	Ribosomal 18s and 28s RNA	75,84
RNR3	12	-	Ribosomal 18s and 28s RNA	75,84
RT1A	20	IX	Major histocompatibility locus	5,23,69,70
RT1B	20	IX	Major histocompatibility locus	7,51,52,67,79
RT1C	20	IX	Major histocompatibility locus	45
RT1D	20	IX	Major histocompatibility locus	51
RT1E	20	IX	Major histocompatibility locus	47
RT2	-	V	Cell surface alloantigen	68,69
RT4	1	I	Cell surface alloantigen	46
RT5	-	VII	Cell surface alloantigen	46
RT6	1	I	Cell surface alloantigen	18,34,102
RT9	-	V	Cell surface alloantigen	48
RTLH1	20	IX	Cell surface alloantigen	102
SAI1	5	II	Transformation suppressor	39
SIS	7	-	Simian sarcoma viral oncogene homologue	21

Table 1. Continued

Symbol	Chr	LG	Gene description	Reference
SP	20	IX	Serum protein	13
SVP1	-	IV	Seminal vesicle protein	24
TAM1	1	I	Tamase	58
TBM	1	I	Tubular basement membrane	60
TF	8	-	Transferrin	92
TG	7	-	Thyroglobulin	6
TK	10	-	Thymidine kinase (soluble)	50
TPI1	4	-	Triosephosphate isomerase 1	50
VDBP	14	-	Vitamin D-binding protein (=GC)	12
YES1	1	I	Avian Y73 SV oncogene homologue 1	50

Fig. 1. Rat linkage map

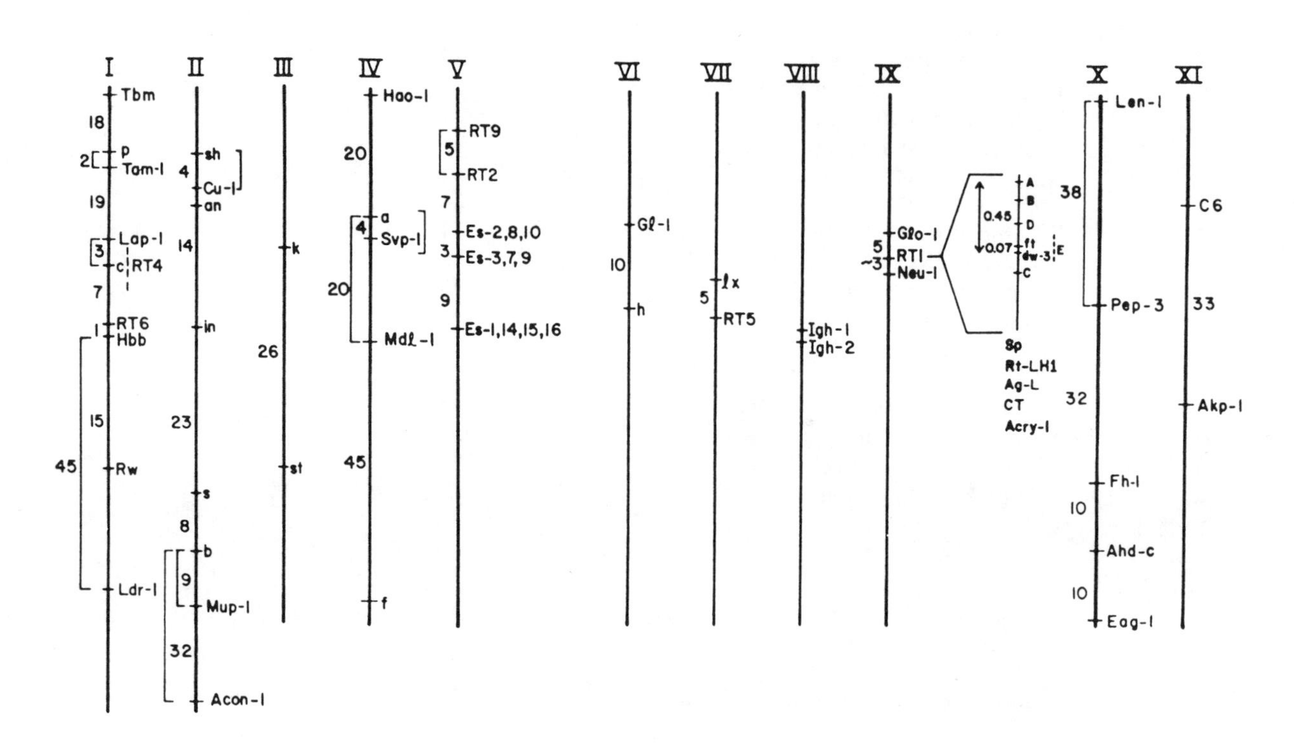

Table 2. Summary of chromosomal assignments of rat genes

Chr.	Gene
1	CALC
	CYP2B1
	CYP2B2
	GPI
	GST
	HBB
	HRAS1
	INS1
	INS2
	LDHA
	PEPD
	PTH
	RT4
	RT6
	TAM1
	TBM
	YES1
2	ACADM
	DHFR1
	FGA
	FGB
	FGG
	GOT1
	MLVI2
3	ABL
	AK1
	FPGS
	IVD
	MYCB
	RNR1
4	DHFR2
	IGK
	KRAS2
	PGLY1
	RAF
	TPI1
5	ACO1
	AK2
	ALDOB
	D4RP
	D5G
	DSI1
	ENO1
	FUCA
	GDH
	GGTB2
	GLUT1
	IFNA
	IFNB
	LCK1
	MOS
	MUP
	MYCL
	ORM
	PGD
	PGM1
	PND
	RBPC1
	RBPC2
	RBPC3
	SAI1
6	IGH1
	IGH2
	IGH
	MYCN
7	LCK2
	MIS1
	MLVI1
	MLVI4
	MYC
	SIS
	TG
8	ME1
	MP1
	PCCB
	TF
9	ALDH
	C3
	CRYG1
	CRYG2
	CRYG3
	CRYG4
	CRYG5
	CRYG6
	EAG1
	FH
	GLS
10	ERBB2
	GALK
	GH
	TK
11	B1
	IGL
	RNR2
12	RN5S1
	RNR3
13	A39
	F5
	RAFAS
14	AFP
	ALB
	EGFR
	VDBP
15	ERBA2
	MLVI3
17	PRL
19	RN5S2
20	AGL
	C4
	CBS
	CRYA1
	CT
	GLO1
	HAO1
	NEU
	RT1A
	RT1B
	RT1C
	RT1D
	RT1E
	RTLH1
	SP
X	G6PD
	GLA
	HPRT
	HRAS2
	HX
	PFK2
	PGK
Y	HY

References

1 ADAMS, M., et al. 1984. Biochem. Genet. 22, 611-629.
2 AUGUSTINSSON, K.B. and HENDRICSON, B. 1979. Biochem. Biophys. Acta 124, 323-331.
3 BAZIN, H., et al. 1974. J. Immunol. 112, 1035-1041.
4 BECKERS, A. and BAZIN, H. 1975. Immunochem. 12, 671-675.
5 BOGDEN, A.E. and APTEKMAN P.M. 1960. Cancer Res. 20, 1372-1382.
6 BROCAS, H., et al. 1985. Cytogenet. Cell Genet. 39, 150-152.
7 BUTCHER, G.W. and HOWARD, J.C. 1977. Nature 266,362-364.
8 CARLEER, J. and ANSAY, M. 1976. Int. J. Biochem. 7,565-567.
9 CARTER, N.D. and PARR, C.W. 1969. Nature 224,1214.
10 COLLARD, J.G., et al. 1982. Cytogenet. Cell Genet. 32, 257-258.
11 COOKE, N., et al. 1986. Endocrinol. 119, 2451-2455.
12 COOKE, N., et al. 1987. Cytogenet. Cell Genet. 44, 98-100.
13 CRAMER, D.V. 1983. Immunogenet. 18, 593-598
14 CRAMER, D.V., et al. 1985. Biochem. Genet. 23, 623-629.
15 CRAMER, D.V., et al. 1986. Biochem. Genet. 24, 217-227.
16 DAHLBÄCK, B., et al. 1988. Somatic Cell Mol. Genet. 14, 509-514.
17 DARVILLE, M.I., et al. 1989. Int. Physiol. Biochim. In press
18 DE WITT, C.W. and McCULLOUGH, M. 1972. Transpl. 19, 310-317.
19 DEN DUNNEN, J., et al. 1987. Exp. Eye Res. 45, 747-750.
20 DONNER, M.E., et al. 1985. Biochem. Genet. 24, 217-227.
21 FANG, X.-E., et al. 1985. Cytogenet. Cell Genet. 40, 627.
22 FRENCH, E.A., et al. 1971. Biochem. Genet. 5, 397-404.
23 FRENZL, B., et al. 1960. Fol. Biol. 6, 121-126.
24 GASSER, D.L. 1972. Biochem. Genet. 6, 61-63.
25 GASSER, D.L., et al. 1973. Biochem. Genet. 10, 207-217.
26 GERHARD, D.S., et al. 1984. DNA 3, 139-145.
27 GILL, T.J., et al. 1982. J. Immunogenet. 9,281-293.
28 GRANADOS, J., et al. 1984. J. Immunolog. 133, 405-407.
29 HANSON, C., et al. 1988. BioScience, Malmö, Sweden
30 HANSON, C., et al. Hereditas In press
31 HEDRICH, H.J. and DEIMLING, O.V. J. Heredity In press.
32 HEDRICH, H.J., et al. 1987. Biochem. Genet. In press.
33 HESLOP, B.F. 1973. Transpl. 15, 31-35.
34 HOWARD, J.C. and SCOTT, D.W. 1974. Immunol. 27, 903-922.
35 INGVARSSON, S., et al. 1987a. Cytogenet. Cell Genet. 45, 174-176.
36 INGVARSSON, S., et al. 1987b. Somatic Cell Mol. Genet. 13, 335-339.
37 INGVARSSON, S., et al. 1988a. Somatic Cell Mol. Genet. 14, 401-405.
38 INGVARSSON, S., et al. 1988b. Mol. Cell. Biol. 8, 3168-3174
39 ISLAM, M.Q., et al. 1989. J. Cell Sci. 92, 147-162.
40 JIMENEZ-MARIN, D. 1974. J. Heredity 65, 235-237.
41 JIMENEZ-MARIN, D. and DESSAUER, H.C. 1973. Comp. Biochem. Physiol. 46B, 487-492.
42 KLINGA, K., et al. 1989. Hereditas In press.
43 KOEHNE, C., et al. 1989. J. Virol. 63, 2366-2369
44 KOGA, A., et al. 1972. Jap. J. Genet. 47, 335-338.
45 KOHOUTOVA, M., et al. 1980. Immunogenet. 11, 499-506.
46 KREN, V., et al. 1973. Transpl. Proc. 5, 1463-1466.
47 KUNZ, H.W., et al. 1982. J. Immunol. 128, 402-408.
48 KUNZ, H.W., et al. 1985. J. Immunol. 12, 75-78.
49 KURTZ, D. 1981. J. Mol. Applied Genet. 1, 29-38.
50 LEVAN, G., et al. 1986. Rat News Letter 17, 3-8
51 LOBEL, S.A. and CRAMER, D.V. 1981. Immunogenet. 13, 465-473.
52 LOCKER, J., et al. 1989. Manuscript in preparation.
53 LYNCH, D.H. and DE WITT, C.W. 1980. J. Immunol. 124, 2247-2253.

54 MARINO, N.W., et al. 1986. Cytogenet. Cell Genet. 42, 36-41.
55 MARSHAK, A., et al. 1977. J. Exp. Med. 146, 1773-1790.
56 MASUDA, R., et al. 1986. Jpn. J. Cancer Res. 77, 1055-1058.
57 MATSUMOTO, K. 1980. Biochem. Genet. 18, 879-887.
58 MATSUMOTO, K. and GASSER, D.L. 1983. Biochem. Genet. 21, 1209-1215.
59 MATSUMOTO, K., et al. 1982. Genetics 20, 443-448.
60 MATSUMOTO, K., et al. 1984. Immunogenet. 20, 117-123.
61 MOCK, B., et al. 1989. Genomics In press.
62 MOUTIER, R., et al. 1973a. Biochem. Genet. 8, 321-328.
63 MOUTIER, R., et al. 1973b. Biochem. Genet. 9, 109-115.
64 MOUTIER, R., et al. 1973c. Biochem. Genet. 10, 395-398.
65 MULLEN, Y. and HILDERMAN, W.H. 1972. Transpl. 13, 521-529
66 NIKAIDO, H., et al. 1982. J. Hered. 73, 119-122.
67 OIKAWA, T., et al. 1984. Cytogenet. Cell Genet. 37, 558.
68 OWEN, R.D. 1962. Ann. N.Y. Acad. Sci. 97, 37-42.
69 PALM, J. 1962. Ann. N.Y. Acad. Sci. 97, 57-58.
70 PALM, J. 1971. Transpl. 11,175-183.
71 PEAR, W., et al. 1986. Immunogenet. 23, 293-295.
72 PERLMANN, C., et al. 1985. Immunogenet.. 22, 97-100.
73 PERROTEZ, C. et al. 1989. Cytogenet. Cell Genet. In press.
74 RAMPERSAUD, A., et al. 1987. J. Biol. Chem. 262, 5649-5663.
75 SASAKI, M., et al. 1986. Cytogenet. Cell Genet. 41, 83-88.
76 SCHRÖDER, J., et al. 1980. Immunogenet. 10,125-131.
77 SKOW, L.C., et al. 1985. Immunogenet. 22, 291-293.
78 SOARES, M.B., et al. 1985. Mol. Cell Biol. 5, 2090-2103.
79 STARK, O., et al. 1977. Immunogenet. 5, 183-187.
80 STOLC, V. and GILL, T.J. 1983. Biochem. Genet. 21, 933-941.
81 STOLC, V., et al. 1980. Immunol. 125, 1167-1170.
82 SUGIYAMA, K., et al. 1984. Jpn. J. Genet. 59, 577-583.
83 SÜMEGI, J., et al. 1983. Nature 306, 497-498.
84 SZABO, P., et al. 1978. Chromosoma 65, 161-172.
85 SZPIRER, C., et al. 1988a. Current Topics Microbiol. Immunol. 137, 33-38.
86 SZPIRER, C., et al. 1989a. Cytogenet. Cell Genet. In press
87 SZPIRER, C., et al. 1989b. Cytogenet. Cell Genet. In press
88 SZPIRER, C., et al. 1989c. Submitted.
89 SZPIRER, J., et al. 1984. Cytogenet. Cell Genet. 38, 142-149.
90 SZPIRER, J., et al. 1985. Somatic Cell Mol. Genet. 11, 93-97.
91 SZPIRER, J., et al. 1987. Cytogenet. Cell Genet. 46, 701.
92 SZPIRER, J., et al. 1988b. Cytogenet. Cell Genet. 47, 42-45.
93 TAKAHASHI, R., et al. 1986. Proc. Natl. Acad. Sci. 83, 1079-1083.
94 TSICHLIS, P., et al. 1985. J. Virol. 56, 938-943.
95 VAN DE BERG, J.L., et al. 1983. J. Immunogenet. 8, 239-242.
96 VAN ZUTPHEN, L.F.M., et al. 1981. Biochem. Genet. 19, 173-186.
97 WATTERS, J.W.F., et al. 1987. Immunogenet. 25, 204-206.
98 WOMACK, J.E. 1972a. J. Heredity 63, 41-42.
99 WOMACK, J.E. 1972b. Experientia 28, 1372.
100 WOMACK, J.E. 1973. Biochem. Genet. 9, 13-24.
101 WOMACK, J.E. and CRAMER, D.V. 1980. Biochem. Genet. 18, 1019-1026.
102 WONIGEIT, K. 1979. Transpl. Proc. 15, 1687-1688.
103 YAMADA, J., et al. 1980. Biochem. Genet. 18, 433-438.
104 YAMORI, Y. and OKAMOTU, K. 1970. Lab. Invest. 22, 206-211.
105 YOSHIDA, M.C. 1978. Cytogenet. Cell Genet. 22, 606-609.
106 YOSHIDA, M.C. 1979. Proc. Jpn Acad. 55, 403-406.
107 YOSHIDA, M.C. 1982. Cytogenet. Cell Genet. 32, 330.
108 YOSHIDA, M.C. 1984a. Cytogenet. Cell Genet. 37, 613.
109 YOSHIDA, M.C. 1984b. Cytogenet. Cell Genet. 37, 613.
110 ZHANG, J., et al. 1988. Cytogenet. Cell Genet. 48, 121-123

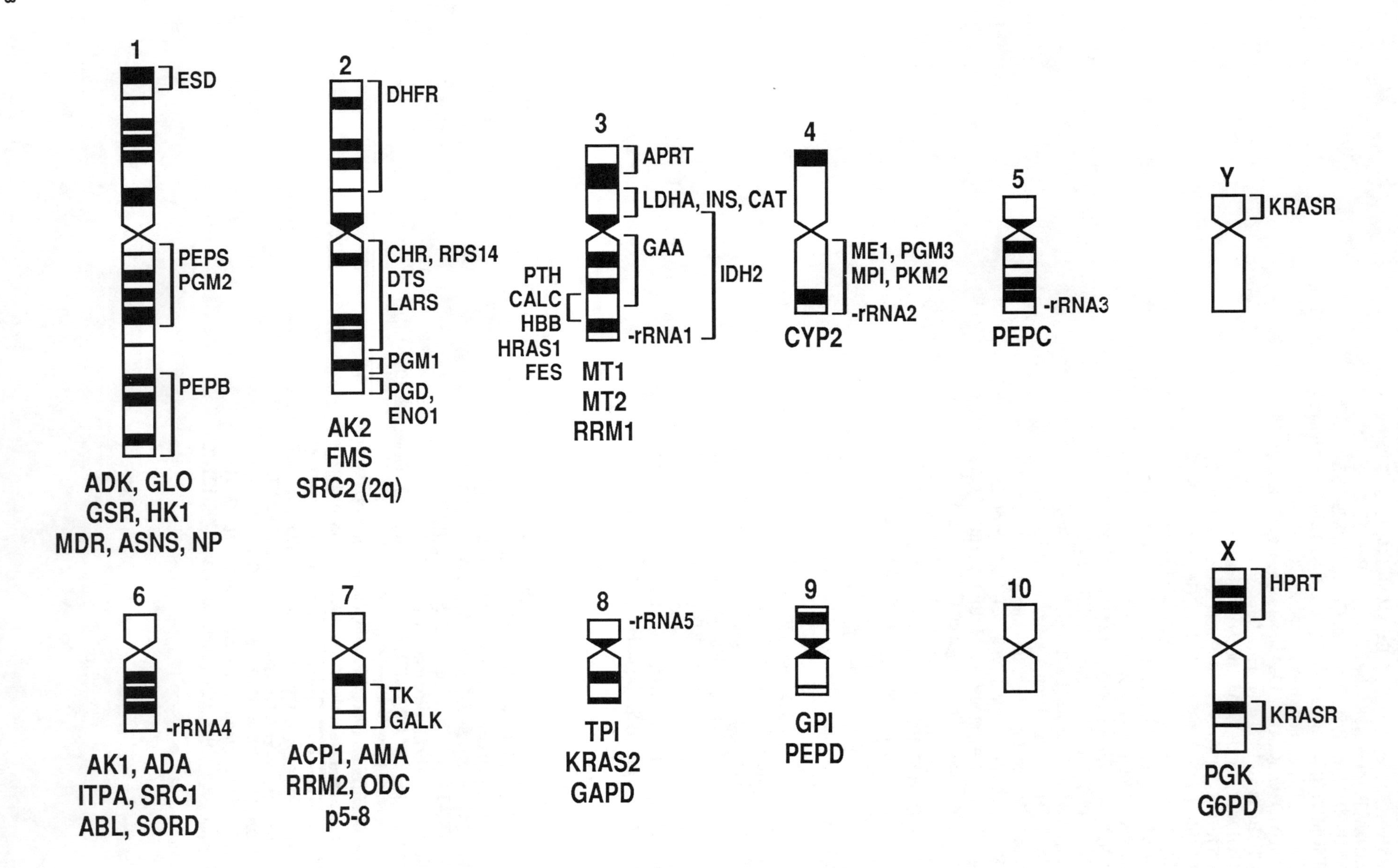

CHINESE HAMSTER GENE MAP

The Chinese Hamster Gene Map (Cricetulus griseus) (2N = 22)

Date: August 1989

Raymond L. Stallings[1], Gerald M. Adair[3], and Michael J. Siciliano[2]
[1]Los Alamos National Laboratory, Genetics Group, Los Alamos, NM 87545,
[2]Department of Genetics, Houston, TX 77030, [3]Science Park Research Division, Smithville, TX 77897.

Nomenclature and abbreviations for Chinese hamster genes follow the recommendations for an international system for human gene nomenclature (Shows et al., 1979 Cytogenet. Cell Genet.).

GENE	PROTEIN OR GENE PRODUCT	CHROMOSOME	REFERENCE
ABL	Abelson murine leukemia viral oncogene homolog	6	26
ADA	adenosine deaminase	6	19, 22
ADK	adenosine kinase	1	7, 18
ACP1	acid phosphatase 1	7	18
AK1	adenylate kinase 1	6	16, 19
AK2	adenylate kinase 2	2	19
AMA	RNA polymerase II	7	14
APRT	adenine phosphoribosyl transferase	3p	1, 2, 3
ASNS	aspargine synthetase	1	4
CALC	calcitonin	3q	33
CAT	catalase	3p	33
CHR	chromate resistance	2q	5
CYP2	cytochrome p450, dioxin inducible	4	10
DHFR	dihydrofolate reductase	2p	31
DTS	diptheria toxin sensitivity	2q	13, 21
ESD	esterase D	1p	7, 18
ENO1	enolase 1	2q	13, 18
FES	feline sarcoma viral oncogene	3q	26
FMS	McDonough feline sarcoma viral oncogene homolog	2q	26
GAA	α-glucosidase	3q	1, 3
GALK	galactokinase	7q	14, 18, 22
GAPD	glyceraldehyde-3-phosphate dehydrogenase	8	9
GALT	galactose-1-phosphate uridyl transferase	2q	18
GLO	glyoxalase	1	18
G6PD	glucose-6-phosphate dehydrogenase	X	15, 30
GPI	glucose phosphate isomerase	9	16, 17
GSR	gluthathione reductase	1	7, 18
HBB	β-globin	3q	33
HPRT	hypoxanthine guanine phosphoribosyl transferase	X	6, 15, 30
HRAS1	Harvey rat sarcoma 1 viral oncogene homolog	3q	27
HK1	hexokinase 1	1	16
IDH2	isocitrate dehydrogenase 2	3	1, 3

GENE	PROTEIN OR GENE PRODUCT	CHROMOSOME	REFERENCE
INS	Insulin	3P	33
ITPA	inosine triosephosphatase	6	19, 22
KRAS2	Kirsten rat sarcoma 2 viral oncogene homolog	8	27
KRASR	Kirsten rat sarcoma viral oncogene related	Y, X	32
LARS	leucyl-tRNA-synthetase	2q	29
LDHA	lactate dehydrogenase A	3p	1, 3
MDR	multi-drug resistant phenotype	1	28
ME1	malic enzyme 1	4q	16, 24
MPI	mannose phosphate isomerase	4q	16, 24
MT1	metallothionein 1	3	25
MT2	metallothionein 2	3	25
NP	nucleoside phosphorylase	1	7, 18
ODC	ornithine decarboxylase	7	12
PEPB	peptidase B	1	7, 18
PEPC	peptidase C	5	16
PEPD	peptidase D	9	16, 17
PEPS	peptidase S	1	7, 18
PGD	6-phosphogluconate dehydrogenase	2q	13, 18
PGK	phosphoglycerate kinase	X	15, 30
PGM1	phosphoglucomutase 1	2q	13, 18
PGM2	phosphoglucomutase 2	1	7, 18
PGM3	phosphoglucomutase 3	4q	24
PKM2	pyruvate kinase 2	4q	24
PTH	parathyroid hormone	3q	33
P5-8	anonymous DNA sequence	7	12
RPS14	small ribosomal protein 14 (EMT resistance)	2q	5
RRM1	ribonucleotide reductase M1 subunit	3	12
RRM2	ribonucleotide reductase M2 subunit	7	12
SORD	sorbital dehydrogenase	6	16
SRC1	Avian sarcoma viral oncogene homolog 1	6	26
SRC2	Avian sarcoma viral oncogene homolog 2	2q	26
RNR1	ribosomal RNA 1	3q	8, 11
RNR2	ribosomal RNA 2	4q	8, 11
RNR3	ribosomal RNA 3	5q	8, 11
RNR4	ribosomal RNA 4	6q	8, 11
RNR5	ribosomal RNA 5	8p	8, 11
TK	thymidine kinase	7q	14, 18, 22
TPI	triosephosphate isomerase	8	17

References

1. Adair, GM, Stallings, RL, Friend, K, and Siciliano, MJ 1983. Somat. Cell Genet. 9:477-487.
2. Adair, GM, Stallings, RL, Nairn, R, and Siciliano, MJ 1983. Proc. Natl. Acad. Sci. USA 80:5961-5964.
3. Adair, GM, Stallings, RL, and Siciliano, MJ 1984. Somat. Cell Genet. 10:283-295.
4. Andrulius, IL, Duff, C, Evans-Blacker, S, Worton, RG, and Siminovitch, L 1983. Molec. Cell. Biol. 3:391-398.
5. Campbell, CE and Worton, RG 1980. Somat. Cell Genet. 6:215-225.
6. Farrell, SA and Worton, RG 1977. Somat. Cell Genet. 3:539-551.
7. Dhar, V, Searle, BM,, and Athwal, RS 1984. Somat. Cell Molec. Genet.

10:547-559.
8. Goodpasture, C and Bloom, SE 1975. Chromosoma 53:37-50.
9. Greenspan, J 1984. Doctoral Dissertation. The University of Texas Graduate School of Biomedical Sicence, Houston, Texas.
10. Hildebrand, CE, Stallings, RL, Gonzalez, FJ, and Nevert, DW 1985. Somat. Cell Molec. Genet. 11:391-395.
11. Hsu, TC, Spirito, SE, and Pardue, ML 1975. Chromosoma 53: 25-26.
12. Tonin, PN, Stallings, RL, Carmon, MD, Bertino, JR, Srinvasan, PR, and Lewis, WH 1987. Cytogenet. Cell Genet. 45:102-108.
13. Roberts, M and Ruddle, F 1980. Expl. Cell Res. 127:47-54.
14. Roberts, M, Scangos, A, Hart, JT, and Ruddle, FH 1983. Somatic Cell Genet. 9:235-248.
15. Rosenstraus, M and Chasen, LA 1975. Proc. Natl. Acad. Sci. USA 72:493-497.
16. Satoh, H and Yoshida, MC 1985. Cytogenet. Cell Genet. 39:285-291.
17. Siciliano, MJ, Stallings, RL, Adiar, GM, Humphrey, RM, and Siciliano, MJ 1983. Cytogenet. Cell Genet. 35:15-20.
18. Stallings, RL and Siciliano, MJ 1981. Somat. Cell Genet. 7:683-698.
19. Stallings, RL and Siciliano, MJ 1982. J. Hered. 73:399-404.
20. Stallings, RL and Siciliano, MJ 1983. Isozymes: Current Topics in Biological and Medical Research 10:313-321.
21. Stallings, RL, Siciliano MJ, Adiar, GM, and Humphrey, RM 1982. Somat. Cell Genet. 8:413-422.
22. Stallings, RL, Adair, GM, Siciliano, J, Greenspan, J, and Siciliano, MJ 1983. Molec. Cell Biol. 3:1967-1974.
23. Stallings, RL, Adair, GM, Lin, J, and Siciliano, MJ 1984. Cytogenet. Cell Genet. 38:132-137.
24. Stallings, RL, Adair, GM, and Siciliano, MJ 1984. Somat. Cell Genet. 10:109-112.
25. Stallings, RL, Munk, AC, Longmire, JL, Hildebrand, CE, and Crawford, BD 1984. Molec. Cell. Biol. 4:2932-2936.
26. Stallings, RL, Munk, AC, Longmire, JL, Jett, JH, Wilder, ME, Siciliano, MJ, Adair, GM, and Crawford, BD 1985. Chromosoma 92:156-163.
27. Stallings, RL, Crawford, BD, Black, RJ, and Chang, EH 1986. Cytogenet. Cell Genet. in press.
28. Teeter, LK, Sanford, JA, Sen, S, Stallings, RL, Siciliano, MJ, and Kuo, MT 1986. Molec. Cell. Biol. in press.
29. Wasmuth, JJ and Chu, LY 1980. J. Cell Biol. 87:697-702.
30. Westerveld, A, Visser, R, Freeke, MA, and Bootsma, D 1972. Biochem. Genet. 7:33-40.
31. Worton, RG, Duff, C, and Flintoff, W 1981. Molec. Cell. Biol. 1:330-335.
32. Stenman, G, Anisowicz, A, and Sager, R 1988. Somat. Cell Molec. Genet. 14:639-644.
33. Stenman, G, and Sager, R 1987. Proc. Natl. Acad. Sci. 84:9909-9102.

CHO CELL GENE MAP

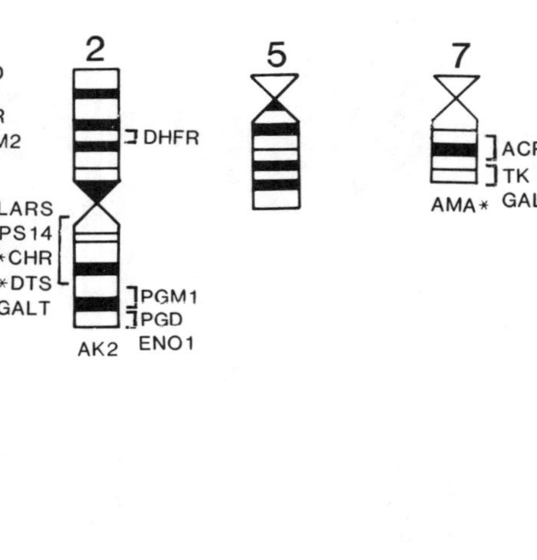
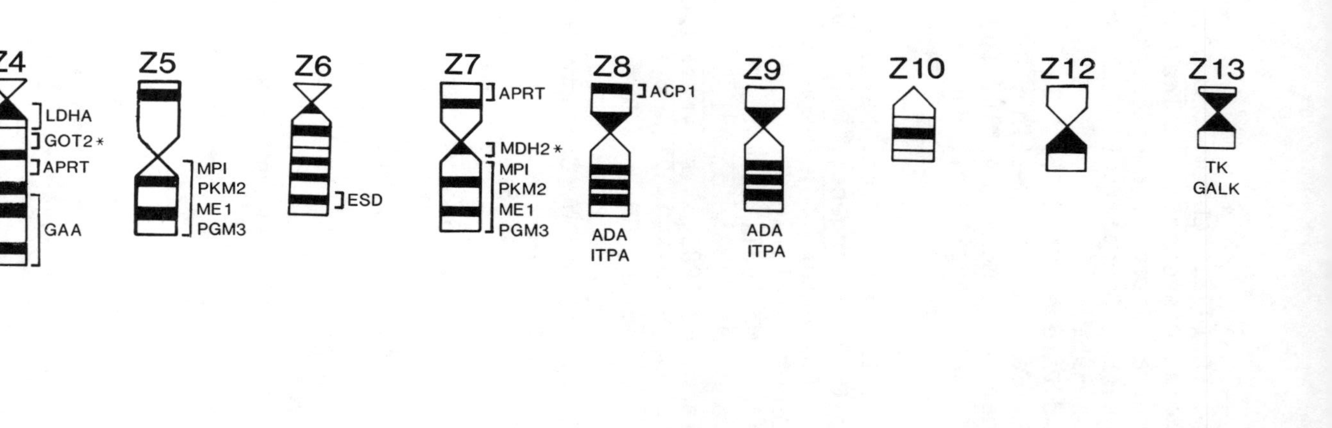
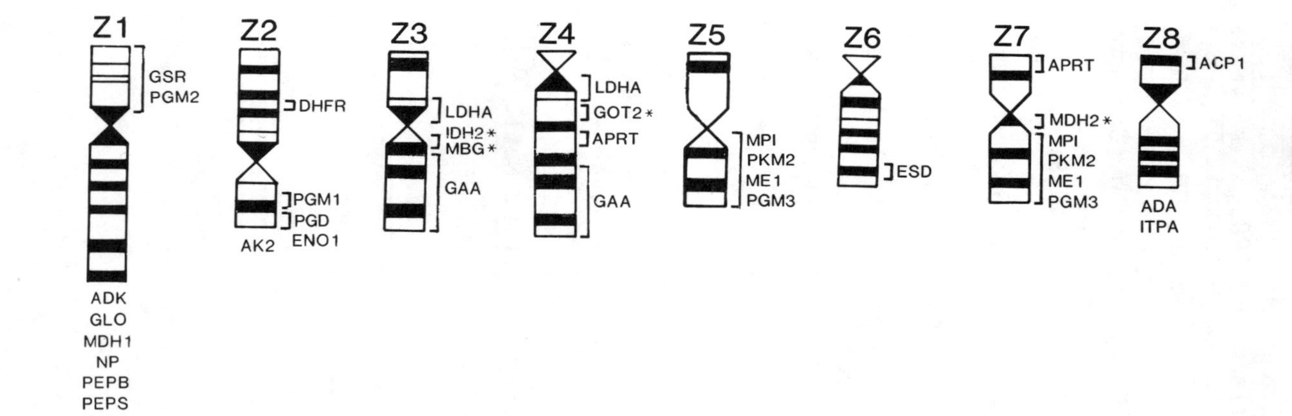

Gene Map of the Chinese Hamster (*Cricetulus griseus*) CHO Cell Line

August, 1989

Gerald M. Adair[1], Raymond L. Stallings[2], and Michael J. Siciliano[3]

[1] The University of Texas M.D. Anderson Cancer Center, Science Park-Research Division, Smithville, TX 78957,
[2] Life Sciences Divison, Los Alamos National Laboratory, Los Alamos, NM 87545,
[3] Department of Genetics, The University of Texas M.D. Anderson Cancer Center, Houston, TX 77030.

The CHO (Chinese hamster ovary) cell line has been used for a wide range of somatic cell genetic and mutation research since its establishment over thirty years ago in the laboratory of Dr. Theodore Puck (14). CHO cells have a near-diploid modal chromosome number of 21 (2n=22). Although many of the CHO chromosomes reflect deletions, translocations, and other chromosomal rearrangements that occurred during evolution of this cell line, the present karyotype appears to be stable and the origins and derivations of most of the rearranged Z-group chromosomes can be traced back to the original Chinese hamster karyotype on the basis of their G-banding patterns (20). Over 60 gene loci have been mapped in Chinese hamster (*Cricetulus griseus*). At least 43 genes have been localized to specific chromosomes in CHO cells; twenty-seven have been regionally mapped. Fifteen of these gene loci (indicated by *) appear to be physically or functionally hemizygous in CHO cells. Much of this work was supported by U.S. Public Health Service Grants CA28711 and CA04484 from the National Cancer Institute. Nomenclature and abbreviations follow the recommendations for an international system for human gene nomenclature (18, 20).

Gene Symbol	Gene Product or Genetic Marker	Chromosome(s)	References
ACP1	Acid phosphatase 1	7q, Z8p	19, 24
ADA	Adenosine deaminase	Z8, Z9	20, 24
ADK	Adenosine kinase	1, Z1	21, 26
AK2	Adenylate kinase 2	2	22
AMA	α-Amanitin resistance (RNA polymerase II)	7	16
APRT	Adenine phosphoribosyl transferase	Z4q, Z7p	1-3, 7, 8
CHR	Chromate resistance	2q	10, 27
DHFR	Dihydrofolate reductase	2p, Z2p	12, 29
DTS	Diphtheria toxin sensitivity	2q	15, 23
ENO1	Enolase 1	2q, Z2p	15, 21, 23
ESD	Esterase D	1p, Z6q	21, 25
GAA	α-Glucosidase	Z3q, Z4q	1, 3
GALK	Galactokinase	7q, Z13	16, 21, 24
GALT	Galactase-1-phosphate uridyl transferase	2q	21, 23
GAPD	Glyceraldehyde-3-phosphate dehydrogenase	8	13
GLO1	Glyoxylase 1	1, Z1	21, 25
G6PD	Glucose-6-phosphate dehydrogenase	X	11, 17, 28
GOT2	Glutamic-oxaloacetic transaminase 2	Z4	4, 9
GPI	Glucose phosphate isomerase	9	14, 19
GSR	Glutathione reductase	1p, Z1p	9, 21, 25
HPRT	Hypoxanthine phosphoribosyl transferase	Xp	11, 17, 28
IDH2	Isocitrate dehydrogenase 2	Z3q	1, 3
ITPA	Inosine triose phosphatase	Z8, Z9	22, 24
LARS	Leucyl-tRNA synthetase	2q	27
LDHA	Lactate dehydrogenase A	Z3q, Z4q	1, 3, 8
MBG	Methyl glyoxylbisguanyl hydrazone resistance	Z3	5
MDH1	Malate dehydrogenase 1	1, Z1	9
MDH2	Malate dehydrogenase 2 (mitochondrial)	Z7	6-8
ME1	Malic enzyme 1	Z5q, Z7q	3, 26
MPI	Mannose phosphate isomerase	Z5q, Z7q	3, 26
NP	Nucleoside phosphorylase	1, Z1	21, 25
PEPB	Peptidase B	1, Z1	21, 25
PEPD	Peptidase D	9	19
PEPS	Peptidase S	1, Z1	21, 25
PGD	6-Phosphogluconate dehydrogenase	2q, Z2q	15, 21, 23
PGK	Phosphoglycerate kinase	X	11, 17, 28
PGM1	Phosphoglucomutase-1	2q, Z2q	15, 21, 23
PGM2	Phosphoglucomutase-2	1, Z1	21, 25
PGM3	Phosphoglucomutase-3	Z5q, Z7q	2, 3, 26
PKM2	Pyruvate kinase-2	Z5q, Z7q	3, 26
RPS14	Emetine resistance (Small ribosomal protein 14)	2q	10, 27
TK	Thymidine kinase	7q, Z13	16, 21, 24
TPI	Triose phosphate isomerase	8	19

REFERENCES

1. ADAIR, G.M., et al. 1983. *Somat. Cell Genet.* 9: 477-487.
2. ADAIR, G.M., et al. 1983. *Proc. Natl. Acad. Sci. USA* 80: 5961-5964.
3. ADAIR, G.M., et al. 1984. *Somat. Cell Molec. Genet.* 10: 283-295.
4. ADAIR, G.M., et al. 1984. *Human Gene Mapping* 7: 399-400.
5. ADAIR, G.M. and SICILIANO, M.J. 1985. *Mol. Cell. Biol.* 5: 109-113.
6. ADAIR, G.M. and SICILIANO, M.J. 1986. *Somat. Cell Molec. Genet.* 12: 111-119.
7. ADAIR, G.M., et al. 1989. *Somat. Cell Molec. Genet.* 15: 271-277.
8. ADAIR, G.M., et al. 1989. *Somat. Cell Molec. Genet.* 15: (in press).
9. ADAIR, G.M. and SICILIANO, M.J. (unpublished results).
10. CAMBELL, C.E. and WORTON, R.G. 1980. *Somat. Cell Genet.* 6: 215-225.
11. FARRELL, S.A. and WORTON, R.G. 1977. *Somat. Cell Genet.* 3: 539-551.
12. FUNAGE, V.L. and MYODA, T.T. 1986. *Somat. Cell Molec. Genet.* 12: 649-655.
13. GREENSPAN, J. 1984. Dissertation, The University of Texas Graduate School of Biomedical Sciences, Houston, TX.
14. PUCK, T.T. 1985. In: *Molecular Cell Genetics*, (ed. M.M. Gottesman), pp. 37-64, John Wiley & Sons, NY.
15. ROBERTS, M. and RUDDLE, F. 1980. *Exptl. Cell Res.* 127: 47-54.
16. ROBERTS, M., et al. 1983. *Somat. Cell Genet.* 9: 235-248.
17. ROSENSTRAUSS, M. and CHASIN, L.A. 1975. *Proc. Natl. Acad. Sci. USA* 72: 493-497.
18. SHOWS, T.B., et al. 1979. *Cytogenet. Cell. Genet.* 25: 96-116.
19. SICILIANO, M.J., et al. 1983. *Cytogenet. Cell. Genet.* 35: 15-20.
20. SICILIANO, M.J., et al. 1985. In: *Molecular Cell Genetics*, (ed. M.M. Gottesman), pp. 95-135, John Wiley & Sons, NY.
21. STALLINGS, R.L. and SICILIANO, M.J. 1981. *Somat. Cell Genet.* 7: 683-698.
22. STALLINGS, R.L. and SICILIANO, M.J. 1982. *J. Hered.* 73: 399-404.
23. STALLINGS, R.L., et al. 1982. *Somat. Cell Genet.* 8: 413-422.
24. STALLINGS, R.L., et al. 1983. *Mol. Cell. Biol.* 3: 1967-1974.
25. STALLINGS, R.L., et al. 1984. *Somat. Cell Molec. Genet.* 10: 109-113.
26. STALLINGS, R.L., et al. 1984. *Cytogenet. Cell. Genet.* 38: 132-137.
27. WASMUTH, J.J. and CHU, L.Y. 1980. *J. Cell Biol.* 87: 697-702.
28. WESTERVELD, A., et al. 1972. *Biochem. Genet.* 7: 33-40.
29. WORTON, R.G., et al. 1981. *Mol. Cell. Biol.* 1: 330-335.

LINKAGE MAP OF THE DEERMOUSE

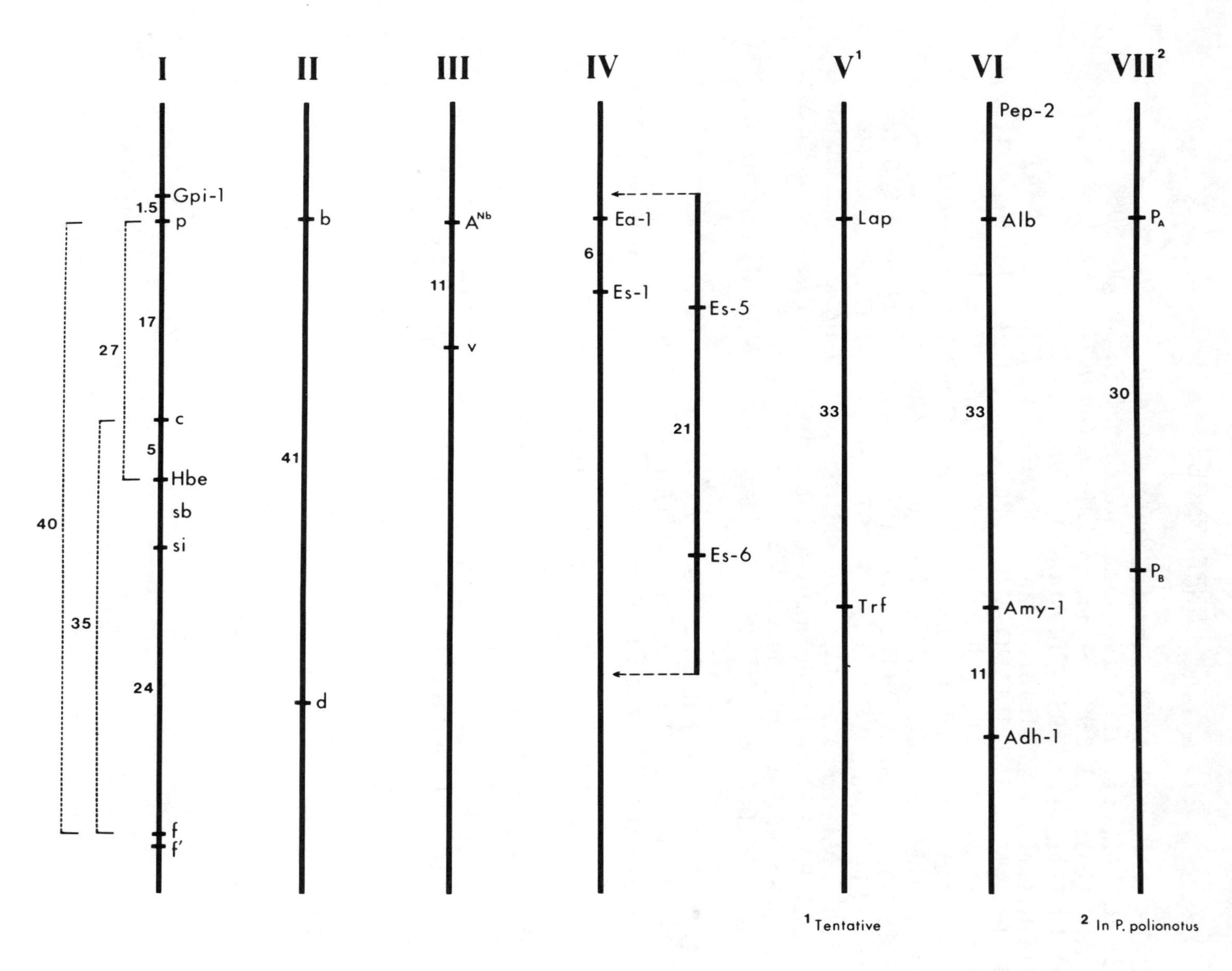

[1] Tentative

[2] In P. polionotus

Genetic Linkage Map of Deer Mouse (*Peromyscus maniculatus*) 2N = 48

July 1989

Wallace D. Dawson
Department of Biological Sciences
University of South Carolina
Columbia SC 29208

Linkage data for the deermouse (*Peromyscus maniculatus*) collected before 1972 are summarized by Robinson (15, 16). The system of assigning linkage groups on the basis of a single marker employed during the 1940's and 50's (2,13) is no longer used. "Group IV" in the earlier system is now Group II, and old Groups "II" and "III" have been abandoned. In the interim since Robinson's review several additional linkages have been added (3, 8, 9, 17). The current status of the linkage map for the deermouse and its sibling species *P. polionotus* is represented in the accompanying figure. Six linkage groups are now established by formal genetics and another is tentative. An additional linkage, *Es-5* - *Es-6*, by homology with *Mus*, will probably map to Group IV (8), and is designated IVa in the table.

The order of loci in Group I was reported informally by Huestis and Silliman in an unpublished communication, according to Robinson (15), and has been partially confirmed by Dodson (unpub.). Linkage of *Trf* and *Lap* is tentative (8), but is homologous with a similar linkage in *Mus*. The *Pep-2* locus is provisionally assigned to Group VI proximal to *Alb*, but has not been mapped further (9).

Positive, but not significant, lod scores suggesting possible linkage between the gene pairs *Adh* - *6Pgd*, *Adh* - *Got-1*, *Adh* - *Idh*, *Alb* - *Pept-1*, *Alb* - *Sdh* and *Est-4* - *Sdh*, respectively, were reported by Baccus *et al.* (1). Subsequent information indicates that *Adh-1* and *Got-1* are independent, as are the *Alb* and *Sdh-1* loci (9).

The *Hbe* locus is part of the triplicated beta globin site (*Hbb*), according to Snyder (17). Unpublished data from Snyder maps the position of the *Gpi-1* and *Hbe* loci relative to the albino (*c*) and pink-eyed dilution (*p*) loci. Silliman (unpub.) proposed that there is a duplication, *f'*, closely linked to the *f* locus. The Pm blood group locus, formerly designated *Pm*, is redesignated *Ea-1*.

Two significant markers on the *Peromyscus* linkage map. *d* and *v*, are now extinct in laboratory stocks of deermice. The "flexed tail" trait which occurs in a laboratory stock may not be identical by descent with the original trait used in early linkage studies, but it maps to the same location in Group I.

The chromosome number of all *Peromyscus* species is 2N = 48 (6). None of the linkage groups have been assigned to chromosomes, but partial banding homology between Chromosome 1 of *Rattus* and *Peromyscus* (12) suggests that Linkage Group I is probably located on Chromosome 1 in deer mouse, as is the homologous group in rat (7).

MAPPED LOCI IN *PEROMYSCUS MANICULATUS*

Gene Symbol	Name of Locus	Linkage Group	Reference
A^{Nb}	Agouti	III	13.
Adh-1	Alcohol dehydrogenase (liver)	VI	9.
Alb	Albumin (serum)	VI	9.
Amy-1	Amylase (salivary)	VI	9.
b	Brown	II	13.
c	Albino	I	4, 5, 11, 18.
d	Dilute	II	13.
Ea-1	Pm erythrocytic antigen	IV	14.
Es-1	Esterase-1 (erythrocytic)	IV	14.
Es-5	Esterase-5 (kidney)	IVa	8.
Es-6	Esterase-6 (kidney)	IVa	8.
f	Flexed tail	I	10, 11.
Gpi-1	Glucose phosphate isomerase	I	17.
Hbb	Beta globin (hemoglobin)	I	17.
Lap-1	Leucine aminopeptidase (serum)	V	8.
p	Pink-eyed dilution	I	4, 5, 17, 18.
P_A	Pointed rump pattern A	VII	3.
P_B	Pointed rump pattern B	VII	3.
Pep-2	Tripeptidase (erythrocytic)	VI ?	8.
sb	Snub nose	I	16.
si	Silver	I	10, 11.
Trf	Transferrin (serum)	V	8.
v	Waltzer	III	13.

References:

1. Baccus, R., J. Joule and W.J. Kimberling. (1980). J. Mamm., 61:423-435.
2. Barto, E. (1942). Papers Mich. Acad. Sci., Arts. Lets. 27:195-213
3. Bowen, W.W. and W. D. Dawson. (1977). J. Mamm., 58:521-530.
4. Clark, F.H. (1936). J. Hered., 27:259-260.
5. Clark, F.H. (1938). Contrib. Lab. Vert. Genet., 7:1-11.
6. Committee. (1977). Cytogenetics, 19:38-43.
7. Cramer, D.V. (1988). Rat News Letter, 20:15-20
8. Dawson, W.D. (1982). Acta Theriol., 27:213-230.
9. Dawson, W.D., L.L. Huang, M.R. Felder and J.B. Shaffer (1983). Biocehm. Genet., 21:1101-1114.
10. Huestis, R.R. and V. Piestrak (1942). J. Hered.,33:289-291.
11. Huestis, R.R. and G. Lindstedt (1946). Am. Nat., 80:85-91.
12. Koop, B.F., R.J. Baker, M.W. Haiduk and M.D. Engstrom. (1984) Genetica, 64:199-208.
13. McIntosh, W.B. (1956). Contrib. Lab. Vert. Biol., 73:1-27.
14. Randerson, S. (1973). J. Hered., 64:371-372.
15. Robinson, R. (1964). Heredity, 19:701-709.
16. Robinson, R. (1972). *Gene Mapping in Laboratory Mammals*, pp.431-441.
17. Snyder, L.R.G. (1980). Biochem. Genet., 18:209-220.
18. Sumner, F.F. (1922). Am. Nat., 56:412-417.

Mesocricetus auratus (Syrian Hamster): 2n = 44

June 1989

Roy Robinson
St.Stephens Road Nursery
Ealing, London W13 8HB, England

Linkage group	Loci	Crossover value (%)	References
I	c - p	30.6 $\pm$ 1.8	4
II	e - hy	3.6 $\pm$ 0.9	8
III	Ba - l	8.9 $\pm$ 1.7	6
IV	J - sz	7.4 $\pm$ 2.9	9
V	cm - Lg	13.6 $\pm$ 4.2	10, 11
VI	Sa - U	15.0 $\pm$ 2.2	5
Sex	Mo - To	10.4 $\pm$ 4.4	2

References:

1. Nixon, C. W., Connelly, M. E. 1968. J. Hered., 59:276-278
2. Nixon, C. W., Robinson, R. 1973. Genetica, 44:588-590
3. Petersen, J. S., et al. 1977. J. Hered., 64:131-132
4. Robinson, R. 1973. J. Hered., 64:232
5. Robinson, R. 1988. J. Hered., 79:71
6. Robinson, R. et al. 1978. J. Hered., 69:199
7. Robinson, R. et al. 1987. J. Hered., 78:128-129
8. Yoon, C. H., Petersen, J. S. 1977. J. Hered., 68:418
9. Yoon, C. H., Petersen, J. S. 1979. J. Hered., 70:279-280
10. Yoon, C. H., et al. 1980. J. Hered., 71:61-52
11. Yoon, C. H., et al. 1980 J. Hered., 71:287-288

Notes:

The genes pa and T are known to be on the X chromosome with Mo and To but no linkage data are available (1, 3). The gene p (pink eyed dilution) was formerly known as b (brown) (7).

Meriones unquiculatus (Mongolian Gerbil): 2n = 44

June 1989

Roy Robinson
St. Stephens Road Nursery
Ealing, London W13 8HB, England

Linkage group	Loci	Crossover Value (%)	References
I	$c^h - p$	30.0 ± 5.0	2
II	$a - s^p$	40.2 ± 4.4	1

References:

1. Allan, D., Robinson, R. 1988> J. Hered., 79:386-387
2, Leiper, B. D., Robinson, R. 1986. J. Hered., 77:207

Note:
Descriptions of the mutant coat colour genes of the species and some loci independency data may be found in ref. 1.

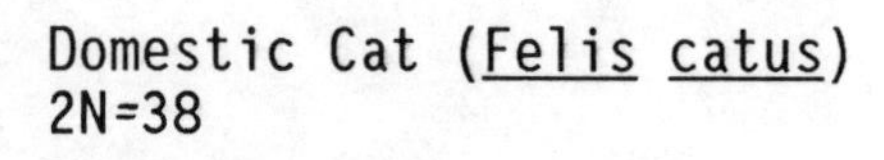
Domestic Cat (Felis catus)
2N=38

December, 1989

Stephen J. O'Brien
Laboratory of Viral Carcinogenesis
National Cancer Institute
Frederick, Maryland 21701-1013

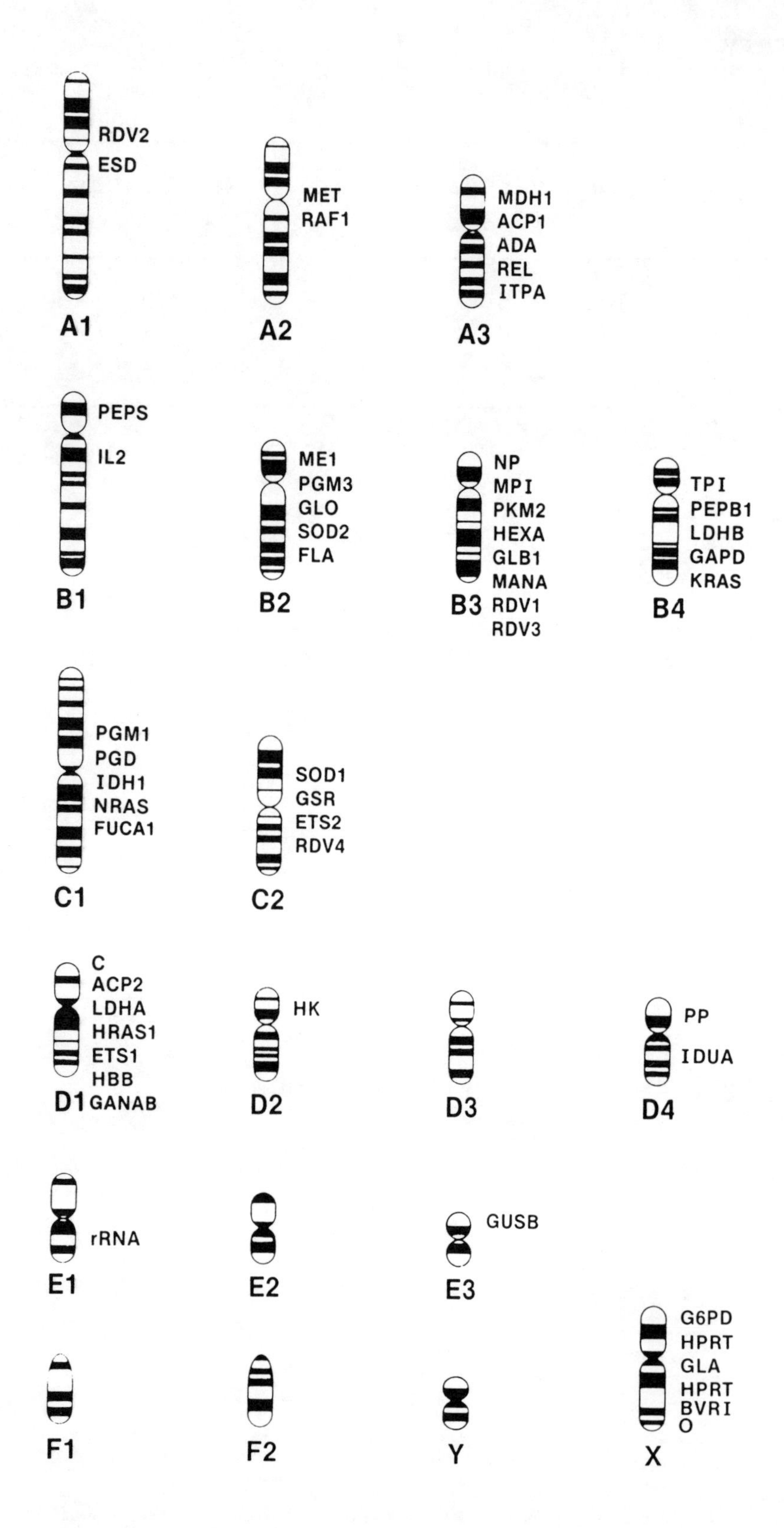

*Regional positions for the cat are not determined

Table I - Genetic Loci Assigned to Chromosomes in the Cat

Chromosome	Gene Symbol	Gene Name	Assay/ Mapping Procedure	References
A1	ESD	Esterase-D	I/H	7
A1	RDV2	RD-114 endogenous retrovirus-2	D/H	10
A2	RAF1	raf-1 proto-oncogene homolog	D/H	4
A2	MET	met proto-oncogene homolog	D/H	4
A3	MDH1	Malate dehydrogenase-1	I/H	7
A3	ACP1	Acid phosphatase-1	I/H	7
A3	ITPA	Inosine triphosphatase	I/H	1
A3	ADA	Adenosine deaminase	I/H	1
A3	REL	rel proto-oncogene homolog	D/H	2
B1	PEPS	Peptidase S	I/H	7
B1	IL2	Interleukin-2	D/H	13
B2	ME1	Malic enzyme-1 (soluble)	I/H	7
B2	PGM3	Phosphoglucomutase-3	I/H	7
B2	GLO	Glyoxylase	I/H	7
B2	SOD2	Superoxide dismutase	I/H	7
B2	FLA-I	Feline leukocyte antigen-1 (major histocompatibility complex, class-I)	D/H	16
B2	FLA-II	Feline leukocyte antigen-2 (major histocompatibility complex, class-II)	D/H	16
B3	NP	Purine nucleoside phosphorylase	I/H	7
B3	MPI	Mannose phosphate isomerase	I/H	7
B3	PKM2	Pyruvic kinase	I/H	7
B3	HEXA	Hexosaminidase-A	I/H	5, 7
B3	GLB1	Galactosidase-beta	I/H	5
B3	MAN	α-mannosidase	I/H	5
B3	RDV1	RD-114 endogenous retrovirus-1	D/H	10
B3	RDV3	RD-114 endogenous retrovirus-3	D/H	10
B4	TPI	Triosephosphate isomerase	I/H	7
B4	PEPB1	Peptidase B1	I/H	7
B4	LDHB	Lactate dehydrogenase-B	I/H	7
B4	GAPD	Glyceraldehyde-3-phosphate dehydrogenase	I/H	7
B4	KRAS	K-ras proto-oncogene homolog	D/H	
C1	PGM1	Phosphoglucomutase-1	I/H	7
C1	PGD	6-Phosphogluconate dehydrogenase	I/H	7
C1	IDH1	Isocitrate dehydrogenase-1 (soluble)	I/H	7
C1	NRAS	N-ras proto-oncogene homolog	D/H	4
C1	FUCA	α-L-Fucosidase	I/H	5
C2	SOD1	Superoxide dismutase-1	I/H	7
C2	GSR	Glutathione reductase	I/H	1
C2	ETS2	ets proto-oncogene homolog	D/H	14
C2	RDV4	RD-114 endogenous retrovirus-4	D/H	10
D1	LDHA	Lactate dehydrogenase-A	I/H	1
D1	ACP2	Acid phosphatase-2	I/H	1
D1	HRAS1	Harvey ras proto-oncogene homolog	D/H	17
D1	ETS1	ets proto-oncogene homolog	D/H	14
D1	HBB	β-globin	I,D/H,C	8
D1	C	Color, albino, siamese	M/C	8
D1	GANAB	Glucosidase-A	I/H	5
D2	HK-1	Hexokinase-1	I/H	7
D4	PP	Inorganic pyrophosphatase	I/H	7
D4	IDUA	α-L-iduronidase	I/H	12
E1	rRNA	Ribosomal DNA	D/A	9, 15
E3	GUSB	β-glucuronidase	I/H	5
X	BVR1	BALB retrovirus restriction-1	H	
X	G6PD	Glucose-6-phosphate dehydrogenase	I/H	7
X	HPRT	Hypoxanthine-guanine phosphoribosyl transferase	I/H	7
X	GLA	α-galactosidase	I/H	7
X	O	Orange	M/C	3
U2	GPI	Glucose phosphate isomerase	I/H	7
U4	PEPA	Peptidase A	I/H	7
U5	AK1	Adenylate kinase-1	I/H	7

Key:
I - isozyme stain development
D - DNA probe Southern blot
M - morphological phenotype
H - somatic cell hybrid
C - sexual cross
A - in situ hybridization

For a general review of cat genetics see:
Robinson, R. Trends Genet. 1: 236-239, 1985.
O'Brien, S. J. Trends Genet. 2: 137-142, 1986.

References:

1. Berman, E.J., Nash, W.G., Seuanez, H.N., and O'Brien, S.J.: Chromosomal mapping of enzyme loci in the domestic cat: GSR to C2, ADA and ITPA to A3, and LDHA-ACP2 to D1. Cytogenet. Cell Genet. 41: 114-120, 1986.

2. Brownell, E., Kozak, C.A., Fowle, J.R., III, Modi, W.S., Rice, N.R., and O'Brien, S.J.: Comparative genetic mapping of cellular rel sequences in man, mouse, and the domestic cat. Am. J. Hum. Genet. 39: 194-202, 1986.

3. Centerwall, W.R., and Benirschke, K. Male tortoiseshell and calico cats. Animal models of sex chromosome mosaics, aneuploids, polyploids, and chimerics. J. Hered. 64: 272-278, 1973.

4. Dean, M., Kozak, C., Robbins, J., Callahan, R., O'Brien, S., and Vande Woude, G.F.: Chromosomal localization of the met proto-oncogene in the mouse and cat genome. Genomics 1: 167-173, 1987.

5. Gilbert, D.A., O'Brien, J.S. and O'Brien, S.J.: Chromosomal mapping of lysosomal enzyme structural genes in the domestic cat. Genomics 2: 329-336, 1988.

6. O'Brien, S.J. Bvr-1: A restriction locus of a type C RNA virus in the feline cellular genome: Identification, location and phenotypic characterization in cat x mouse somatic cell hybrids. Proc. Natl. Acad. Sci. U.S.A. 73: 4618-4622, 1976.

7. O'Brien, S.J. and Nash, W.G.: Genetic mapping in mammals: Chromosome map of the domestic cat. Science 216: 257-265, 1982.

8. O'Brien, S.J., Haskins, M.E., Winkler, C.A., Nash, W.G., and Patterson, D.F.: Chromosomal mapping of beta-globin and albino loci in the domestic cat reveals a mammalian chromosome group. J. Hered., in press.

9. Pearson, M.D., Seabright, M., and Maclean, N. Silver staining of nucleolar organizer regions in the domestic cat, *Felis catus*. Cytogenet. Cell Genet. 24: 245-247, 1979.

10. Reeves, R.H., Nash, W.G., and O'Brien, S.J.: Genetic mapping of endogenous RD-114 retroviral sequences of the domestic cats. J. Virol. 56: 303-306, 1985.

11. Robinson, R. *Genetics for Cat Breeders*. Pergamon Press, New York, 1977.

12. Schuchman, E.H., O'Brien, S.J., and Desnick, R.J.: Assignment of the feline α-L-iduronidase gene to chromosome D4. Genomics 4: 442-444, 1989.

13. Seigel, L.J., Harper, M., Wong-Staal, F., Gallo, R.C., Nash, W.G., and O'Brien, S.J.: Gene for T-cell growth factor: Location on human chromosome 4q and feline chromosome B1. Science 223: 175-178, 1984.

14. Watson, D.K., McWilliams-Smith, M.J., Kozak, C., Reeves, R., Gearhart, J., Nunn, M.F., Nash, W., Fowle, J.R., III, Duesberg, P., Papas, T.S., and O'Brien, S.J.: Conserved chromosomal positions of dual domains of the ets protooncogene in cats, mice and humans. Proc. Natl. Acad. Sci. U.S.A. 83: 1792-1796, 1986.

15. Yu, L., Szabo, P., Hardy, W.D., and Prensky, W. The location, the number, and the activity of ribosomal genes in the domestic cat. J. Cell Biol. 87: 49a abs., 1980.

16. Yuhki, N.Y. and O'Brien, S.J.: Molecular characterization and genetic mapping of class I and class II MHC genes of the domestic cat. Immunogenetics 27: 414-425 1988.

17. Unpublished observations.

DOMESTIC DOG (Canis familiaris)(2n=78)

January, 1987.

P. Meera Khan, C. Brahe & L.M.M. Wijnen
Dept. of Human Genetics, University of Leiden, WW 72, 2333 AL Leiden,
The Netherlands.

Genetic map of biochemical loci in the dog

Canine chromosomes*	Genes**	References
U 1	ENO1, PGD	4
U 2	PGM1	4
U 3	GUK1	4
U 4	PEPC	4
U 5	ACP1	4
U 6	MDH1	2,4
U 7	IDH1	2,4
U 8	ACY1, GPX1	4
U 9	PGM2	4
U10	PGM3, ME1, SOD2, DLA, C4	4,6,7
U11	MDH2	4
U12	AK3	4
U13	PPA	4
U14	LDHA	2,4
U15	ACP2	4
U16	ESA4	4
U17	GAPD, TPI, LDHB	2,4
U18	ESD	4
U19	NP	4
U20	MPI, PKM2	4
U21	PGP	4
U22	GPI	2,4
U23	ITPA	4
U24	SOD1	4
U25	ACO2	4
X	GLA, HPRT, G6PD, F8, F9	2,4,8

*U1 to U25 are tentative designations for unidentified canine autosomes.
**DLA: the MHC genes; C4: genes for complement factor subunits C4A and C4G; F9 and F8: genes for the clotting factors IX and VIII respectively.

Symbol	Enzyme	E.C. No.	Chrom.	Reference
ACP1	Acid phosphatase-1	3.1.3.2	U 5	4
ACP2	Acid phosphatase-2	3.1.3.2	U15	4
ACO2	Aconitase-2	4.2.1.3	U25	4
ACY1	Aminoacylase-1	3.5.1.14	U 8	4
AK3	Adenylate kinase-3	2.7.4.3	U12	4
ENO1	Enolase-1	4.2.1.11	U 1	4
ESA4	Esterase-A4	3.1.1.1	U16	4
ESD	Esterase D	3.1.1.1	U18	4
GAPD	Glyceraldehyde-3-phosphate dehydrogenase	1.2.1.12	U17	2,4
GLA	Alpha-Galactosidase A	3.2.1.22	X	4,8
GPI	Glucose phosphate isomerase	5.3.1.9	U22	2,4
GPX1	Glutathione peroxidase-1	1.11.1.9	U 8	4
G6PD	Glucose-6-phosphate dehydrogenase	1.1.1.49	X	2,4,5,8
GUK1	Guanylate Kinase-1	2.7.4.8	U 3	4
HPRT	Hypoxanthine-G. phosphoribosyl transferase	2.4.2.8	X	4,8
IDH1	Isocitrate dehydrogenase-1	1.1.1.42	U 7	2,4
ITPA	Inosine triphosphatase	3.6.1.19	U23	4
LDHA	Lactate dehydrogenase A	1.1.1.27	U14	2,4
LDHB	Lactate dehydrogenase B	1.1.1.27	U17	2,4
MDH1	Malate dehydrogenase-1	1.1.1.37	U 6	2,4
MDH2	Malate dehydrogenase-2	1.1.1.37	U11	4,5
ME1	Malic enzyme-1	1.1.1.40	U10	4
MPI	Mannose phosphate isomerase	5.3.1.8	U20	4
NP	Purine nucleoside phosphorylase	2.4.2.1	U19	4
PEPC	Peptidase C	3.4.11.*	U 4	4
PGD	6-phosphogluconate dehydrogenase	1.1.1.44	U 1	4
PGM1	Phosphoglucomutase-1	2.7.5.1	U 2	4,5
PGM2	Phosphoglucomutase-2	2.7.5.6	U 9	4,5
PGM3	Phosphoglucomutase-3	2.7.5.1	U10	4,5,6,7
PGP	Phosphoglycolate phosphatase		U21	4
PKM2	Pyruvate kinase M2	2.7.1.40	U20	4
PPA	Inorganic pyrophosphatase	3.6.1.1	U13	4
SOD1	Superoxide dismutase-1	1.15.1.1	U24	4,5
SOD2	Superoxide dismutase-2	1.15.1.1	U10	4
TPI	Triosephosphate isomerase	5.3.1.1	U17	2,4

REFERENCES

1. BAUR, E.W. and SCHORR, R.T. 1969. Science 166: 1524-1525.
2. BRUNS, G.A.P., et al. 1979. Cytogenet. Cell Genet. 22: 547-551.
3. HUTT, F.B. 1979. Genetics for dog breeders. W.H. Freeman and Company, San Francisco.
4. MEERA KHAN, P., et al. 1984. Human Gene Mapping 7.
5. MEERA KHAN, P., et al. 1973. Transplantation 15: 624-628.
6. MEERA KHAN, P., et al. 1978. Cytogenet. Cell Genet. 22: 585-587, 1978.
7. GROSSE-WILDE, H., et al. 1983. Immunogenet. 18:537-540.
8. BRINKHOUS, K.M., et al. 1973. Blood, 41:577-585.

HORSE (Equus Caballus) 2N = 64 June 1989

Lowell R. Weitkamp, Genetic Markers Laboratory, University of Rochester School of Medicine and Dentistry, Rochester, New York 14642.
Kaj Sandberg, Department of Animal Breeding and Genetics, Swedish University of Agricultural Sciences, Uppsala, Sweden, S-750 07.

The standard nomenclature for the equine karyotype has been the Reading system (15). This was improved by high resolution G-banding (21) and more recently has been revised at the Tenth International Chromosome Conference (Uppsala, Sweden, 1989). Two linkage groups have been assigned to autosomes, chromosome 10 (8 in the Reading terminology) and 20.

Table 1. Loci assigned to chromosomes or linkage groups. [1]

Locus	Marker and Source	Chromosome	References
A	A blood group, rbc	20 (U3)	7, 8
A1B	(Xk) A1B glycoprotein, plasma	10 (U4)	2, 16, 33
ALB	albumin, plasma	U2	3, 4, 25, 26, 29, 31
E	chestnut, coat color	U2	3, 28
C4	complement C4, DNA	20 (U3)	17
CYP21	(21OH) cytochrome P450, steroid 21-hydroxylase, DNA	20 (U3)	17
ELA-A	equine lymphocyte alloantigens	20 (U3)	6, 7, 8
ELA-B	equine lymphocyte alloantigens	20 (U3)	1, 9, 19
ES	carboxylesterase, plasma	U2	3, 4, 5, 25
ESCI	class I molecule, plasma	20 (U3)	18
GC	vitamin D binding protein, plasma	U2	3, 26, 31
GOT2	mitochondrial glutamate oxalo-acetate transaminase, wbc	U2	5
G6PD	glucose-6-phosphate dehydroge-nase, cultured cells	X	14, 23, 30
F13A	clotting factor F13A, plasma	20 (U3)	36
HPRT	hypoxanthine-guanine phospho-ribosyl transferase, cultured cells	X	14
HP	haptoglobin, plasma	U2	35
K	K blood group, rbc	U1	4, 24, 25
ME1	soluble malic enzyme, plasma or rbc	10 (U4)	33
PGD	6-phosphogluconate dehydrogenase, rbc	U1	4, 24, 25
PGK	phosphoglycerate kinase, cultured cells	X	14
GPI	phosphohexose isomerase, rbc	10 (U4)	2, 12
PI	protease inhibitor, plasma	U5	10
RN	roan, coat color	U2	3, 28
TO	tobiano spotting, coat color	U2	11, 29
U	U blood group, rbc	U5	10

[1]Locus symbols have been revised to conform to the conventions for human gene nomenclature (27).

ELA was assigned by *in situ* hybridization to chromosome 20 q14-q22 using a cloned DNA sequence derived from a class I gene of the porcine major histocompatibility complex (6, 22). The probable order of loci on this chromosome is F13A: A: ELA-A: ELA-B (9, 36). At least one other ELA locus has been identified in recombinant families (20), but it is unknown whether it is class I or class II. A soluble class I molecule, ESCI, analogous to mouse Q10 is tightly linked to ELA (18). The genes for complement C4 and 21-hydroxylase are linked to ELA by DNA analysis (17). GPI was assigned to 10 p*ter*, using a porcine genomic probe, by *in situ* hybridization (12).

There are several tightly linked loci for hemoglobin alpha chains (13) which are not assigned to a linkage group. Thirteen equine marker genes segregate independently of each other and of the five established autosomal linkage groups (25, 32). Most linkage investigations were conducted on sire families. For two groups in which females were studied, K:PGD and ALB:ES, recombination is more frequent in females than males (25, 34). Recombination frequencies for individual sires differ significantly for the K:PGD segment (4).

Table 2. Recombination frequencies between linked loci.

Linkage Group	Loci	Male Recombination Frequency	References
U1	K:PGD	0.26	4, 24, 25
U2	ALB:E:ES	0.07, 0.29	3, 25, 34
	ALB:HP:ES	0.22, 0.17	35
	ALB:RN	0.03	3
	RN:E	0.035	28
	ALB:GC	0.009	26, 31
	ALB,GC:TO	~0	11, 29
	ES:GOT2	~0	5
20 (U3)	A:ELA-A	0.016	7, 8
	ELA-A:ELA-B	0.02	9
	F13A:ELA-A	0.17	36
10 (U4)	A1B:ME1	~0	33
	A1B:GPI	0.23	2
U5	PI:U	0.13	10

References

1. Alexander A.J. et al. 1987. Immunogenetics 25:47-54.
2. Andersson L. et al. 1983. Anim Bld Grps Biochem Genet 14:45-50.
3. Andersson L. and Sandberg K. 1982. J Heredity 73:91-94.
4. Andersson L. and Sandberg K. 1984. Genetics 106:109-122.
5. Andersson L. et al. 1983. J Heredity 74:361-364.
6. Ansari H.A. et al. 1988. Immunogenetics 28:362-364.
7. Bailey E. et al. 1979. Science 204:1317-1319.
8. Bailey E. 1983. Anim Bld Grps Biochem Genet 14:37-43.
9. Bernoco D. et al. 1987. Anim Genet 18:103-118.
10. Bowling A.T. 1986. Anim Genet 17:217-223.
11. Bowling A.T. 1987. J Heredity 78:248-250.
12. Chowdhary B.P. et al. 1989. Tenth Internat. Chrom. Conf., Uppsala, Sweden (Abstracts, p.24).
13. Clegg J.B. 1987. Molecular Biology and Evolution 4:492-503.
14. Deys B.F. 1972. Ph.D. Thesis, University of Leiden.
15. Ford C.E. et al., 1980. Hereditas 92:145-162.
16. Juneja R.K. et al. 1987. Anim Genet 18:119-124.
17. Kay et al. 1987. J Immunogenetics 14:247-253.
18. Lew A.M. et al., 1986. Immunogenetics 24:128-130.
19. Lazary S. et al., 1986. J Immunogenetics 13:315-325.
20. Lazary S. et al., 1988. Animal Genet 19:447-456.
21. Maciulis A. et al., 1984. J Heredity 75:265-268.
22. Makinen A. et al. 1989. Hereditas 110:93-96.
23. Mathai C.K. et al. 1966. Nature 210:115-116.
24. Sandberg K. 1974. Anim Bld Grps Biochem Genet 5:137-141.
25. Sandberg K. and Andersson L. 1984. Hereditas 100:199-208.
26. Sandberg K. and Juneja R.K. 1978. Anim Bld Grps Biochem Genet 9:169-173.
27. Shows T.B. et al., 1987. Cytogenet Cell Genet. 46:11-28.
28. Sponenberg D.P. et al. 1984. J Heredity 75:413-414.
29. Trommershausen-Smith A. 1978. J Heredity 69:214-216.
30. Trujillo J.M. et al. 1965. Science 148:1603-1604.
31. Weitkamp L.R. and Allen P.Z. 1979. Genetics 92:1347-1354.
32. Weitkamp L.R. and Bailey E. 1985. Anim Bld Grps Biochem Genet 16:61-63.
33. Weitkamp L.R. et al. 1982. Anim Bld Grps Biochem Genet 13:279-284.
34. Weitkamp L.R. et al. 1982. Anim Bld Grps Biochem Genet 13:305-306.
35. Weitkamp L.R. et al. 1985. Anim Bld Grps Biochem Genet 16, Suppl 1:78.
36. Weitkamp L.R. et al. 1988. Anim Genet 20, Suppl 1:10.

THE GENE MAP OF THE PIG (Sus scrofa domestica L.) 2N = 38

August 1989

ECHARD Geneviève
Laboratoire de Génétique Cellulaire - I.N.R.A. -
BP 27 - 31326 CASTANET-TOLOSAN CEDEX , FRANCE

A standardized G-banded karyotype of the pig was published in 1980 (Hereditas 92 : 145-162) following the Reading Conference (1976). The groups working on pig cytogenetics and gene mapping are using the chromosome numbering adopted at that time. A more detailed standard karyotype (Q-G-R bands), proposed by the committee for standardization, was published in Hereditas 1988 109, 151-158. The numbering of individual bands was similar to the system adopted for human chromosomes.
Part of the informations concerning the map of the pig was compiled from : Pig genetics: a review, OLLIVIER L. & SELLIER P. (48).

LINKAGES, SYNTENIES AND ASSIGNMENT TO CHROMOSOMES.

Group I	: K, HPX
Group II	: AM, I
Group III	: PI1, PO1, B, IGH1, IGH2, IGH3, IGH4
Group IV	: CAS fractions
Group V	: ELF, TF, CP
Group VI	: SLB, L
Group VII	: LP-B, LP-T, LP-V
Chromosome 1	: IFNA (1q2.5)
Chromosome 3	: MDH1
Chromosome 5	: LDHB
Chromosome 6	: GPI-HAL-S-H-PO2-PGD
Chromosome 7	: SLA (7p1.4 -> q1.2), J, C, MPI, PKM2, NP
Chromosome 8	: RNR (NOR)
Chromosome 9	: SOD 1
Chromosome 10	: RNR (NOR)
Chromosome 15	: G
Chromosome X	: SPL, TRA III, HPRT, G6PD, PGK, GLA

The linkage groups were established by family studies. Two of them were assigned to chromosomes: SLA, J, C, (chr.7); GPI, HAL, S, H, PO2, PGD (chr.6). For the others the carrier chromosome is unknown

Concerning chromosome assignments, only the results with no controversy are given in the table. For some enzyme markers, mapped by somatic cell hybridization two chromosomes are proposed: for PGMI chr. 10 or 6, PEPB chr. 11 or 5, LDHA chr. 4 or 2 (16, 55). The gene controlling the alpha-globuline A23 allotype was assigned to chromosome 16 (41).

GENETIC LOCI

Gene symbol	Locus	Linkage groups and chromosomes	Ref.
AM	Sérum amylase	II	2
B	Blood group B	III	3,39
C	Blood group C	7	7,30,49
CP	Ceruloplasmin	V	40
CAS	Caseins	IV	24
ELF	Early lethal factor	V	33
G	Blood group G	15	18,29,36
GLA	alpha-galactosidase A	X	44
G6PD	Glucose-6-phosphate dehydrogenase	X	15,22,44
GPI	Glucose phosphate isomerase	6	5,6,10,12,35,53
H	Blood group H	6	4,5,6,12,29,34
HAL	Halothane sensitivity	6	8,12,34,35
HPRT	Hypoxanthine guanine phosphoribosyl-transferase	X	15,22,44
HPX	Hemopexin	I	27,32
I	Blood group I	II	2,47
IFNA	Interferon alpha (leucocyte)	1	56,57
IGH	Immunoglobulin	III	39,50,51
J	Blood group J	7	7,28
K	Blood group K	I	27,32
L	Blood group L	VI	31
LDHA	Lactate dehydrogenase A	4 or 2	16,55
LDHB	Lactate dehydrogenase B	5	16,55
LP-B,T,V	Lipoproteins B,T,V	VII	52
MDH1	Malate dehydrogenase soluble	3	16
MPI	Mannose phosphate isomerase	7	13,17,22,23,55
NP	Nucleoside phosphorylase	7	13,17,23,55
PEPB	Peptidase B	11 or 5	16,55
PGD	6-phosphogluconate dehydrogenase	6	4,6,25,53,54
PGM1	Phosphoglucomutase 1	10 or 6	16,55
PGK	Phosphoglycerate kinase	X	15,22,44
PI1	Protease inhibitor - 1	III	19,20,37,39
PKM2	Pyruvate kinase (M2)	7	22,23
PO1	Postalbumin - 1	III	19,20,39
PO2	Postalbumin - 2	6	38
RNR	Ribosomal RNA (NOR)	10 and 8	45,46
S	Inhibition of A-O blood group	6	9,54
SLA	Pig Major histocompatibility complex	7	21,14,49,28,30

Gene symbol	Locus	Linkage groups and chromosomes	Ref.
SLB	Pig alloantigen	VI	31
SOD1	Super oxyde dismutase soluble	9	17,43
SPL	Splay leg condition	X	42
TF	Transferrin	V	33,40
TRAIII	Paralytic Tremor AIII	X	26

REFERENCES

1 - ANDRESEN E., 1966. Genetics 54, 805-812.
2 - ANDRESEN E., 1966. Science 153, 1660-1661.
3 - ANDRESEN E., 1968. K. Vetrog. Landbohjsk Aarsskr. 1-11.
4 - ANDRESEN E., 1970. Acta Vet. Scand. 11, 136-137.
5 - ANDRESEN E., 1970. Anim. Blood Grps Biochem. Genet. 1, 171-172.
6 - ANDRESEN E., 1971. Anim. Blood Grps Biochem. Genet. 2, 119-120.
7 - ANDRESEN E. and BAKER L.N., 1964. Genetics 49, 379-386.
8 - ANDRESEN E. and JENSEN P., 1977. Nord. Vet. Med. 29, 502-504.
9 - ANDRESEN E. et al., 1981. Z. Tierzücht. ZüchtBiol. 98, 45-54.
10 - BOOSMA A.A., 1988. Pro. 8th European Coll. on Cytogenet of Domestic Animals (Bristol) 99-105.
11 - CHRISTENSEN K. et al., 1985. Hereditas 102, 231-235.
12 - DAVIES W. et al., 1988. Anim. Genet. 19, 203-212.
13 - DOLF G. and STRANZINGER G., 1986. Genet. Sel. Evol. 18, 375-384.
14 - ECHARD G. et al., 1986. Cytogenet. Cell Genet. 41, 126-128.
15 - FÖRSTER M. et al., 1980. Naturwissenschaften 67, 48.
16 - FÖRSTER M. and HECHT W., 1984. Proceedings of the 6th European Colloquium on Cytogenetics of Domestic Animals, 351-355.
17 - FÖRSTER M. and HECHT W., 1985. Züchtungskunde 54, 249-255.
18 - FRIES R. et al., 1983. J. Hered. 74, 426-430.
19 - GAHNE B. and JUNEJA R.K., 1982. In : Proceedings of the 8th International Conference of Animal blood groups and Biochemical polymorphisms, OTTAWA, 111.
20 - GAHNE B. and JUNEJA R.K., 1985. Protides of the Biological Fluids 33, 119-122.
21 - GEFFROTIN C. et al., 1984. Annls Genet. 27, 213-219.
22 - GELLIN J. et al., 1980. Ann. Genet. 23, 15-21.
23 - GELLIN J. et al., 1981. Cytogenet. Cell Genet. 30, 59-62.
24 - GLASNAK V., 1968. Comp. Biochem. Physiol. 25, 355-357.
25 - GUERIN G. et al., 1983. Genet. Sel. Evol. 15, 55-64.
26 - HARDING J.D.L. et al., 1973. Vet. Rec. 92, 527-529.
27 - HESSELHOLT M. and NIELSEN P.B., 1966. In : Polymorphismes Biochimiques des Animaux. 445-448, I.N.R.A. Paris.
28 - HRADECKY J. et al., 1982. Anim. Blood Grps Biochem. Genet. 13, 223-224.
29 - HRADECKY J. et al., 1987. Anim. Genet. 18, 279-282.

30 - HRUBAN V. et al., 1976. Tissue Antigens 7, 267-271.
31 - HRUBAN V. et al., 1978. J. Immunogenetics 5, 173-178.
32 - IMLAH P., 1965. In : Blood groups of Animals 109-122. Matousek J (ed). Publishing house of the Czechoslovak Academy of Sciences, Prague.
33 - IMLAH P., 1970. Anim. Blood Grps Biochem. Genet. 1, 5-13.
34 - IMLAH P., 1980. Anim. Blood Grps Biochem. Genet. 11 (suppl 1) 47.
35 - JORGENSEN P.F. et al., 1976. Acta Vet. Scand. 17, 370-372.
36 - JORGENSEN P.F., 1979. Acta Agrid. Scand. suppl 21, 386-395.
37 - JUNEJA R.K., GAHNE B., 1981. Anim. Blood Grps Biochem. Genet. 12, 47-51.
38 - JUNEJA R.K. et al., 1983. Anim. Blood Grps Biochem. Genet. 14, 27-36.
39 - JUNEJA R.K. et al., 1986. Anim. Genet. 17, 225-233.
40 - JUNEJA R.K. et al., 1989. Anim. Genet. 20, 105-109.
41 - KNYAZEV S.P. and TIKHONOV V.M., 1984. Proceedings of the 6th European Colloquium on Cytogenetics of Domestic Animals 395-399.
42 - LAX T., 1971. J. Hered. 62, 250-251.
43 - LEONG M.M.L. et al., 1983. Can. J. Genet. Cytol. 25, 233-238.
44 - LEONG M.M.L. et al., 1983. Can. J. Genet. Cytol. 25, 239-245.
45 - LIN C.C. et al., 1980. Can. J. Genet. Cytol. 22, 103-116.
46 - MIYAKE Y. et al., 1988. Jpn. J. Vet. Sci. 50, 341-345.
47 - NIELSEN P.B., 1966. In : Polymorphismes Biochimiques des Animaux 449-452. I.N.R.A. Paris.
48 - OLLIVIER L. and SELLIER P., 1982. Ann. Genet. Sel. Anim. 14, 481-544.
49 - RABIN M. et al., 1985. Cytogenet. Cell Genet. 39, 206-209.
50 - RAPACZ J. and HASLER-RAPACZ J., 1974. Anim. Blood Grps Biochem. Genet. 5 (suppl 1) 33-34.
51 - RAPACZ J. and HASLER-RAPACZ J., 1982. In : 2nd World Congress on Genetics applied to Livestock production, VIII 601-606, Editor GARSI, Madrid.
52 - RAPACZ J., 1982. In : 2nd World Congress on Genetics applied to Livestock production, VI, 365-374, Editor GARSI, Madrid.
53 - RASMUSEN B.A. et al., 1980. Anim. Blood Grps Biochem. Genet. 11, 93-107.
54 - RASMUSEN B.A., 1981. Anim. Blood Grps Biochem. Genet. 12, 207-209.
55 - RYTTMAN H. et al., 1986. Anim. Genet. 17, 323-333.
56 - YERLE M. et al., 1986. Cytogenet. Cell Genet. 42, 129-132.
57 - YERLE M. et al., 1989. Genetics, Selection, Evolution (in press).

ESTABLISHED LINKAGE MAP OF THE RABBIT

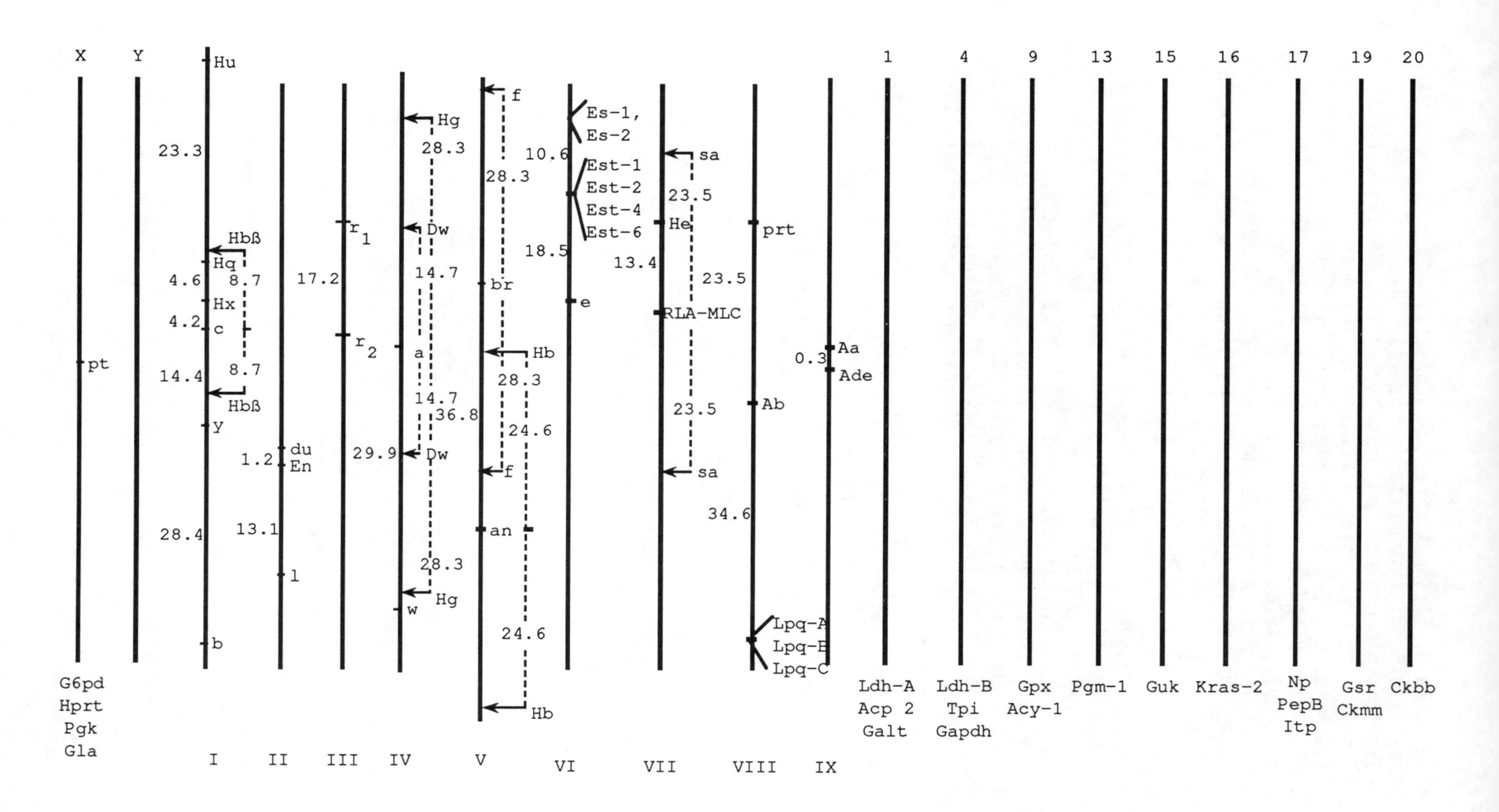

LINKAGE MAP OF THE RABBIT (*Oryctolagus cuniculus*) (2N = 44)

August 1989

DR. RICHARD R. FOX
The Jackson Laboratory
Bar Harbor, Maine 04609

GENE SYMBOL	LOCUS	LINKAGE GROUP	TISSUE	REF
a	Agouti	IV	coat color	38
Aa	Immunoglobulin heavy chain (variable region)	IX	serum	29,30
Ab	Immunoglobulin kappa light chain	VIII	serum	51,52,53, 54
Acp-2	Acid phosphatase-2	Chr 1	cultured cells	22
Acy-1	Aminoacylase-1	Chr 9	cultured cells	46
Ade	Immunoglobulin heavy chain (constant region)	IX	serum	29,30
an	A-antigen (Anti-human A cell)	V	RBC	43
b	Brown	I	coat color	8,9,12
br	Brachydactyly	V	gross morphology	15,43
c	Albino	I	coat color	8,9,12
Ckbb	Creatine kinase, brain form	Chr 20	brain	31
Ckmm	Creatine kinase, muscle form	Chr 19	muscle	32
du	Recessive white spotting (Dutch pattern)	II	coat color	7,9
Dw	Dwarf	IV	gross morphology	15
e	Yellow (extension)	VI	coat color	27,42
En	Dominant white spotting (English pattern)	II	coat color	6,7,9
Es-1	Esterase (erythrocyte)	VI	RBC	27,44,55
Es-2	Esterase (platelet)	VI	RBC	27,44
Est-1	Cocainesterase	VI	serum	27,55
Est-2	Atropinesterase	VI	serum	25,40,51
Est-4	Esterase	VI	serum	56
Est-6	Carboxylesterase	VI	liver	57
f	Furless	V	fur	14,41
G6pd	Glucose-6-phosphate dehydrogenase	X-chr	cultured cells	5,17,19, 20,46
Galt	Galactose-1-phosphate uridyl transferase	Chr 1	cultured cells	47
Gapdh	Glyceraldehyde-3-phosphate dehydrogenase	Chr 4	cultured cells	46,47
Gla	α-galactosidase	X-chr	cultured cells	19,20,46
Gpx	Glutathion peroxydase	Chr 9	cultured cells	46,47
Gsr	Glutathion reductase	Chr 19	cultured cells	46,47
Guk	Guanylate kinase	Chr 15	cultured cells	46,47
Hb	Hb blood group	V	RBC	49
Hbβ	Hemoglobin β-chain	I	RBC	39
He	He blood group	VII	RBC	49,50
Hg	Hg blood group	IV	RBC	49

GENE SYMBOL	LOCUS	LINKAGE GROUP	TISSUE	REF
Hprt	Hypoxanthine phosphoribosyl transferase	X-chr	cultured cells	17,19,20, 46
Hq	Hq blood group	I	RBC	49
Hu	Hu blood group	I	RBC	49
Hx	Hemopexin	I	serum	28,49
Itp	Inosine triposphatase	Chr 17	cultured cells	46
Kras-2	Ki-*ras* proto-oncogene	Chr 16	cultured cells	33
l	Angora	II	fur	6,7,9
Ldh-A	Lactate dehydrogenase-A	Chr 1	cultured cells	22,45,46
Ldh-B	Lactate dehydrogenase-B	Chr 4	cultured cells	16,21,22, 45
Lpq-A	Low-density lipoprotein	VIII	serum	52,53,54
Lpq-B	Low-density lipoprotein	VIII	serum	52,53,54
Lpq-C	Low-density lipoprotein	VIII	serum	52,53,54
Np	Nucleoside phosphorylase-1	Chr 17	cultured cells	45,46
Pep-B	Peptidase B	Chr 17	cultured cells	46
Pgk	Phosphoglycerate kinase	X-chr	cultured cells	17,19,46
Pgm-1	Phosphoglucomutase 1	Chr 13	cultured cells	47
prt	Serum protein	VIII	serum	51,52,53, 54
pt	Paralytic tremor	X-chr	CNS	34
r_1	Rex 1 (French rex)	III	fur	12,14
r_2	Rex 2 (German rex)	III	fur	12,14
RL-A	Major histocompatability complex	VII	serum	49,50
sa	Satin	VII	coat color	49
Tpi	Triosephosphate isomerase-1	Chr 4	cultured cells	21,22,45, 46
w	Wide band agouti	IV	coat color	40
y	Yellow fat	I	fat	11,12,35

Additional review articles on the linkages in the rabbit, including maps at various time periods and some negative data, include the following (10, 13, 23, 24, 36, 37, 38, 41, 48). Formerly it was believed that the *bu* locus was also in Linkage Group I, 16 cM from the *c* locus (2), but more complete data (3) suggest that, if linkage is present, it is a loose linkage of about 41cM. Other very likely linkages, for which crossovers have not yet been observed, include those of the three *Lpq* loci (1, 53, 54), *Mtz* with *Mt* (4), *Es-1* with *Es-2* (44), *Est-1* with *Est-2*, *Est-4* and *Est-6* (55,56,57) and *RLA* with *MLC* (50). Based on pedigree information, there is a high probability that the *mst* locus is located in L.G. IV (26).

A standardized karyotype for *Oryctolagus cuniculis* has been agreed upon by representatives of most groups known to have published banded chromosome analyses of laboratory stocks and has been published in Cytogenetics and Cell Genetics (18). A comparison of the standardized system with the various previously reported numbering systems is given. The chromosome numbering is according to proposed standard (18).

REFERENCES

1. ALBERS, J. J. and S. DRAY. 1969. J. Immunol. 103:155-162.
2. BAUER, E. J., Jr. and J. BENNETT. 1964. Genetics 50:234
3. BENNETT, J., et al. 1973. J. Hered. 64:363-364.
4. BERNE, B. H., et al. 1970. J. Immunol. 105:856-864.
5. BOMSEL-HELMREICH, O. 1970. Adv. Biosci. 6:381-403.
6. CASTLE, W. E. 1921. Science 54:634-636.
7. CASTLE, W. E. 1924. Proc. Nat. Acad. Sci. (USA) 10:107-108.
8. CASTLE, W. E. 1924. Proc. Nat. Acad. Sci. (USA) 10:486-488.
9. CASTLE, W. E. 1926. Carnegie Inst. Wash. Pub. 337:1-47.
10. CASTLE, W. E. 1929. Z. indukt. Abstamm. Vererbungsl. 52:53-60.
11. CASTLE, W. E. 1933. Proc. Nat. Acad. Sci. (USA) 19:947-950.
12. CASTLE, W. E. 1936. Proc. Nat. Acad. Sci. (USA) 22:222-225.
13. CASTLE, W. E. 1940. Mammalian Genetics. Harvard Univ. Press, Cambridge.
14. CASTLE, W. E. and H. NACHTSHEIM. 1933. Proc. Nat. Acad. Sci. (USA).
15. CASTLE, W. E. and P. B. SAWIN. 1941. Proc. Nat. Acad. Sci. (USA) 27:519-523.
16. CIANFRIGLIA, M., et al. 1979. Cytogenet. Cell Genet. 25:141.
17. CIANFRIGLIA, M., et al. 1979. Cytogenet. Cell Genet. 25:142.
18. COMMITTEE for Standardized Karyotype of *Ocyctolagus cuniculus*. 1981. Cytogenet. Cell Genet. 31:240-248.
19. ECHARD, G., and M. GILLOIS. 1979. Cytogenet. Cell Genet. 25:148-149.
20. ECHARD, G., et al., 1981. Cytogenet. Cell Genet. 29:176-183.
21. ECHARD, G., et al., 1982. Cytogenet. Cell Genet. 32:269.
22. ECHARD, G., et al., 1982. Cytogenet. Cell Genet. 34:289-295.
23. FOX, R. R. 1974. Chapter 1. In S. H. Weisbroth, R. E. Flatt, and A. L. Kraus (eds.) The Biology of the Laboratory Rabbit. pp. 1-22. Academic Press, New York.
24. FOX, R. R. 1975. Chapter 14. In R. C. King (ed.) Handbook of Genetics. Vol. 4. Vertebrates of Genetic Interest. pp. 309-328. Pleunum Publ. Corp., New York.
25. FOX, R. R. 1981. In K. Myers and C. D. MacInnes (eds.), Proceedings of World Lagomorph Conference, University of Guelph, Ontario, held 13-17 August 1979. University Guelph, Guelph, pp. 4-16.
26. FOX, R. R., and D. D. CRARY. 1980. J. Hered. 70:369-372.
27. FOX, R. R., and L. F. M. VAN ZUTPHEN. 1979. Genetics 93:183-188.
28. HAGEN, K. L., et al. 1978. Anim. Blood Grps. Biochem. Genet. 9:151-159.
29. MAGE, R. G. 1979. Ann. Immunol. (Inst. Pasteur) 130C:105-114
30. MAGE, R. G. 1982. Immunogenetics 15:287-297.
31. MAHONEY, C. E., et al. 1988. Cytogenet. Cell Genet. 48:160-163.
32. MARTIN-DeLEON, P. A., et al. 198__. Mapping of the creatine kinase M gene to 19 q11-q12 in the rabbit genome. Cytogenet. Cell Genet. (in press).
33. MARTIN-DeLEON, P. A. and S. R. PICCIANO. 1988. Cytogenet. Cell Genet. 48:201-204.
34. OSETOWSKA, E. 1967. Acta Neuropathologica 8:331-344.
35. PEASE, M. S. 1930. Verhandl. I. Internat. Kanichenzuchter-kongr. Leipzig 91-95.
36. RIFAAT, O. M. 1954. Heredity 8:107-116.
37. ROBINSON, R. 1956. J. Genet. 54:358-369.
38. ROBINSON, R. 1958. Bibliogr. Genet. (Hague) 17:229-588 (454-460).
39. SANDBERG, K. and L. ANDERSON 1987. J. Hered. 78:124-125.
40. SAWIN, P. B. 1934. J. Hered. 25:477-481.
41. SAWIN, P. B. 1955. Adv. Genet. 7:183-226.
42. SAWIN, P. B., and D. GLICK. 1943. Proc. Nat. Acad. Sci. (USA) 29:55-59.
43. SAWIN, P. B., et al. 1944. Proc. Nat. Acad. Sci. (USA) 30:217-221.
44. SCHIFF, R., and C. STORMONT. 1970. Biochem. Genet. 4:11-23.
45. SOULIE, J., et al. 1982. Cytogenet. Cell Genet. 32:319.
46. SOULIE, J. and J. de GROUCHY. 1982. Human Genet. 60:172-175.
47. SOULIE, J. and J. de GROUCHY. 1983. Human Genet. 63:48-52.
48. SPENDLOVE, W. H. and R. ROBINSON. 1970. Genetica 41:635-637.
49. TISSOT, R. G. 1978. Personal communication to R. R. Fox.
50. TISSOT, R. G., and C. COHEN. 1974. Transplantation 18:142-149.
51. USHER, D. C., et al. 1978. J. Immunol. 120:1832-1835.
52. USHER, D. C., et al. 1978. Immunogenetics 7:117-124.
53. USHER, D. C., et al. 1982. J. Hered. 73:286-290.
54. USHER, D. C., et al. 1983. Biochem. Genet. 21:511-526.
55. ZUTPHEN, L. F. M. VAN, et al. 1977. Biochem. Genet. 15:989-1000.
56. ZUTPHEN, L. F. M. VAN, et al. 1983. Biochem. Genet. 21:772-780.
57. ZUTPHEN, L. F. M. VAN, et al. 1987. Biochem. Genet. 25:335-344.

THE GENE MAP OF THE SHEEP (Ovis aries) 2N = 54

August 1989

ECHARD Geneviève
Laboratoire de Génétique Cellulaire - I.N.R.A. -
BP 27 - 31326 CASTANET-TOLOSAN CEDEX , FRANCE

The committee for standardization of the sheep karyotype published in 1985 a standard nomenclature for the G-bands (Hereditas 103 : 165-170) following the system adopted for human chromosomes. The numbering of the chromosomes was the one proposed at the Reading Conference (1980 Hereditas 92 : 145-162).

LINKAGES, SYNTENIES AND ASSIGNMENT TO CHROMOSOMES.

C,I	Linkage Group I
PGD-ENO1-PEPC	U1
PGM1 or 2	U2
NP-PKM2	U3
PGM3-ME1	U8
MP1	U9
SOD1-PRGS-PAIS	U10
IDH1	U11
GPI	U12
LDHA	U13
MDH2	U14
ADA-ITPA	U15
GSR	U16
HBG	U17
ACO1	U18
TPI-LDHB-PEPB-GAPD	Chromosome 3
SHMT-KRTB (3q14-22)	
KRTA	Chromosome 11q24-28
OLA	Chromosome 20q12-23
RNR (NOR)	Chromosomes 1,2,3,4,25
G6PD-HPRT-GLA-PGK	Chromosome X
OTC-RCP-ARAF1-PLP-SYN	

* U for unknown chromosome

The syntenies were established by somatic cell hybridization. The RNR localization were obtained using silver staining, and the regional mapping of genes by in situ hybridization.

GENETIC LOCI

Gene symbol	Locus	Chromosome	Ref.
ACO1	Aconitase 1	U18	13
ADA	Adenosine deaminase	U15	10
ARAF1	v-raf oncogene homolog 1	X	13
C	C blood group	I	8,9
ENO1	Enolase 1	U1	10
GAPD	Glyceraldehyde-3-phosphate dehydrogenase	3	13
G6PD	Glucose-6-phosphate dehydrogenase	X	10
GLA	Alpha-galactosidase	X	10
GPI	Glucose phosphate isomerase	U12	10
GSR	Glutathione reductase	U16	5
HBG	Hemoglobin gamma	U17	12
HPRT	Hypoxanthine guanine phosphoribosyl transferase	X	10
I	I blood group	I	8,9
IDH1	Isocitrate dehydrogenase 1	U11	10
ITPA	Inosine triphosphate	U15	13
KRTA	Keratin A	11	4
KRTB	Keratin B	3	4
LDHA	Lactate dehydrogenase A	U13	10
LDHB	Lactate dehydrogenase B	3	10,6
MDH2	Malate dehydrogenase 2	U14	10
ME1	Malic enzyme 1	U8	10
MPI	Mannose phosphate isomerase	U9	11
NP	Nucleoside phosphorylase	U3	11
OLA	Ovine major histocompatibility complex	20	7,3
OTC	Ornithine carbamoyl transferase	X	13
PEPB	Peptidase B	3	10,6
PEPC	Peptidase C	U1	12
PGD	6-Phosphogluconate dehydrogenase	U1	10
PGK	Phosphoglycerate kinase		10
PGM1	Phosphoglucomutase 1	U2	10
PGM3	Phosphoglucomutase 3	U8	10
PKM2	Pyruvate kinase 2	U3	11,6
PAIS	Phosphoribosylaminoimidazole synthetase	U10	1
PLP	Proteolipid protein	X	13
PRGS	Phosphoribosyl glycinamide synthetase	U10	1
RCP	Red pigment, color vision	X	13
RNR	Ribosomal RNA (NOR)	1,2,3,4,25	2

Gene symbol	Locus	Chromosome	Ref.
SHMT	Serine hydroxymethyl transferase	3	6
SOD1	Superoxyde dismutase soluble	U10	1
SYN	Synapsin	X	13
TPI	Triose phosphate isomerase	3	10,6

REFERENCES

1 - BROAD T.E. et al 1984, (HGM7) Cytogenet.Cell Genet. 37 427.
2 - HENDERSON L.M., BRUERE A.N. 1977, Cytogenet. Cell Genet. 19 326-334.
3 - HEDIGER R. et al 1988, Proceeding 8th Europ. Coll. on Cytogenet. Dom. Anim. (Bristol) 143.
4 - HEDIGER R. et al 1989, Anim. Genet. (21st Int. Conf. Anim. Blood Groups and Biochem. Polym. (Turin) 20 95.
5 - IMAN-GHALI M. et al 1987, (HGM9) Cytogenet. Cell Genet. 46 632-633.
6 - JONES C. et al 1985, HGM8, Cytogenet. Cell Genet. 40 662.
7 - MILLOT P. et al 1981, Ann. Genet. 24 82-88.
8 - NGUYEN T.C. 1985, Anim.Blood Groups & Bioch.Genet. 16 13-17
9 - RASMUSEN B.A. 1966, Genetics 54 356.
10 - SAIDI-MEHTAR N. et al 1981, Cytogenet.Cell Genet. 30 193-204
11 - SAIDI-MEHTAR N., HORS-CAYLA M.C. 1981, Ann.Genet. 24 148-151
12 - SAIDI-MEHTAR N. et al 1987, (HGM9) Cytogenet. Cell Genet. 46 686.
13 - SAIDI-MEHTAR N. et al 1989, (HGM10) Cytogenet. Cell Genet. (in press).

Gene Map of the Cow (*Bos taurus*) 2N=60

August 1989

James E. Womack
Department of Veterinary Pathology
Texas A&M University
College Station, Texas 77843

Syntenic Group	Chromosome	Gene Locus
U1		PGD, ENO1, AT3, ABLL
U2		SOD2, ME1, PGM3
U3	5	GAPD, LDHB, TPI1, PEPB, IFNG, A2M, INT1, HOX3, LALBA, KRAS2, GLI, PAH, NKNB, KRT2, GDH, LYS
U4	21	MPI, CYP11A, FES
U5		PKM2, NP, HEXA, FOS, KRT8L
U6		PGM1
U7		LDHA, TYR
U8		MDH2, ASL, PRM, GUSB, HBA
U9	18	GPI, DIA4
U10		SOD1, IFREC, PRGS, PAIS, CRYA1, SST, APP, ETS2, S100B, COL6A1, COL6A2, CBS, GAP43, PFKL, CD18, TF, CP
U11		ITPA, ADA, VIM
U12		ACY1, RHO
U13		HOX1, MET, COL1A2
U14		GSR, PLAT
U15	6	PGM2, PEPS, CASAS1, CASAS2, CASB, CASK
U16		ABL, ASS, GRP78, LGB, J
U17	8	IDH1, FN1, CRYG
U18		ACO1, IFNA, IFNB, GSN, GGTB2, ALDOB, ALDH1, C5, ITI, NEFM
U19	15	CAT, A, PTH, HBB,CRYA2, FSHB
U20	23	GLO1, CYP21, BOLA-A, BOLA-B, BOLA-D, PRL, TCP-1, M, HSPA1
U21	19	GH, HOX2, KRT1, TK
U22		AMH, SPARC
U23		ALDH2, IL2, IGL
U24	14	TG, MOS, CA2, MYC, CYP11B
U25		PEPA
U26		GOT1, CYP17, ADRA2R
U27		POLR2, UMPH2
U28		MBP
U29		–
X	X	G6PD, HPRT, PGK1, GLA, F9, DMD
Y	Y	DYZB, DYZ1

Linkage Groups Not Assigned to Syntenic Groups or Chromosomes:

LG VI	ALB, GC
LG VII	S, PI2

Gene Symbol	Location	Description	Reference
A2M	Chr 5	Alpha-2-macroglobulin	33
A	Chr 15	A blood group	12, 25
ABL	U16	Abelson oncogene homologue	33
ABLL	U1	Abelson oncogene homologue-like	33
ACO1	U18	Aconitase 1, soluble	38
ACY1	U12	Aminoacylase 1	38
ADA	U11	Adenosine deaminase	38
ADRA2R	U26	Adrenergic, alpha-2-, receptor	40
ALB	LG VI	Albumin	5
ALDH1	U18	Aldehyde dehydrogenase 1, soluble	33
ALDH2	U23	Aldehyde dehydrogenase 2, mitochondrial	33
ALDOB	U18	Aldolase B	33
AMH	U22	Anti-Mullerian hormone	39
APP	U10	Amyloid beta (A4) precursor protein	32
ASL	U8	Argininosuccinate lyase	39
ASS	U16	Argininosuccinate synthetase	33
AT3	U1	Antithrombin III	39
BOLA-A	Chr 23	Major histocompatibility complex - class I	3, 11, 31, 36
BOLA-B	Chr 23	Major histocompatibility complex - class I	11
BOLA-D	Chr 23	Major histocompatibility complex - class II	3, 11, 31, 36
C5	U18	Complement component 5	33
CA2	Chr 14	Carbonic anhydrase II	34
CASAS1	Chr 6	Casein, alpha S1	23, 39
CASAS2	Chr 6	Casein, alpha S2	23, 39
CASB	Chr 6	Casein, beta	23, 39
CASK	Chr 6	Casein, kappa	23, 39
CAT	Chr 15	Catalase	33
CBS	U10	Cystathionine-beta-synthase	32
CD18	U10	Antigen CD18	32
COL1A2	U13	Collagen, type 1, alpha 2	39
COL6A1	U10	Collagen, type VI, alpha 1	32
COL6A2	U10	Collagen, type VI, alpha 2	32
CP	U10	Ceruloplasmin	22
CRYA1	U10	Crystallin, alpha, polypeptide 1	29
CRYA2	Chr 15	Crystallin, alpha polypeptide 2	39
CRYG	Chr 8	Crystallin, gamma	2, 41
CYP11A	Chr 21	Cytochrome P450, subfamily XIA	30
CYP11B	Chr 14	Cytochrome P450, subfamily XIB	30
CYP17	U26	Cytochrome P40, family XVII	30
CYP21	Chr 23	Cytochrome P40, family XXI	11, 29
DIA4	Chr 18	Diaphorase (NADH NADPH)	6, 39
DMD	X	Dystrophin	25
DYZB	Y	DNA segment	7
DYZ1	Y	DNA segment	17
ENO1	U1	Enolase 1	20
ETS2	U10	Avian erythroblastosis virus E26 oncogene homologue	32
F9	X	Coagulation factor IX	25

FES	Chr 21	Feline sarcoma viral oncogene homologue	35
FN1	Chr 8	Fibronectin 1	2, 41
FOS	U5	Murine FBJ osteosarcoma viral oncogene homologue	35, 39
FSHB	Chr 15	Follicle stimulating hormone, beta peptide	13
GAP43	U10	Growth associated protein (43 kD)	32
GAPD	Chr 5	Glyceraldehyde-3-phosphate dehydrogenase	38
GC	LG VI	Vitamin D binding protein	5
GDH	Chr 5	Glucose dehydrogenase	38
GGTB2	U18	Glycoprotein-4-beta galactosyl transferase 2	33
GH	Chr 18	Growth hormone	39
GLA	X	Galactosidase, alpha	19, 28
GLI	Chr 5	Glioma-associated oncogene homologue	33
GLO1	Chr 23	Glyoxalase 1	29, 38
GOT1	U26	Glutamic-oxaloacetic transaminase 1, soluble	26
GPI	Chr 18	Glucose phosphate isomerase	6,10, 38
GRP78	U16	Glucose related protein (78 kD)	33
GSN	U18	Gelsolin	33
GSR	U14	Glutathione reductase	38
GUSB	U8	Glucuronidase, beta	39
G6PD	X	Glucose-6-phosphate dehydrogenase	19, 28, 38
HBA	U8	Hemoglobin, alpha	40
HBB	Chr 15	Hemoglobin, beta	12, 22
HEXA	U 5	Hexosaminidase A	39
HOX1	U13	Homeo box region 1	39
HOX2	Chr 19	Homeo box region 2	39
HOX3	Chr 5	Homeo box region 3	33
HPRT	X	Hypoxanthine phosphoribosyltransferase	19, 28
HSPA1	Chr 23	Heat shock 70 kD protein 1	30
IDH1	Chr 8	Isocitrate dehydrogenase 1, soluble	20, 38, 41
IFNA	U18	Interferon, alpha	1
IFNB	U18	Interferon, beta	1
IFNG	Chr 5	Interferon, gamma	1, 39
IFREC	U10	Interferon receptor	37
IGLC	U23	Immunoglobulin lambda polypeptide, constant region	35, 39
IL2	U23	Interleukin 2	39
INT1	Chr 5	MMTV integration site oncogene homologue	33
ITI	U18	Inter-alpha-trypsin inhibitor, protein HC	33
ITPA	U11	Inosine triphosphatase	38
J	U16	J blood group	21, 22
KRAS2	Chr 5	Kirsten rat sarcoma 2 viral oncogene homologue	33
KRT1	Chr 19	Keratin (Type I)	9, 39
KRT2	Chr 5	Keratin (Type II)	9,39
KRT8L	U5	Keratin 8-like	40
LALBA	Chr 5	Lactalbumin, alpha	33
LDHA	U7	Lactate dehydrogenase A	8, 10, 20, 38

LDHB	Chr 5	Lactate dehydrogenase B	8, 10, 20, 38
LGB	U16	Lactoglobulin, beta	19, 20, 39
LYS	Chr 5	Lysozyme	40
M	Chr 23	M blood group	3, 11, 24
MBP	U28	Myelin basic protein	40
MDH2	U8	Malate dehydrogenase, NAD, mitochondrial	10, 20, 38
ME1	U2	Malic enzyme 1, soluble	10, 20, 38
MET	U13	Met proto-oncogene	39
MOS	Chr 14	Molony murine sarcoma viral oncogene homologue	34
MPI	Chr 21	Mannose phosphate isomerase	10, 20, 38
MYC	Chr 14	Avian myelocytomastis viral oncogene homologue	34
NEFM	U18	Neurofilament, medium peptide	39
NKNB	Chr 5	Neurokinin B	33
NP	U5	Nucleoside phosphorylase	38
PAH	Chr 5	Phenylalanin hydroxylase	33
PAIS	U10	Phosphoribosylaminoimidazole synthetase	27
PEPA	U25	Peptidase A	26, 39
PEPB	Chr 5	Peptidase B	8, 10, 20, 38
PEPS	Chr 6	Peptidase S	26, 39
PFKL	U10	Phosphofructokinase, liver type	32
PGD	U1	Phosphogluconate dehydrogenase	10, 20, 38
PGK1	X	Phosphoglycerate kinase 1	19
PGM1	U6	Phosphoglucomutase 1	10, 20, 38
PGM2	Chr 6	Phosphoglucomutase 2	38
PGM3	U2	Phosphoglucomutase 3	8, 10, 20, 38
PI2	LG VII	Protease inhibitor	16
PKM2	U5	Pyruvate kinase	10, 20, 26, 38
PLAT	U14	Plasminogen activator, tissue	39
POLR2	U27	Polymerase (RNA) II (DNA directed) large polypeptide	40
PRGS	U10	Phosphoribosylglycinamide synthetase	27
PRL	Chr 23	Prolactin	11, 18, 29
PRM	U8	Protamine	32
PTH	Chr 15	Parathyroid hormone	12, 14
RHO	U12	Rhodopsin	18
S	LG VII	S blood group	16
S100B	U10	S100 protein, beta polypeptide	32
SOD1	U10	Superoxide dismutase 1, soluble	8, 20, 38
SOD2	U2	Superoxide dismutase 2, mitochondrial	10, 20, 38
SPARC	U22	Secreted protein, acidic, cystein-rich	39
SST	U10	Somatostatin	9
TCP1	Chr 23	T complex protein	4
TF	U10	Transferrin	23, 32
TG	Chr 14	Thyroglobulin	34
TK	U 21	Thymidine kinase	39
TPI1	Chr 5	Triosephosphate isomerase 1	10, 20, 38

TYR	U7	Tyrosinase	15, 39
UMPH2	U27	Uridine 5'-monophosphate phosphohydrolase 2	39
VIM	U11	Vimentin	40

REFERENCES

1. Adkison, L. R., et al., 1988a. *Cytogenet. Cell Genet.* 47:62-65.
2. Adkison, L.R. et al., 1988b. *Cytogenet. Cell Genet.* 47:155-159.
3. Andersson, L., et al. 1986. *Anim. Genet.* 17:95-112.
4. Andersson, L. 1988. *J. Hered.* 79:1-5.
5. Bouquet, Y., et al., 1986. *Anim. Genet.* 17:175-182.
6. Chowdhary, B.P., et al. 1989. *Sixth North American Colloquium on Cytogenet. of Domestic Anim.*, Purdue.
7. Cotinot, C., et al. 1987. HGM9. *Cytogenet. Cell Genet.* 46:598.
8. Dain, A. R., et al. 1984. *Biochem. Genet.* 22:429-439.
9. Dietz, A.B. and J.E. Womack. 1989. *J. Hered.*, in press.
10. Echard, G., et al. 1984. *Cytogenet. Cell Genet.* 37:458-459.
11. Fries, R., et al. 1986. *Anim. Genet.* 17:287-294.
12. Fries, R., et al. 1988. *Genomics* 3:302-307.
13. Fries, R., et al. 1989. HGM10. *Cytogenet. Cell Genet.*, in press.
14. Foreman, M.E. and J.E. Womack. 1989. *Biochem. Genet.*, in press.
15. Foreman, M.E., et al. 1989, submitted.
16. Georges, M., et. al. *Anim. Genet.*18:311-316.
17. Georges, M. et al. 1989. HGM10. *Cytogenet. Cell Genet.*, in press.
18. Hallerman, E.M., et al. 1988. *Anim. Genet.* 19:123-131.
19. Heuertz, S. and M.-C. Hors-Cayla. 1978. *Ann. Genet.* 21:197-202.
20. Heuertz, S. and M.-C. Hors-Cayla 1981. *Cytogenet. Cell Genet.* 30:137-145.
21. Hines, H. C., et al. 1969. *Genetics* 62:401-412.
22. Larsen, B. 1970. *Arsberetn. Inst. Sterilitesforsk., Copenhagen* 13:165-194.
23. Larsen, B. 1977. *Anim. Blood Grps. biochem. Genet.* 8:111-113.
24. Leveziel, H. and H. C. Hines. 1984. *Genetique, Selection, Evolution* 16:405-416.
25. Livingston, R. and J.E. Womack, 1989. submitted.
26. Moll, Y.D. and J.E. Womack. 1989., submitted.
27. McAvin, J. C., et al. 1988. *Biochem. Genet.* 26:9-18.
28. Shimizu, N., et al. 1981. *Cytogenet. Cell Genet.* 29:26-31.
29. Skow, L. C., et al. 1988. *DNA* 7:143-149.
30. Skow, L.C., et al. unpublished.
31. Spooner, R. L., et al. 1978. *J. Immunogenet.* 5:335-346.
32. Threadgill, D.S. et al. 1989. HGM10. *Cytogenet. Cell Genet.*, in press.
33. Threadgill, D.W. and J.E. Womack. 1989. HGM10. *Cytogenet. Cell Genet.*, in press.
34. Threadgill, D.W., et al. 1989. *Cytogenet. Cell Genet.*, in press.
35. Tobin, T. and J.E. Womack. 1989. *Second Symposium on the Genetic Engineering of Animals*, Cornell.
36. Usinger, W. R., et al. 1981. *Immunogenet.* 14:423-428.
37. Womack, J. E. and J. M. Cummins. 1984. *Cytogenet. Cell Genet.* 37:612.
38. Womack, J. E. and Y. D. Moll. 1986. *J. Hered.* 77:2-7.
39. Womack, J.E. et al.1989. HGM10. *Cytogenet. Cell Genet.*, in press.
40. Womack, J.E., et al. unpublished.
41. Zneimer, S.M. and Womack, J.E. 1989. *Genomics* 5:215-220.

AMERICAN MINK (MUSTELA VISON) (2N = 30)

May, 1989

O.L.Serov, S.D.Pack, Institute of Cytology & Genetics, Academy of Sciences of the USSR, Novosibirsk-90.

Chromosome localization of 55 mink genes was done by means of mink-mouse and mink-Chinese hamster cell hybrids; 30 localizations were carried out by use of both hybrids.

Regional assignments of the genes for ENO1, PGM1, PGD, ADK, PP, GOT1, HK1 and NP on chromosome 2 (13,16), those for ESD, ALDC and UMPH2 on chromosome 8 (8), and that for G6PD on the X chromosome were done by use of spontaneous chromosome rearrangements appearing either in mink cultured cells or hybrid cells. Gene transfer procedures were used for regional assignment of the GALK, TK1, ALDC, UMPH2 and HOX2 genes on chromosome 8 (2,8,13).

Three linkage groups were identified by breeding test: 1)PEPD--19 cM--(ES1, ES2, ESR)--14 cM--ES3 on chromosome 7 (13); 2) LPM--11 cM--PEPB on chromosome 9 (18); 3) the dominant mutation Ebony (gene Eb) affecting underfur, lightening it to white, and the recessive royal pastel (gene b) are 26 cM apart; their chromosome localization is unknown (15).

Gene order on chromosome 8 was determined as GALK--TK1--ALDC--(UMPH2, HOX2)--pter (2,8), and that on the X chromosome as GLA--PGK1--HPRT--G6PD--qter (17).

Mink chromosome karyotyped by analysis of pro- and metaphase chromosomes stained with the G-banded method (not less than 300 bands were resolved). Chromosome numeration and nomenclature were as described elsewhere (2.5.13).

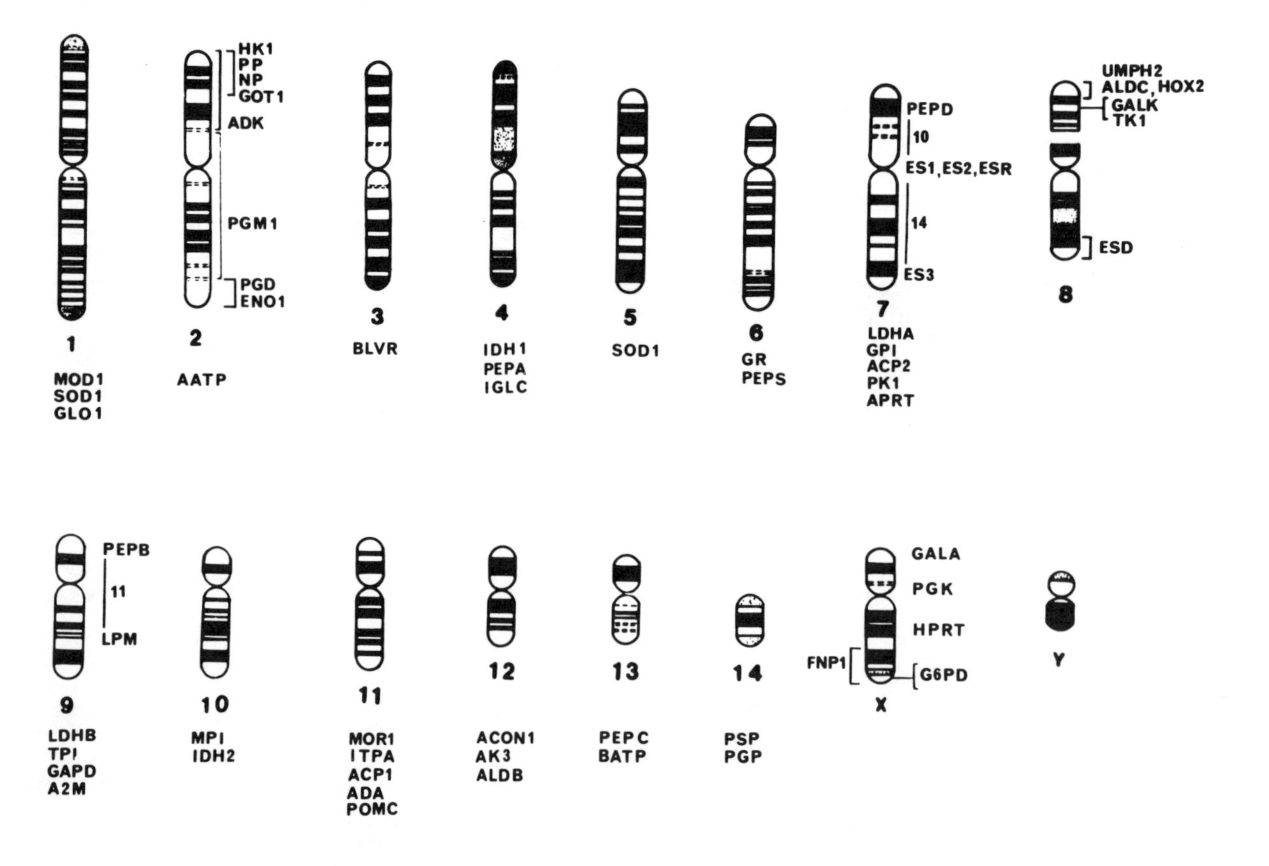

Gene	Locus	Chromosome	Reference
AATP	ATP-ase, alpha-subunit	2	6
ACON1	Aconitase-1	12	1
ACP1	Acid phosphatase-1	11	13
ACP2	Acid phosphatase-2	7	1
ADA	Adenosine deaminase	11	13
ADK	Adenosine kinase	2pter--p11.1	1, 16
AK3	Adenylate kinase-3	12	13
ALDB	Aldolase B	12	19
ALDC	Aldolase C	8pter--p25	2
A2M	α_2-Macroglobulin	9	18
APRT	Adenine phosphoribosyl - transferase	7	8
BATP	ATP-ase, beta-subunit	13	4
BLVR	Biliverdin reductase	3	13
ENO1	Enolase-1	2q24.4--qter	1, 16
ES1	Esterase-1	7	13
ES2	Esterase-2 (presumed)	7	13
ES3	Esterase-3	7	13
ESD	Esterase D	8q24--qter	2, 8
ESR	Esterase regulator	7	13
FNP1	Fibronectin pseudogene-1	Xq14--pter	14
GALK	Galactokinase	8p24	2
GAPD	Glyceraldehyde-3-phosphate dehydrogenase	9	13
GLA	α-Galactosidase	X	1, 17
GLO1	Glyoxalase-1	1	19
GOT1	Glutamate-oxaloacetate transaminase-1	2pter--p22	11, 16
GPI	Glucosephosphate isomerase	7	10
G6PD	Glucose-6-phosphate dehydrogenase	Xq15.22--qter	17
GR	Glutathione reductase	6	11
HK1	Hexokinase-1	2pter--p22	1, 16
HOX2	Homeo box-2	8pter--p25	8
HPRT	Hypoxanthine phosphoribosyl-transferase	X	9
IDH1	Isocitrate dehydrogenase-1	4	11
IDH2	Isocitrate dehydrogenase-2	10	11
IGLC	Immunoglobulin lambda polypeptide, constant region	4	19
ITPA	Inosine triphosphatase	11	1
LDHA	Lactate dehydrogenase A	7	10
LDHB	Lactate dehydrogenase B	9	10
LPM	Lipoprotein of mink	9	18
ME1	Malic enzyme-1	1	9
MDH1	Malate dehydrogenase-1 (NAD dependent)	11	9
MPI	Mannose phosphate isomerase	10	1
NP	Nucleoside phosphorylase	2pter--p22	13
PEPA	Peptidase A	4	12
PEPB	Peptidase B	9	7
PEPC	Peptidase C	13	12
PEPD	Peptidase D	7	8, 13

Gene	Locus	Chromosome	Reference
PEPS	Peptidase S	6	8
PGD	6-Phosphogluconate dehydrogenase	2q24.4--qter	10, 16
PGM1	Phosphoglucomutase-1	2p22--q24.4	10, 13
PGK1	Phosphoglycerate kinase-1	X	11, 17
PGP	Phosphoglycolate phosphatase	14	8
PSP	Phosphoserin phosphatase	14	8
POMC	Proopiomelanocortin	11	3
PK1	Pyruvate kinase-1	7	1, 13
PP	Inorganic pyrophosphatase	2pter--p22	12, 16
SOD1	Superoxide dismutase-1	5	12
SOD2	Superoxide dismutase-2	1	13
TK1	Thymidine kinase-1	8p24	2
TPI	Triosephosphate isomerase	9	10
UMPH2	Uridine 5'-monophosphate phosphohydrolase-2	8pter--p25	8

REFERENCES

1. Gradov,A.A., et al. 1983. Theor.Appl.Genet. 67:59-65.
2. Gradov,A.A., et al. 1985. Mol.Gen.Genet. 200:433-438.
3. Khlebodarova,T.M., et al. 1988. Genomics 2:185-188.
4. Khlebodarova,T.M., et al. 1988. FEBS Lett. 236:240-242.
5. Mandahl,N. and K.Fredga. 1975. Hereditas 81:211-220.
6. Matveeva,N.M., et al. 1987. FEBS Lett. 217:42-44.
7. Mullakandov,M.R., et al. 1986. Theor.Appl.Genet. 73:272-277.
8. Pack,S.D., et al. 1989. Cytogenet.Cell Genet. in press.
9. Rubtsov,N.B., et al. 1981. Theor.Appl.Genet. 60:99-106.
10. Rubtsov,N.B., et al. 1981. Cytogenet.Cell Genet. 31:184-187.
11. Rubtsov,N.B., et al. 1982.Cytogenet. Cell Genet. 33:256-260.
12. Rubtsov,N.B., et al. 1982. Theor.Appl.Genet. 63:331-336.
13. Serov,O.L., et al. 1987. In M.C.Rattazzi, J.G.Scandalios, G.S.Whitt, and C.L.Markert (eds). Isozymes: Current Topics in Biological and Medical Research. Vol.15. pp.179-215. A.R.Liss, New York.
14. Serov,O.L., et al. in preparation.
15. Shackelford,R.M. 1949. Amer.Natur. 83:49-67.
16. Zhdanova,N.S., et al. 1985. Cytogenet.Cell Genet. 39:296-298.
17. Zhdanova,N.S., et al. 1988. Cytogenet.Cell Genet. 48:2-5.
18. Yermolaev,V.P., et al. 1989. Theor.Appl.Genet. in press.
19. Khlebodarova,T.M. unpublished.

Gene maps of marsupials (mammalian Infraclass Metatheria) and monotremes (mammalian Subclass Prototheria)

August 1989

Jennifer A. Marshall Graves
Department of Genetics and Human Variation
La Trobe University
Bundoora, Victoria 3083
Australia

Marsupials constitute a separate infraclass (Metatheria) of mammals, which diverged from placentals (Eutheria) 130-150 million years ago (1). Monotremes (egg-laying mammals) are a subclass (Prototheria) which diverged from Subclass Theria (Metatheria and Eutheria) probably even earlier.

Marsupials are a diverse group, comprising over 250 living species in 16 families, in at least 3 orders (27). The extraordinary conservation of a 2n = 14 "basic karyotype" (21,37) leads us to anticipate conservation of synteny; it may be possible to construct a "basic gene map" of a marsupial ancestor.

Marsupial gene mapping has been hampered by the difficulties of captive breeding, and the problems of obtaining and analysing marsupial cell hybrids (10,26). Data were fragmentary, and derived from studies of many diverse species, using different approaches. Recently there have been efforts to concentrate attention on model marsupial species, which can be laboratory bred (2,22,40). Gene mapping in marsupials is advancing very rapidly now, with the refinement of techniques for using heterologous DNA probes to detect marsupial sequences in cell hybrids, and to localize genes by _in situ_ hybridization to marsupial chromosomes (39,40).

There are only 3 extant species of monotremes, in 2 families; Tachyglossidae (the 2 echidna species _Tachyglossus aculeatus_ and _Zaglossus bruijnii_) and Ornithorhynchidae (the platypus, _Ornithorhynchus anatinus_). Their karyotypes bear obvious homology to each other; the platypus has 2n = 52, while the two echidna species (which are G-band identical) have 2n = 63 ♂ , 64 ♀ (34,54,55). Monotremes are unique among vertebrates in possessing a translocation complex of 6 or 8 unpaired chromosomes which form a chain at meiosis. In all three species, the X and Y chromosomes are G-band homologous over most of their length (53).

Monotremes will not breed in captivity and are difficult to study in the wild. The only classic genetic data is the observation of males which appear to be heterozygous for PGK alleles (47), suggesting autosomal inheritance of this gene. Rodent-monotreme cell hybrids have proved intractable for chromosome assignment, and many years' work has produced only two syntenic relationships, HPRT-PGK and ENO1-6PGD (52).

The most interesting features of the marsupial and monotreme gene maps are the localizations of genes which are on the X or Y in man and other eutherian mammals. Most human Xq markers are located on the X also in all marsupials and in platypus suggesting that this region represents an ancient conserved X. This observation seems not to depend on X chromosome inactivation, however, since the platypus Xq and Yq are G-band homologous and do not appear to be inactivated (53). Surprisingly, human Xp markers are excluded from the X in cell hybrids and are located to autosomes by in situ hybridization in marsupials and monotremes (9,11,17,20, 39,41,42), suggesting that the human Xp is a rather recent recruit to the eutherian sex chromosomes. The exclusion from the Y and the X of sequences homologous to the human ZFY gene (40) makes this gene an unlikely candidate for a sex determining role.

I have tabulated data from classic genetic studies, somatic cell genetics and in situ hybridization for species of the two best studied marsupial families (Family Macropodidae, Order Diprotodontia and Family Dasyuridae, Order Polyprotodontia), as well as for the best studied monotreme, the platypus. Fragmentary data for a variety of other marsupial species are listed in ref. 18.

		Marsupials							Monotremes
Gene Symbol	Human chr	Family Macropodidae			Family Dasyuridae				
		M.e 2n=16	M.rb 2n=16	M.r 2n=20	P.m 2n=14	S.c 2n=14	D.r 2n=14	D.v 2n=14	O.a 2n=54
STS	Xp		A*	A	A	A			
POLA	Xp		A						
DMD	Xp	5p					3q		
OTC	Xp	1p					3p		2p
CYBB	Xp	5p	X	X					1p,4q
MAOA	Xp	5p					3q		1,2
SYN1	Xp	1p					3p		
EPA	Xp			X					1q
ARAF1	Xp	X	X	X	X				2q
CCG1	Xq		X	X	X				
PGK1	Xq	X	X	X	X	X			U2
AR	Xq		X	X	X				Xp
PLP	Xq	X	X	X	X				Xp
GLA	Xq	X	X	X	X				Xp
SCAR14	Xq	X	X	X	X				
TBG	Xq	1q							3q
MCF2	Xq								Xq
HPRT	Xq	X	X	X	X	X			U2
F9	Xq		X	X					Xp
F8	Xq		X						Xp
RCP	Xq	X	X	X	X				Xp
G6PD	Xq	X	X	X	X	X			Xq
GDX	Xq	X	X	X	X				Xq
P3	Xq	X	X	X	X				Xq
ZFY	Yp	1p,5p,7				3p,3q,7			

* A designates exclusion from X

Gene Assignments in Mammalian Infraclass Metatheria (Marsupials) and Subclass Prototheria (Monotremes)

Species names: M.e = Macropus eugenii; M.rb = M. robustus; M.r = M. rufus; P.m = Planigale maculata; S.c = Sminthopsis crassicaudata; D.r = Dasykaluta rosamondae; D.v = Dasyurus viverrinus; O.n = Ornithorhynchus anatinus

		Marsupials							Monotremes
Gene Symbol	Human chr	Family Macropodidae			Family Dasyuridae				
		M.e 2n=16	M.rb 2n=16	M.r 2n=20	P.m. 2n=14	S.c 2n=14	D.r 2n=14	D.v 2n=14	O.a 2n=52
ENO1	1p								U1
PGD	1p					U2			U1
RAF1	3p		4						
TF	3q					U2			
HSPA1	6p	1q,2q,5p							
MDH2	7p		1						
MYC	8q		4						
LDHA	11p			5					
HBB	11p	3q						4	
HRAS	11p	3q					4q		
CAT	11p	3q					4q		
PI	14q	U1				U1			
IGHE	14q	3q							5
HBA	16q							2	
PEPA	18q		4						
GPI	19q	U1				U1			
ADA	20q					U1			
SOD1	21q	7				U2			
ETS2	21q	3q							

References

1. Air, G.M., Thompson, E.O.P., Richardson, B.J. and Sharman, G.B. (1971) Nature 229: 391-394.
2. Bennett, J.H., Breed, W.G., Hayman, D.L. and Hope, R.M. (1989) "Mammals from pouches and eggs: Genetics, breeding and evolution of marsupials and monotremes", (eds.) J.A.M. Graves, R.M. Hope and D.W. Cooper. CSIRO (Melbourne). In press.
3. Bennett, J.H., Hayman, D.L. and Hope, R.M. (1986) Nature 323: 59-60.
4. Briscoe, D.A., Murray, J.D. and Sharman, G.B. (1981) Aust. J. Biol. Sci. 34: 341-346.
5. Cooper, D.W. and Sharman, G.B. (1964) Nature 203: 1094.
6. Cooper, D.W. and Hope, R.M. (1971) Biochem. Genet. 5: 765-68.
7. Cooper, D.W., VandeBerg, J.L., Sharman, G.B. and Poole, W.E. (1971) Nature 230: 155.
8. Cooper, D.W., Woolley, P.A., Maynes, G.M., Sherman, F.S. and Poole, W.E. (1983) Aust. J. Biol. Sci. 36: 511-517.
9. Cooper, D.W., McAllan, B.M., Donald, J.A., Dawson, G.W., Dobrovic, A. and Graves, J.A.M. (1984) Human Gene Mapping 7: 439.
10. Dawson, G.W. and Graves, J.A.M. (1984) Chromosoma 91: 20-27.
11. Dawson, G.W. and Graves, J.A.M. (1986) Cytogenet. Cell Genet. 42: 80-84.
12. Dawson, G.W., Johnston, P.G. and Graves, J.A.M. (1986) Cytogenet. Cell Genet. 42: 80-84.
13. Dobrovic, A. (1984) PhD Thesis, La Trobe University, Melbourne, Australia.
14. Dobrovic, A. and Graves, J.A.M. (1986) Cytogenet. Cell Genet. 41: 9-13.
15. Donald, J.A. and Adams, M.A. (1981) Biochem. Genet. 19: 901-908.
16. Donald, J.A. and Hope, R.M. (1981) Cytogenet. Cell Genet. 29: 127-137.
17. Graves, J.A.M. (1987) Trends in Genetics 3: 252-256.
18. Graves, J.A.M.)(1987) Genetic Maps, S.J. O'Brien (ed.) Cold Spring Harbor 4: 504-507.
19. Graves, J.A.M., Chew, G.K., Cooper, D.W. and Johnston, P.G. (1979) Somat. Cell. Genet. 5: 481-489.
20. Graves, J.A.M., Sinclair, A.H. and Spencer, J.A. (1989) "Mammals from pouches and eggs: Genetics, breeding and evolution of marsupials and monotremes", (eds.) J.A.M. Graves, R.M. Hope and D.W. Cooper. CSIRO (Melbourne). In press.
21. Hayman, D.L. and Martin, P.G (1969) Comparative Mammalian Cytogenetics, Springer, N.Y.
22. Hinds, L.A., Poole, W.E., Tyndale-Biscoe, C.H., van Oorschot, R.A.H. and Cooper, D.W. (1989) "Mammals from pouces and eggs: Genetics, breeding and evolution of marsupials and monotremes", (eds.) J.A.M. Graves, R.M. Hope and D.W. Cooper. CSIRO (Melbourne). In press.
23. Hope, R.M. (1972) Biochem. Genet. 3: 95-99.
24. Hope, R.M. Personal Communication.
25. Hope, R.M. and Godfrey, G.K. (1968) Aust. J. Biol. Sci. 21: 587-591.
26. Hope, R.M. and Graves, J.A.M. (1978) Aust. J. Biol. Sci. 31: 527-543.
27. Hope, R.M., Bennett, J.H., Chesson, C.M. and Adams, M. (1984) Biochem. Genet. 22: 221-229.

28. Johnston, P.G. and Sharman, G.B. (1975) Aust. J. Biol. Sci. 28: 567-574.
29. Johnston, P.G., VandeBerg, J.L. and Sharman, G.B. (1975) Biochem. Genet. 13: 235-242.
30. Johnston, P.G., Sharman, G.B., James, E. and Cooper, D.W. (1978) Aust. J. Biol. Sci. 31: 415-424.
31. Kirsch, J.A.W. (1977) "The Biology of Marsupials", (eds.) B. Stonehouse and D. Gilmore. Macmillian, London, 9-26.
32. Kirsch, J.A.W. and Poole, W.E. (1967) Nature New Biol. 215: 1097-1098.
33. Kola, I. Personal communication.
34. Murtagh, C.E. (1977) Chromosoma 65: 37-57.
35. Page, D.C., Mosher, R., Simpson, E.M., Fisher, E.M.C., Mardon, G., Pollack, J., McGillivray, B., de la Chapelle, A. and Brown, L.G. (1987) Cell 20: 555-566.
36. Richardson, B.J., Czuppon, A.B. and Sharman, G.B. (1971) Nature New Biol. 230: 154.
37. Rofe, R. and Hayman, D.L. (1985) Cytogenet. Cell Genet. 39: 40-50.
38. Samollow, P.B., Ford, A.L. and VandeBerg, J.L. (1986) Genetics (in press).
39. Sinclair, A.H., Wrigley, J.M. and Graves, J.A.M. (1987) Genet. Res. 50: 131-136.
40. Sinclair, A.H., Foster, J.W., Spencer, J.A., Page, D.C., Palmer, M., Goodfellow, P.N. and Graves, J.A.M. (1988) Nature 336: 780-783.
41. Sinclair, A.H. (1989) PhD Thesis, La Trobe University, Melbourne, Australia.
42. Spencer, J.A. (1989) PhD Thesis, La Trobe University, Melbourne, Australia.
43. Spencer, J.A, Watson, J.M. and Graves, J.A.M. (1989) Human Gene Mapping 10 (in press).
44. Sykes, P.J. and Hope, R.M. (1978) Aust. J. Expt. Biol. and Med. Sci. 56: 703-711.
45. Sykes, P.J. and Hope, R.M. (1985) Aust. J. Biol. Sci. 38: 365-376.
46. VandeBerg, J.L., Cooper, D.W., Sharman, G.B. and Poole, W.E. (1977) Aust. J. Biol. Sci. 30: 115-125.
47. VandeBerg, J.L. and Cooper, D.W. (1978) Biochem. Genet. 16: 1031-1034.
48. VanderBerg, J.L., Thiel, J.E., Hope, R.M. and Cooper, D.W. (1979) Biochem. Genet. 17: 325-332.
49. VandeBerg, J.L., Cooper, D.W., Sharman, G.B. and Poole, W.E. (1980) Genetics 95: 413-424.
50. VandeBerg, J.L. (1989) "Mammals from pouches and eggs: Genetics, breeding and evolution of marsupials and monotremes", (eds.) J.A.M. Graves, Hope, R.M. and Cooper, D.W. CSIRO (Melbourne). In press.
51. Wainwright, B. and Hope, R.M. (1985) Proc. Natl. Acad. Sci. USA 82 8105-1808.
52. Watson, J.M. and Graves, J.A.M. (1988) Aust. J. Biol. Sci. 41: 231-237.
53. Watson, J.M. and Graves, J.A.M. (1988) J. Hered. 79: 115-118.
54. Watson, J.M. (1989) "Mammals from pouches and eggs: Genetics, breeding and evolution of marsupials and monotremes" (eds. J.A.M. Graves, R.M. Hope and Cooper, D.W. CSIRO (Melbourne). In press.
55. Wrigley, J.M. and Graves, J.A.M. (1988) Chromosoma 96: 231-247.
56. Young, G.J., Graves, J.A.M., Barbieri, I., Woolley, P.A., Cooper, D.W. and Westerman, M. (1982) "Carnivorous marsupials" (ed.) M. Archer. Roy. Zool. Sec. NSW, Sydney, Australia.

Primate genetic maps
August 1989

N.CREAU-GOLDBERG, C. COCHET, C. TURLEAU and J. de GROUCHY
INSERM U.173, Hôpital Necker-Enfants Malades,
149, rue de Sèvres, 75743 Paris Cedex 15,FRANCE.

The maps of the following primates were compiled with a unique bibliography : chimpanzee, gorilla, orangutan (Pongidae); gibbon (Hylobatidae); rhesus monkey, baboons,African green monkey (Cercopithecidae); capuchin monkey (Cebidae); mouse lemur (Lemuridae). For the Pongidae, two types of chromosome nomenclature are presented (8) : the arabic nomenclature allocated on the basis of centromere position and chromosome length and the roman nomenclature allocated on homologies with the human karyotype. For the other species, only arabic nomenclatures exist. Since several nomenclatures have been proposed for each species, the nomenclature used in the map is referenced, for ex. : HCO (46). The gene nomenclature used is that of the HGM10 Nomenclature Committee for man (78). Unassigned syntenic groups are designated by U. The localizations underlined are confirmed, the others are provisionnal (i.e. determined by one group only). For the baboons and rhesus monkey, a different nomenclature was used, though the karyotypes were described as homologous (11).

Main progresses on these maps, since the last edition, were the followings : - the use of in situ hybridization for mapping which gives more precise informations on chromosome evolution. For example, the localizations of AFP and ALB in the Pongidae have demonstrated that a pericentric inversion has occurred specifically in the chimpanzee lineage (67,75).-localization of genes having been recently duplicated in primate evolution which demonstrates intrachromosomic modifications in spite of the conservation of chromosome banding homology, as it was shown for the glycophorins (GYPA, GYPB) genes(72).-confirmation of gene localizations in the gibbon, which strengthens the observation that numerous rearrangements have occurred in the Hylobatidae, and that only few syntenic groups have been conserved (77, 71).

Chimpanzee , Pan troglodytes , PTR (2n = 48)

PTR (8)			Ref.
1	I	PGD, ENO1 ,FUCA1, AK2, PGM1 ,PEPC, RN5S, TRE(p)	4,5,20,28 9,74
2	III	GPX1, ACY1(p), UMPS(q)	16,58
3	IV	PGM2, PPAT, ALB(p), AFP(p), GYPA, GYPB	16,58,67,72
4	V		
5	VI	MHC, GLO1, PGM3, ME1, SOD2	20,34,39
6	VII	GUSB , COL1A2	15,16,57
7	VIII	GSR	16
8	X	GOT1	16
9	XI	LDHA ,ACP2,HBB, MER1, MDU1	3,4,5,20,44, 58
10	XII	GAPD, TPI1 , LDHB , PEPB ,ENO2	3,4,15,20
11	IX	AK3,ACO1,AK1	23
12	IIp	MDH1,ACP1	22,23
13	IIq	IDH1	22
14	XIII	RNR ,RB1	2,14,71
15	XIV	NP,ITPA, RNR ,MTHFC	2,14,16,23,73
16	XV	MPI, PKM2, HEXA	23
17	XVIII	RNR	2,14
18	XVI		
19	XVII	TK , GALK ,COL1A1, A12M4	3,4,5,7,57
20	XIX	GPI	16
21	XX		
22	XXI	SOD1 ,PAIS, RNR , APP, ETS2	2,4,5,14,16,58 71
23	XXII	RNR ,ADSL	2,14,66
X		G6PD , PGK ,GLA,HPRT,STS	4,5,15,20,65
Y		ZFY,TSPY	68,69

August 1989

N.CREAU-GOLDBERG, C. COCHET, C. TURLEAU and J. de GROUCHY
INSERM U.173, Hôpital Necker-Enfants Malades,
149, rue de Sèvres, 75743 Paris Cedex 15,FRANCE.

Gorilla, <u>Gorilla</u> <u>gorilla</u>, GGO (2n = 48)

GGO (8)			Ref.
1	I	PGD, ENO1, PGM1, PEPC, FH, RN5S, TRE(p)	20,9,70
2	III	GPX1	16
3	IV	PGM2, ALB(q),AFP(q),	16,75
4	V		
5	VI	MHC, PGM3, ME1, SOD2	20,34,39
6	VII	GUSB	16
7	VIII	GSR	16
8	X	GOT1	16
9	XI	LDHA	20
10	XII	GAPD, TPI1, LDHB, PEPB	20
11	IIq	IDH1	23
12	IIp	MDH1	23
13	IX	ACO1, AK3	23
14	XIII	ESD	23
15	XV	PKM2, HEXA	23
16	XVIII	PEPA	16
17	XVI		
18	XIV	NP, ITPA	6,23
19	XVII	TK	16
20	XIX	GPI	16
21	XX		
22	XXI	SOD1, <u>RNR</u>	16,13,27
23	XXII	<u>RNR</u>	13,27
X		G6PD, PGK, GLA, STS	20,65
Y		ZFY,TSPY	68,69

August 1989

N.CREAU-GOLDBERG, C. COCHET, C. TURLEAU and J. de GROUCHY
INSERM U.173, Hôpital Necker-Enfants Malades,
149, rue de Sèvres, 75743 Paris Cedex 15,FRANCE.

Orangutan, Pongo pygmaeus, PPY (2n = 48)

PPY (8)			Ref.
1	I	PGD- ENO1 - PGM1 (p), FH(q), PEPC, RN5S, TRE(p)	9,17,20 74
2	III		
3	IV	PGM2, ALB(q), AFP(q)	16
4	V	HEXB	19
5	VI	MHC, PGM3, ME1, SOD2	20,39
6	VIII	GSR	16
7	X	GOT1	16
8	XI	LDHA	20
9	XII	GAPD, TPI1, LDHB, PEPB	20
10	VII		
11	IIq	IDH1, RNR	23,27,35
12	IIp	MDH1, ACP1, RNR	23,27,35
13	IX	RNR	27,35
14	XIII	RNR	27,35
15	XIV	RNR	27,35
16	XV	MPI, PKM2, HEXA, RNR	23,27,35
17	XVIII	RNR	27,35
18	XVI		
19	XVII	TRK1-TRL2-TRQ1(p)	76
20	XIX	GPI	16
21	XX		
22	XXI	SOD1, PAIS, RNR	27,35,58
23	XXII	RNR	27,35
X		G6PD,GLA,STS	20,65
Y		TSPY	69

August 1989

N.CREAU-GOLDBERG, C. COCHET, C. TURLEAU and J. de GROUCHY
INSERM U.173, Hôpital Necker-Enfants Malades,
149, rue de Sèvres, 75743 Paris Cedex 15,FRANCE.

Gibbon, Hylobates (Nomascus) concolor ,HCO (2n = 52)

HCO (46)		Ref.
1	MHC(p), IDH1(q), COL3A1-COL5A2(q)	63,77,71
2		
3	FUCA1, GOT1, SOD2	53
4	GPX1, PEPA	53,77
5	PGM1 ,GUK1, RB1	53,77,63
6	MPI, PKM2, HEXA, SORD	53
7	GYPA	72
8	ACO1, AK1	53
9		
10	COL1A2, TK1	63,77
11	PEPB	77
12		
13		
14		
15	LDHA, ACP2	53
16		
17	PGM3,ME1	53,77
18	GUSB	63
19	ACP1	53
20		
21		
22		
23	SOD1	77
24	PGD ,ENO1,RNR	53,77,54
25	RNR	47,54
X	PGK, GLA, HPRT ,G6PD	53,77
Y		
U1	TPI1 , LDHB ,ENO2,GAPD	53,77,63
U2	UGP2	53
U3	NP, CKBB	53
U4	AK3	53
U5	COL1A1	63

August 1989
N.CREAU-GOLDBERG, C. COCHET, C. TURLEAU and J. de GROUCHY
INSERM U.173, Hôpital Necker-Enfants Malades,
149, rue de Sèvres, 75743 Paris Cedex 15,FRANCE.

Rhesus monkey, _Macaca mulatta_ ,MML (2n = 42)

MML (30)		Ref.
1	PGD, ENO1, PGM1, AK2, GUK1, FH	20
2	MHC, PGM3, SOD2, GUSB, MDH2	20,32,34,39,51
3	GPX1	16
4		
5	HEXB	16
6	PGM2	16
7	NP, HEXA, PKM2	32
8	GSR	16
9	IDH1	32
10		
11	LDHA, ACP2, SV40I	20,33
12	GAPD, TPI1, LDHB, PEPB, CS	20
13	ITP, NAGA, _RNR_	32,1,11
14		
15	MDH1	32
16		
17		
18		
19	GPI	16
20		
X	G6PD, GLA, STS	20,65
Y	ZFY,TSFY*	68,69
U1	ACO1, AK3	23

* TSFY has been localized in _Macaca sylvanus_ .

August 1989

N.CREAU-GOLDBERG, C. COCHET, C. TURLEAU and J. de GROUCHY
INSERM U.173, Hôpital Necker-Enfants Malades,
149, rue de Sèvres, 75743 Paris Cedex 15,FRANCE.

Baboon, Papio papio, hamadryas, cynocephalus ,
PPA-PHA-PCY (2n=42)

(10)		Ref.
1	PGD, ENO1, PGM1 ,PEPC, RN5S	18,36-59,62-9
2	ACY1(pter-q1)	59,61-59,62
3	GUSB, COL1A2, MDH2 (p), SOD1 (p), APP, ETS2	48,49,57,59,61-59,62 71
4	MHC, GLO1 , ME1 , SOD2	49,52,64,61-59,62
5	PGM2 ,PEPS,GYPA	49,52,61-59,62,72
6		
7	NP ,CKBB, MPI , PKM2 , IDH2 , SORD , HEXA	48,49,52,61-59,62
8		
9	PP	59,62
10	ADA , ITPA , RNR	48,52,11,30-59,62,21
11	LDHB ,TPI1	49-59,62
12	IDH1,COL3A1	48-49,62,71
13	MDH1	59-61,62
14	LDHA	49-59,62
15		
16	TK,COL1A1,GAA	55
17	ESD,RB1	59,62,71
18		
19		
20	GPI,PEPD	59,62
X	TBG	56
Y		
U1	PEPB	59,61
U2	ACP1	36
U3	GSR	59,62

For the sake of simplification the three Papio species have been considered together because of their karyotypic identity. Gene localizations underlined are confirmed in the same or other baboon species.

August 1989
N.CREAU-GOLDBERG, C. COCHET, C. TURLEAU and J. de GROUCHY
INSERM U.173, Hôpital Necker-Enfants Malades,
149, rue de Sèvres, 75743 Paris Cedex 15,FRANCE.

African green monkey, Cercopithecus aethiops CAE (2n = 60)

CAE (11)		Ref.
1		
2	GSR	16
3	GAPD,TPI1,LDHB,PEPB,CS	20
4	PGD, ENO1 , PGM1 ,FUCA1	18,20,28
5	COL1A2(q)	64
6		
7		
8		
9		
10		
11		
12		
13	PEPC ,FH	18,20
14	LDHA,SV40I	20,33,40
15		
16	ME1	57
17		
18		
19	TK1, GALK, COL1A1, GAA	7,55
20		
21	MHC, GLO1	64,57
22	PEPA	57
23		
24	GPI	16
25		
26	RNR	25,26
27		
28		
29	GUSB	57
X	G6PD, PGK, GLA	20
Y		
U1	ACO1, AK1	23
U2	NP, MPI, PKM2 , HEXA	23,57
U3	CKBB	57

August 1989

N.CREAU-GOLDBERG, C. COCHET, C. TURLEAU and J. de GROUCHY
INSERM U.173, Hôpital Necker-Enfants Malades,
149, rue de Sèvres, 75743 Paris Cedex 15,FRANCE.

Capuchin monkey, <u>Cebus</u> <u>capucinus</u> ,CCA (2n = 54)

CCA (29)		Ref.
1	PGP, GUSB	50
2	PGM2, GYPA	42,43,72
3	PGM3,ME1	42,43
4	MDH1, ACP1, UGP2	42,43,50
5		
6		
7		
8	GPI	42,43
9	SOD1	42,43
10	TPI1, LDHB, PEPB	42,43
11		
12	AK3	42,43
13		
14	IDH1	42,43
15	PGD, ENO1, PGM1	42,43
16	LDHA, ACP2	50
17		
18	ACY1	50
19		
20		
21		
22		
23		
24	RNR	29
25		
26		
X		
Y		
U1	MPI, PKM2	42,43
U2	GUK1, FH	42,43

August 1989
N.CREAU-GOLDBERG, C. COCHET, C. TURLEAU and J. de GROUCHY
INSERM U.173, Hôpital Necker-Enfants Malades,
149, rue de Sèvres, 75743 Paris Cedex 15,FRANCE.

Mouse lemur, *Microcebus murinus*, MIM (2n = 66)

MIM (37)		Ref.
1	GPX1	50
2	NP, CKBB, MPI, PKM2, HEXA, SORD	41,45,50,57
3	PGD, ENO1, FUCA1, PGM1	41,45,50
4	ACP1, MDH1, UGP2	45,50
5	LDHA	41,45
6	GLO1, PGM3, ME1	41,45,50
7	GAPD, TPI1, LDHB, PEPB, ENO2, CS	41,45,60
8		
9		
10	AK1,AK3	41,45
11	COL1A2	64
12		
13		
14		
15	GOT1	41,45
16		
17		
18		
19		
20		
21		
22		
23		
24		
25		
26		
27		
28		
29		
30		
31		
32		
X	G6PD, GLA, PGK, HPRT	41,45
Y		
U1	PEPC	41,45
U2	GUK1	41,45
U3	PEPD, GPI	41,45

August 1989

N.CREAU-GOLDBERG, C. COCHET, C. TURLEAU and J. de GROUCHY
INSERM U.173, Hôpital Necker-Enfants Malades,
149, rue de Sèvres, 75743 Paris Cedex 15,FRANCE.

REFERENCES (compilation by year order)

1. HENDERSON, A.S., WARBURTON, D., and ATWOOD, K.C. 1974. Chromosoma 44:367-370.
2. HENDERSON, A.S., WARBURTON, D.,and ATWOOD, K.C. 1974. Chromosoma 46:435-441.
3. CHEN, S., McDOUGALL, J.,CREGAN, R., LEWIS, V., and RUDDLE,F. HGM3 1975. 1976. Cytogenet. Cell. Genet. 16:412-415.
4. FINAZ, C., COCHET, C., GROUCHY, J. de, NGUYEN, V.C., REBOURCET, R. and FREZAL, J.1975. Ann. Génét. 18:169-177.
5. GROUCHY, J. de, FINAZ, C., NGUYEN, V.C., REBOURCET, R.,and FREZAL, J. HGM3. 1975. 1976. Cytogenet. Cell. Genet. 16:416-419.
6. MEERA KHAN, P., PEARSON, P.L., WIJNEN, L.L.L., DOPPERT,B.A., WESTERVELD, A., and BOOTSMA, D. HGM3 1975. 1976. Cytogenet. Cell. Genet. 16:420-421.
7. ORKWISZEWSKI, R.G., TEDESCO, T., MELLMAN, W., and CROCE, C. HGM3 1975. 1976. Cytogenet. Cell. Genet. 16:427-429.
8. PARIS CONFERENCE 1971, SUPPLEMENT 1975. Birth Defects, Original Series. 1975. 11:14-29.
9. WARBURTON, D., YU, M., ATWOOD, K.C.and HENDERSON, A.S. HGM3 1975. 1976. Cytogenet. Cell. Genet. 16:440-442.
10. CAMBEFORT, Y., MOUNIE, C. and COLOMBIES, P.1976. Ann. Génét. 19:5-9.
11. FINAZ, C., COCHET, C. and GROUCHY, J.de. 1976. Ann. Génét. 21:149-151.
12. FINAZ, C., DUBOIS, M.F., COCHET, C., VIGNAL, M. and GROUCHY, J. de. 1976. Ann. Génét. 19:213-216.
13. HENDERSON, A.S., ATWOOD, K.C.and WARBURTON, D. 1976. Chromosoma 59:147-155.
14. TANTRAVAHI, R., MILLER, D.A., DEV, V.G. and MILLER, O. 1976. Chromosoma 56:15-27.
15. COCHET, C., FINAZ, C. NGUYEN, V.C.,FREZAL, J. de.1977. Ann. Génét. 20:255-257.
16. ESTOP, A., GARVER, J.,PEARSON, P.L.,DIJKSMAN, T., WIJNEN, L.and MEERA KHAN P. HGM4 1977. 1978. Cytogenet. Cell. Genet. 22:558-563.
17. FINAZ, C;, NGUYEN, V.C., COCHET, C., FREZAL, J. and GROUCHY, J. de. 1977. Ann. Génét. 20:85-92
18. FINAZ, C., NGUYEN, V.C.,COCHET, C., FREZAL, J. and GROUCHY, J. de.1977. Cytogenet. Cell. Genet. 18:160-164.
19. FINAZ, C., NGUYEN, V.C. and GROUCHY, J.de. Personnal communication.
20. GARVER, J.,PEARSON, P.L., ESTOP, A., DIJKSMAN, T., WIJNEN, L., WESTERVELD, A. and MEERA KHAN, P. HGM4 1977. 1978. Cytogenet. Cell. Genet. 22:564-569.
21. HENDERSON, A.S., WARBURTON, D., McGRAW-RIPLEY, S. and ATWOOD, K.C. 1977. Cytogenet. Cell. Genet. 19:281-302.
22. PEARSON, P.L., ESTOP, A., GARVER, J.J., DIJKSMAN, T.M., WIJNEN, L. and MEERA KHAN, P. 1977. Chromosomes Today 6: 201-207.
23. PEARSON, P.L.,GARVER,J., ESTOP,A., DIJKSMAN, T;, WIJNEN, L. and MEERA KHAN, P. 1977; Chromosomes Today 6:201-207.

24. SUN, N.C., SUN, C.R.Y. and HO, T. HGM4 1977. 1978. Cytogenet. Cell. Genet. 22:598-601.
25. DUTRILLAUX, B., VIEGAS-PEQUIGNOT, E., COUTURIER, J. and CHAUVIER, G. 1978. Hum. Genet. 45:283-296.
26. ESTOP, A., GARVER, J.J. and PEARSON, P.L. 1978. Genetica 49:131-138.
27. GOSDEN, J., LAURIE, S. and SUANEZ, H. 1978. Cytogenet. Cell. Genet. 21:1-10.
28. NGUYEN, V.C., WEIL, D., FINAZ, C., REBOURCET, R., COCHET, C., GROUCHY, J. de and FREZAL, J. 1978 Ann. Génét. 21:41-46.
29. DUTRILLAUX, B. 1979. Cytogenet. Cell. Gnent. 24:84-94.
30. DUTRILLAUX, B., BIEMONT, M.C., VIEGAS-PEQUIGNOT, E. and LAURENT, C. 1979. Cytogenet. Cell. Genet. 23:77-83.
31. ESTOP, A. HGM5 1979. 1979. Cytogenet. Cell. Genet. 25:92.
32. ESTOP, A., GARVER, J., MEERA KHAN, P. and PEARSON, P.L. HGM5 1979. 1979. Cytogenet. Cell. Genet. 25:150-151.
33. GARVER, J.J., ABRAHAMS, P., EB, A.V.D. and PEARSON, P.L. HGM5 1979. 1979. Cytogenet Cell. Genet. 25:157.
34. GARVER, J.J., BALNER, H. and PEARSON, P.L. HGM5 1979. 1979 Cytogenet. Cell. Genet. 25:157.
35. HENDERSON, A.S., McGRAW-RIPLEY, S. and WARBURTON, D. 1979. Cytogenet. Cell. Genet. 23:213-216.
36. LALLEY, P.A., DIXON, D. and HELD, M.B. HGM5 1979. Cytogenet. Cell. Genet. 25:178.
37. RUMPLER, Y; and DUTRILLAUX, B. 1979. Cytogenet. Cell. Genet. 24:224-232.
39. GARVER, J., ESTOP, A., BALNER, H. and PEARSON, P.L. 1980. Cytogenet. Cell. Genet. 27:238-245.
40. GARVER, J., PEARSON, P.L., ABRAHAMS, P. and EB, V.D.A. 1980. Som. Cell. Genet. 6:443-450
41. COCHET, C. CREAU-GOLDBERG, N., TURLEAU, C. and GROUCHY, J.de. HGM6 1981. 1982. Cytogenet. Cell. Genet. 32:257.
42. CREAU-GOLDBERG, N., COCHET, C., TURLEAU, C, and GROUCHY, J. de. 1981. Cytogenet. Cell. Genet. 31:228-239.
43. CREAU-GOLDBERG, N., COCHET, C., TURLEAU, C. and GROUCHY, J.de HGM6 1981. 1982. Cytogenet. Cell. Genet. 32:261.
44. PEARSON, P.L. and BAKKER, E. HGM6 1981. 1982. Cytogenet. Cell. Genet. 32:307.
45. COCHET, C., CREAU-GOLDBERG, N., TURLEAU, C. and GROUCHY, J.de 1982. Cytogenet. Cell. Genet. 33:213-221.
46. COUTURIER, J., DUTRILLAUX, B., TURLEAU, C. and GROUCHY J.de. Ann. Génét. 25:5-10.
47. COUTURIER, J. Personnal communication.
48. CREAU-GOLDBERG, N., TURLEAU, C., COCHET, C. and GROUCHY, J.de. 1982. Ann. Génét. 25:14-18.
49. CREAU-GOLDBERG, N., TURLEAU, C., COCHET, C. and GROUCHY, J. de 1983. Ann. Génét. 26:75-78.
50. CREAU-GOLDBERG, N., TURLEAU, C., COCHET, C. and GROUCHY, J. de HGM7 1983. 1984. Cytogenet. Cell. Genet. 37:444.
51. ESTOP, A.M., GARVER, J.J., EGOZCUE, J., MEERA KHAN, P. and PEARSON, P.L. 1983. Cytogenet. Cell. Genet. 35:46-50
52. MOORE, K.L. and LALLEY, P.A. HGM7 1983. 1984. Cytogenet. Cell. Genet. 37:542-543.
53. TURLEAU, C., CREAU-GOLDBERG, N., COCHET, C. and GROUCHY, J. de. 1983. Hum. Genet. 64:65-72.

54. VAN TUINEN, P. and LEDBETTER, D.H. 1983. Am. J. Phys. Anthropol. 61:453-466.
55. CREAU-GOLDBERG, N., TURLEAU, C., COCHET, C., HUERRE, C., JUNIEN, C. and GROUCHY, J. de. 1984. Hum. Genet. 68:333-336.
56. LOCKWOOD, D.H., COPPENHAVER, D.H., FERRELL, R.E. and DAIGER, S.P. 1984. Biochem. Genet. 22:81-88.
57. CREAU-GOLDBERG, N., COCHET, C., TURLEAU,C., HUERRE, C., JUNIEN, C. and GROUCHY, J. de. HGM8 1985. 1985. Cytogenet. Cell. Genet. 40:610.
58. JONES, C., MORSE, H.G. and PALMER, D.K.HGM8 1985. 1985. Cytogenet. Cell. Genet.40:662.
59. MOORE, K.L. and LALLEY, P.A. HGM8 1985. 1985. Cytogenet. Cell. Genet. 40:702.
60. CREAU-GOLDBERG, N. and COCHET, C. Unpublished results.
61. THIESSEN, K.L. and LALLEY, P.A. 1986. Cytogenet.Cell. Genet. 42:19-23.
62. THIESSEN, K.L. and LALLEY, P.A. 1987. Cytogenet.Cell. Genet. 44:82-88.
63. COCHET, C., CREAU-GOLDBERG, N., TURLEAU, C. and GROUCHY, J. de. HGM9 1987. 1987. Cytogenet. Cell. Genet. 46:594.
64. CREAU-GOLDBERG, N., COCHET, C., TURLEAU, C., GROUCHY, J. de. HGM9 1987. 1987. Cytogenet. Cell. Genet. 46:600.
65. FRASER,N., BUCKLE, V., LEVY, E., BALLABIO, A., ROLLO, M., PERSICO, G., CRAIG, I. HGM9 1987. 1987. Cytogenet. Cell. Genet. 46:615-616.
66. JONES, C., MORSE, H.G., GEYER, D. and PATTERSON, D. HGM9 1987. 1987. Cytogenet. Cell. Genet.46:635.
67. MAGENIS, R.E., SHEEHY, R., DUZAICZYK, A. and GIBBS, P.E.M. HGM9 1987. 1987. Cytogenet. Cell. Genet. 46:654.
68. PAGE, D.C., MOSHER, R., SIMPSON, E.M.,FISHER, E.M.C., MARDON, G., POLLACK, J., McGILLIVRAY, G., CHAPELLE, A. de la. and BROWN, L.G. 1987. Cell 51:1091-1104.
69. ARNEMANN,J., THURING, S., KAHLER, S., JAKUBICZKA, S., BERG, L.P.,BURFEIND, P. and SCHMIDTKE,J. HGM10 1989. Cytogenet. Cell. Genet. (in press).
70. BOYD, E., THERIAULT, A., GODDARD, J.P., KALAITSIDAKI, M., SPATHAS, D.H. and CONNOR, J.M. 1989. Hum. Genet. 81:153-156.
71. COCHET, C., CREAU-GOLDBERG, N., GOLDGABER, D., STEHELIN, D., TURLEAU, C, and GROUCHY, J. de. HGM10 1989. Cytogenet. Cell. Genet. (in press).
72. CREAU-GOLDBERG, N., LONDON, J., COCHET, C., RAHUEL, C., CARTRON, J.P., TURLEAU, C. and GROUCHY, J. de. HGM10 1989. Cytogenet. Cell. Genet. (in press).
73. JONES, C., MORSE, H.G., GEYER, D. and PATTERSON, D. HGM10 1989. Cytogenet. Cell. Genet. (in press).
74. KALAITSIDAKI, M;, GODDARD, J.P. and BOYD, E. HGM10 1989. Cytogenet. Cell. Genet. (in press)
75. MAGENIS, R.E., LUO, X.Y., DUGAICZYK, A., RYAN, S.C. and OOSTERHUIS,J.E. HGM10 1989. Cytogenet. Cell. Genet.(in press)
76. MORRISON, N., GODDARD, J.P., BOURN, D., CONNOR, J.M. and BOYD, E. HGM10 1989. Cytogenet. Cell. Genet. (in press).
77. VAN TUINEN, P. and LEDBETTER, D.H. HGM10 1989. Cytogenet. Cell. Genet. (in press).
78. McALPINE, P.J., SHOWS, T.B., BOUCHEIX, C. and PAKSTIS, A.J. HGM10 1989. Nomenclature committee. Cytogenet. Cell. Genet. (in press).

Gene Map of the Cotton-topped Marmoset (*Saquinus oedipus*, 2N=46)

October, 1986

Peter A. Lalley
Institute for Medical Research
110 Hospital Drive
Bennington, Vermont 05201

Chromosome or Syntenic Group*	Gene Locus**			
U1	PEPC			
U2	PGM1	PGD		
U3	IDH1			
U4	MDH1	ACP1		
U5	PGM2	PEPS		
U6	NP	MPI	PKM2	HEXA
U7	GLO	ME		
U8	GR			
U9	HK1	PP		
U10	LDHA	ACP2		
U11	LDHB			
U12	PEPA			
U13	GPI	PEPD		
U14	ADA			
U15	ITPA			
U16	SOD1			

*The autosomal syntenic groups have not been assigned to chromosomes.
**Human gene nomenclature is used for homologous loci.

References:

Hink, L.A. and Lalley, P.A. 1985. Cytogenet. Cell Genet. 40:651.

Gene Map of the Owl Monkey
(Aotus trivirgatus)
Karyotype I(2n=54)
Karyotype II,III,IV(2n=54,53,52)
Karyotype V(2n=46)
Karyotype VI(2n=49/50)
Karyotype VII(2n=51/52)

Nancy Shui-Fong Ma
New England Regional Primate Research Center
Harvard Medical School
Southborough, MA 01772

September, 1989

		OWL MONKEY KARYOTYPE				
	Human	(I)	(II,III,IV)	(V)	(VI)	(VII)
ENO1	1p	12	10		12	12
PGD	1p	12		2	12	
AK2	1p	12	10	2	12	
PGM1	1p	12		2	12	12
NRAS	1p				12	
LMYC	1p				12	
SRCII	1p				12	
FH*	1q		4	8	6	
MDH1	2p	2	14	1	2	2
UGP2*	2p	2	14		2	
IDH1*	2q		16	15	16	
ALB	4q		9	12	1	
PGM2*	4				14	
HLA	6p		7	10	9	9
GLO1	6p		7	10	9	
ME1	6q	11	7	10	9	9
SOD2	6q		7	10	9	
PGM3	6q		7		9	
MYB	6q				9	
MDH2	7q	4	3	5	4	4
COLIA2	7q				14q	
MET	7q				14q	
D7S13	7q				14q	
D7S8	7q				14q	
MOS	8q			15	16	
MYC	8q			15	16	

Cont.

	Human	OWL MONKEY KARYOTYPE (I)	(II,III,IV)	(V)	(VI)	(VII)
ACON1*	9				15	
LDHA	11p	15	2	4	19	19
CAT	11p		2q	4	19	
HBB	11p		2q	4	19	
INS	11p			4	19	
PTH	11p			4	19	
HRAS1	11p				19	
ETS1	11q		3	5	4	
THY1	11q		3	5	4	
LDHB	12p	2	13	1	10	10
KRAS2	12p		13	1	10	
CD4	12		13	1	10	
IGF1	12				10	
A2M	12				10	
TPI	12				10	
FOS	14q				11	
NP	14q	16	12	11	11	11
SORD	15q	16	12	11	11	11
PKM2	15q	16	12	11	11	11
MPI	15q	16	12	11	11	
B2M*	15q		12	11	11	
FES	15q		9	12	1	
GPI	19q	9		2	25	
SIS	22q		6	3	3	
IGLC	22q		6	3	3	
PGK	Xq			X	X	
GLA	Xq				X	
G6PD	Xq				X	
ZFX	Xp		X		Xp	Xp
ZFY	Y		Y		14/y	17/y

*Unpublished assignment.

The numbers in the table refer to chromosomes to which the loci are assigned. Aotus chromosome nomenclature follows Ma et al., Lab. Anim. Sci. 26:1022, 1976.

REFERENCES:

1. Ma, N.S.F., et al. 1982. Immunogenetics 15:1-16.
2. Ma, N.S.F. 1983. J. Hered. 74:27-33.
3. Ma, N.S.F. 1983. Cytogenet. Cell Genet. 35:117-123.
4. Ma, N.S.F. 1984. Cytogenet. Cell Genet. 38:248-256.
5. Ma, N.S.F., Kurnit, D. 1984. Cytogenet. Cell Genet. 38:272-277.
6. Ma, N.S.F., et al. 1985. Cytogenet. Cell Genet. 40:684.
7. Ma, N.S.F. et al. 1986. Cytogenet. Cell Genet. 43:57-68.
8. Ma, N.S.F. 1986. Cytogenet. Cell Genet. 43:211-214.
9. Ma, N.S.F. 1987. J. Hered. 78:87-91.
10. Ma, N.S.F., et al. 1986. Am. J. Hum. Genet. 39(3):A161.
11. Ma, N.S.F., Erickson, J.M. 1988. Cytogenet. Cell Genet. 47:167-169
12. Ma, N.S.F., Gerhard, D.S. 1988. Cytogenet. Cell Genet. 48:170-173.
13. Ma, N.S.F., Harris, T.S. 1989. Cytogenet. Cell Genet. 50:34-39.
14. Ma, N.S.F., et al. 1989. J. Hered. 80:259-263.
15. Ma, N.S.F., et al. 1989. Genomic. in press.

(Supported by NIH grant RR00168 from the Division of Research Resources).

COMPOSITE LINKAGE MAP OF SALMONID FISHES (Salvelinus, Salmo, Oncorhynchus)

July, 1989

Bernie May, Cornell Laboratory for Ecological and Evolutionary Genetics,
Dept. of Natural Resources, Cornell Univ., Ithaca, NY, 14853

Kenneth R. Johnson, Jackson Laboratory, Bar Harbor, ME, 04609.

The following composite linkage map of isozyme loci is based on classical linkages (Table I), pseudolinkages (Table II), and gene-centromere map distances (Table III) in a limited number of species and fertile F_1 and backcross hybrids of the tetraploid derivative salmonid fishes. The charrs studied include brook or speckled trout (*Salvelinus fontinalis*; 2n=84), lake trout (*S. namaycush*; 2n=84), and arctic charr (*S. alpinus*; 2n=80), plus splake (brook x lake), sparctic (brook x arctic), and spleu (brook x japanese charr, *S. leucomaenis*). The trouts studied have been cutthroat (*Salmo clarki*; 2n=64-68), rainbow (*S. gairdneri*; 2n=58-62) plus cutbows (rainbow x cutthroat), and brown (*S. trutta*; 2n=80). Limited studies of salmon have involved pink (*Oncorhynchus gorbuscha*; 2n=52), chum (*O. keta*; 2n=74), and Atlantic (*Salmo salar*; 2n=54-60). Tentative assignment of loci, relative to one another, to chromosomes of this group of fishes is shown in Figures 1 and 2.

Eighteen classical linkage groups, some putative, have been established among salmonid species and hybrids; since the linkage relationships appear to be highly conserved among species, these are grouped and shown in Table I. Note that recombination values are generally greater in females than in males and that members of duplicated loci (relics of tetraploidization) belong to different classical linkage groups. Correct order of linked loci is difficult to determine when data are available only from male parents (for example, Idh-1, Ldh-1 and Ldh-5 data from male cutbow hybrids). Presumably, this is because of structural constraints imposed on crossing over by multivalent pairing involving metacentric (centrically fused) chromosomes (26). Therefore, only gene-gene recombination values in females are given to the left of each chromosome in Figures 1 and 2. Gene-centromere mapping in females (Table III) is beginning to permit ordering of loci relative to each other and to the centromere. Gene-centromere map distances are given to the right of each chromosome in Figures 1 and 2. The variability for gene centromere distances between species and species hybrids may affect some of the proposed ordering in Figures 1 and 2 and may reflect species' intrachromosomal rearrangements. Genes listed in Table III which are unassigned to a linkage group are not presented on the map.

In hybridized genomes (strain or species hybrids) duplicated loci (both isoloci and structurally diploidized duplicated loci), as well as diploidized single loci, belonging to different linkage groups, often show an exceptional pattern of nonrandom assortment (pseudolinkage) in males. That is, nonparental (recombinant) gametes are formed in excess of parental ones. Six pseudolinkage groups have been found; since most are conserved among species, each is grouped and shown in Table II. Pseudolinkage has been proposed (26) to be based on facultative multivalent pairing of metacentric chromosome arms with homeologous arms of other chromosomes coupled with alternate centromeric segregation in male salmonids, as contrasted to strictly bivalent pairing in females (4, 12, 26).

The eighteen linkage groups and six pseudolinkage groups assume the map configurations shown in Figures 1 and 2. Note that arms of metacentrics are labeled L and R and that the six pseudolinkages are identified and the relevant homeologous chromosomes are joined vertically by their respective roman numerals from Table II. Species and hybrids examined (symbols in lower right of Figure 2) to establish linked loci are indicated in the regions between loci. While both types of linkage relations appear to be highly conserved among species, arm 13R is fused to 13L is some species and to 18L in S. alpinus. Such results are not unexpected since all salmonids have relatively the same chromosome arm numbers but quite different 2n numbers, reflecting different numbers and kinds of centric fusions during chromosomal (structural) diploidization.

Although not included in any of the tables or on the map, additional linkage associations may also be detected by studies of linkage disequilibrium. The first such study by Forbes and Allendorf (5) detected the previously reported Idh-3 with Me-2 linkage as well as new associations for Ldh-2 with Me-4 and Gpi-3 with Me-4.

Acknowledgement: We are indebted to Catherine G. Schenck for editorial assistance.

Figure 1. Classical (Arabic) and Pseudolinkage (Roman) Groups in Salmonids

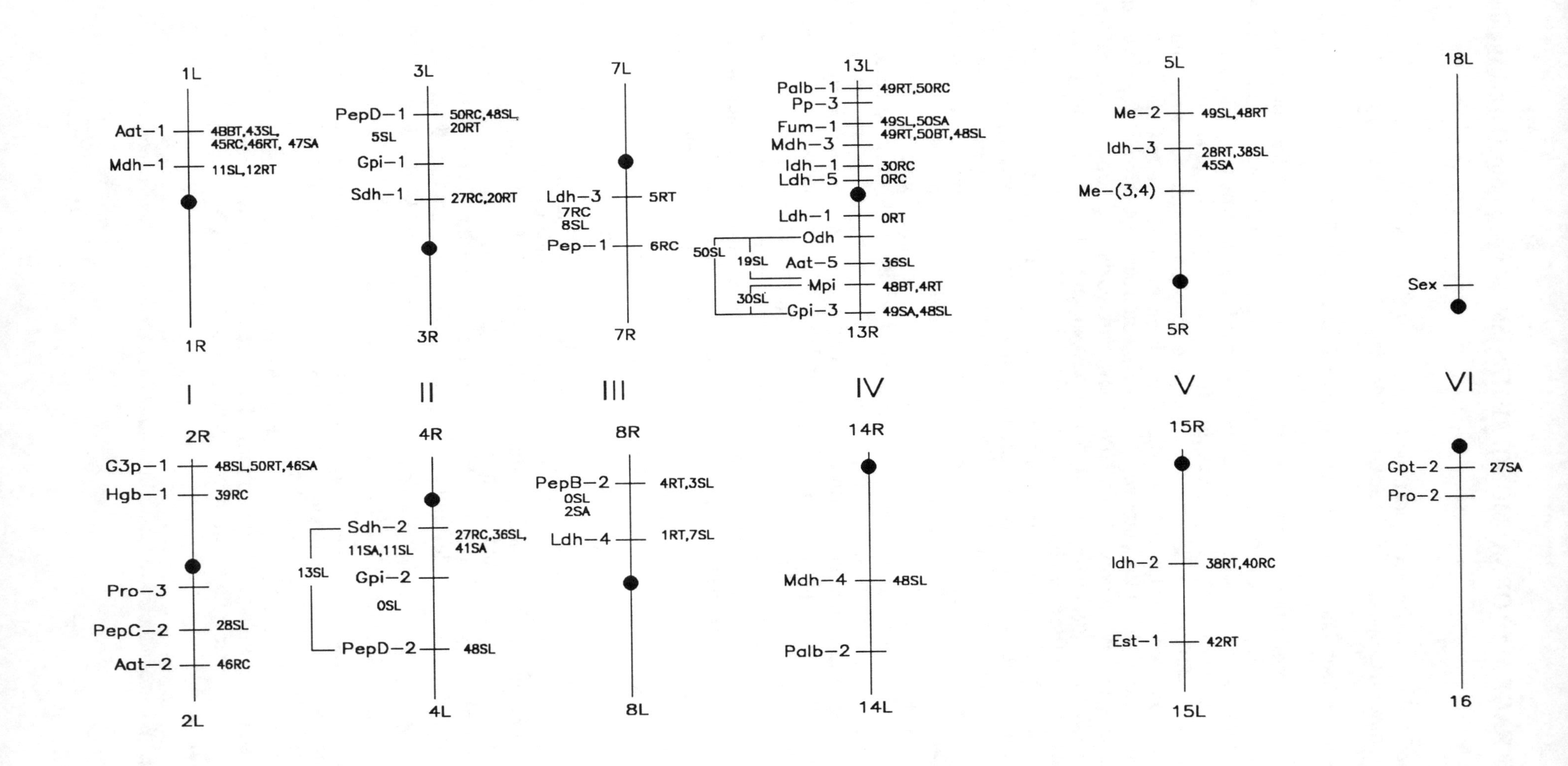

Figure 2. Classical Linkage Groups 6, 9, 10–12, and 16–18 in Salmonids

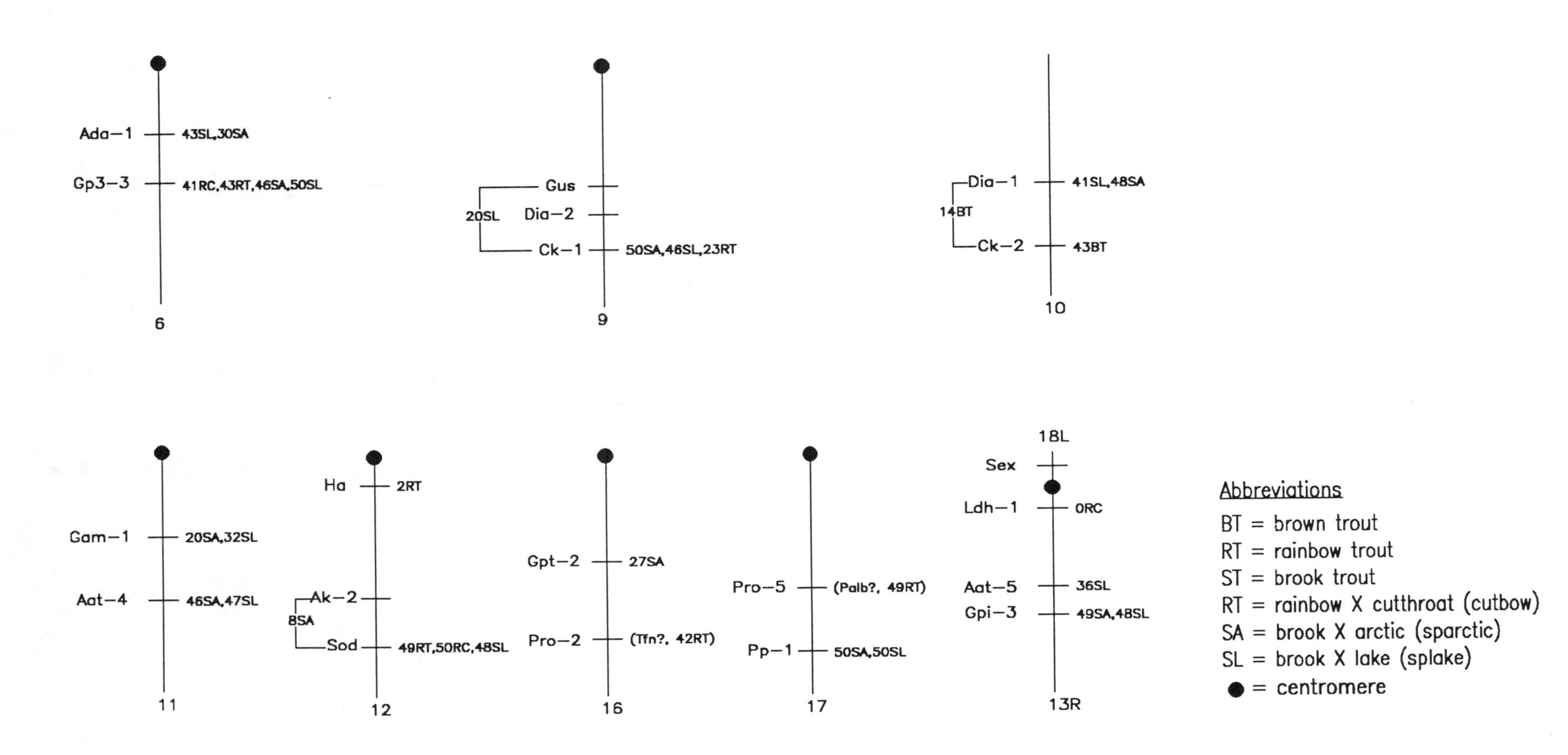

Abbreviations

BT = brown trout
RT = rainbow trout
ST = brook trout
RT = rainbow X cutthroat (cutbow)
SA = brook X arctic (sparctic)
SL = brook X lake (splake)
● = centromere

Table I. Classical linkage groups (chromosomes) in Salmoninae.

Linkage Group	$\hat{r}$	Sex	Species	Reference
1. <u>Aat-1 - Mdh-(1,2)</u>				
Aat-1 with Mdh-(1,2)	.04	M	Splake	15,16
Aat-1 with Mdh-(1,2)	.34	F	Splake	11
Aat-1 with Mdh-(1,2)	.04	M	Brown	9,22
Aat-1 with Mdh-(1,2)	.20	F	Brown	22
Aat-1 with Mdh-(1,2)	.09	M	Sparctic bc	16
2. <u>Aat-2 - Pro-3 - G3p-1 - Hgb-1</u>				
Aat-2 with G3p-1	.11	M	Brook	15,22
Aat-2 with G3p-1	.09	M	Splake	15,16
Aat-2 with G3p-1	~.5	F	Splake	11
Aat-2 with G3p-1	.20	M	Cutbow	11
Aat-2 with G3p-1	.16	M	Sparctic bc	16
Aat-2 with Hgb-1	~.5	F	Cutbow	11
Aat-2 with Hgb-1	.17	M	Cutbow	11
Aat-2 with PepC-2	.05	M	Splake	16
Aat-2 with PepC-2	.03	M	Sparctic bc	16
Aat-2 with Pro-3	.05	M	Cutbow	16
G3p-1 with PepC-2	.19	M	Sparctic bc	16
G3p-1 with PepC-2	.17	M	Splake	16
3. <u>Gpi-1 - PepD-1 - Sdh-1</u>				
Gpi-1 with PepD-1	.04	M	Splake	10
Gpi-1 with PepD-1	.05	F	Splake	10
PepD-1 with Sdh-1	.02	M	Cutbow	11
PepD-1 with Sdh-1	.01	M	Splake	16
4. <u>Gpi-2 - PepD-2 - Sdh-2</u>				
Gpi-2 with PepD-2	.04	M	Splake	10
Gpi-2 with PepD-2	.00	F	Splake	10
Gpi-2 with Sdh-2	.04	M	Splake	10
Gpi-2 with Sdh-2	.11	F	Splake	10
Gpi-2 with Sdh-2	.11	F	Sparctic	11
PepD-2 with Sdh-2	.01	M	Splake	10
PepD-2 with Sdh-2	.13	F	Splake	10
PepD-2 with Sdh-2	.01	M	Sparctic bc	16
5. <u>Idh-3 - Me-2 - Me-(3,4)</u>				
Idh-3 with Me-2	.04	M	Splake	15,16,21
Idh-3 with Me-2	.08	F	Splake	15
Idh-3 with Me-2	.02	M	Brook	16
Idh-3 with Me-2	.10	F	Brown	11
Idh-3 with Me-2	.02	M	Rainbow	18
Idh-3 with Me-2	.32	F	Cutbow	11
Idh-3 with Me-2	.06	M	Cutbow	11
Idh-(3,4) with Me-(3,4)	.00	M	Rainbow	11

	Linkage Group	$\hat{r}$	Sex	Species	Reference
6.	Ada-1 - G3p-3				
	Ada-1 with G3p-3	.01	M	Splake	15,16
	Ada-1 with G3p-3	.15	F	Splake	15
	Ada-1 with G3p-3	.03	M	Sparctic	11
	Ada-1 with G3p-3	.12	F	Sparctic	11
	Ada-1 with G3p-3	.04	M	Rainbow	18
	Ada-1 with G3p-3	.23	M	Cutbow	16
7.	Ldh-3 - PepB-1				
	Ldh-3 with PepB-1	.00	M	Splake	11,16
	Ldh-3 with PepB-1	.08	F	Splake	11
	Ldh-3 with PepB-1	.07	F	Cutbow	11
	Ldh-3 with PepB-1	.04	M	Cutbow	11
8.	Ldh-4 - PepB-2				
	Ldh-4 with PepB-2	.06	M	Splake	11,16
	Ldh-4 with PepB-2	.00	F	Splake	11
	Ldh-4 with PepB-2	.04	M	Sparctic	11
	Ldh-4 with PepB-2	.02	F	Sparctic	11
	Ldh-4 with PepB-2	.03	M	Sparctic bc	16
	Ldh-4 with PepB-2	.00	F	Sparctic bc	16
9.	Gus - Dia-2 - Ck-1				
	Gus with Ck-1	.09	M	Splake	15,16
	Gus with Ck-1	.20	F	Splake	15
	Gus with Dia-2	.07	M	Splake	16
	Ck-1 with Dia-2	.03	M	Splake	16
10.	Dia-1 - Ck-2				
	Dia-1 with Ck-2	.14	F	Brown	11
	Dia-1 with Ck-2	.00	M	Brown	11
11.	Aat-4 - Gam-1				
	Aat-4 with Gam-1	.09	M	Splake	13,16
	Aat-4 with Gam-1	.10	M	Sparctic bc	16
	Aat-4 with Gam-1	.25	F	sparctic bc	16
12.	Ak-2 - Sod - Ha				
	Ak-2 with Sod	.32	M	Sparctic	11
	Ak-2 with Sod	.08	F	Sparctic	11
	Ha with Sod	.15	M	Splake	16
	Ha with Sod	.27	M	Sparctic bc	16
13.	Mdh-3 - Fum-1 - Pp-3 -Palb-1 - Idh-1 - Ldh-1 - Ldh-5 - Odh - Mpi - Gpi-3				
	Mdh-3 with Palb-1	.02	M	Rainbow	11
	Mdh-3 with Palb-1	.03	F	Rainbow	11
	Palb-1 with Idh-1	.08	M	Cutbow	11

	Linkage Group	$\hat{r}$	Sex	Species	Reference
	Palb-1 with Ldh-1	.07	M	Cutbow	11
	Idh-1 with Ldh-1	.31	F	Cutbow	11
	Idh-1 with Ldh-1	.01	M	Cutbow	11
	Idh-1 with Ldh-5	.01	M	Cutbow	11
	Idh-1 with Mdh-3	.09	M	Cutbow	11
	Ldh-1 with Mdh-3	.10	M	Cutbow	11
	Mpi with Gpi-3	.05	M	Brown	9,11
	Mpi with Gpi-3	.10	M	Sparctic	11
	Mpi with Gpi-3	.01	M	Splake	15
	Mpi with Gpi-3	.30	F	Splake	15
	Mpi with Odh	.19	F	Splake	15
	Odh with Gpi-3	~.5	F	Splake	15
	Mpi with Ldh-1	.01	M	Sparctic	11
	Gpi-3 with Ldh-1	.06	M	Sparctic	11
	Gpi-3 with Ldh-1	~.5	F	Sparctic	11
	Gpi-3 with Ldh-1	.02	M	Sparctic bc	16
	Mdh-3 with Mpi	~.5	F	Splake	15
	Gpi-3 with Ldh-5	~.5	M	Brown	22
	Ldh-1 with Ldh-5	.00	M	Cutbow	11
	Ldh-5 with Mdh-3	.05	M	Rainbow	11
	Ldh-5 with Palb-1	.02	M	Rainbow	11
	Ldh-5 with Palb-1	.11	M	Cutbow	11
	Mdh-(3,4) with Fum-(1,2)	.00	M	Cutbow	16
	Mdh-(3,4) with Pp-3	.05	M	Cutbow	16
	Fum-(1,2) with Pp-3	.03	M	Cutbow	16
	Gpi-3 with Aat-5	.00	M	Sparctic bc	16
	Ldh-1 with Aat-5	.04	M	Sparctic bc	16
14.	Mdh-4 - Palb-2				
	Mdh-4 with Palb-2	.03	M	Rainbow	11
15.	Idh-2 - Est-1				
	Idh-2 with Est-1	.09	F	Rainbow	24
6.	Gpt-2 - Pro-2				
	Gpt-2 with Pro-2	.01	M	Splake	16
	Gpt-2 with Pro-2	.05	M	Spractic	16
	(Pro-2 maybe Tfn)				
17.	Pp-1 - Pro-5				
	Pp-1 with Pro-5	.09	M	Sparctic bc	16
	(Pro-5 maybe Palb)				
18.	Sex				
	Sex with Ldh-1	.01	M	Sparctic	14
	Sex with Ldh-1	.01	M	Sparctic bc	14
	Sex with Aat-5	.01	M	Sparctic bc	14
	Sex with Gpi-3	.04	M	Sparctic	14

Table II. Pseudolinkages in Salmoninae.

	Pseudolinkage	$\hat{r}$[1]	Sexes Tested[2]	Species	Reference
I.	<u>Linkage Group 1 Pseudolinked with Linkage Group 2</u>				
	Aat-1 with G3p-1	.79	F+M	Splake	15, 16
	Aat-1 with G3p-1	.79	M	Sparctic bc	16
	Aat-1 with G3p-1	.68	F+M	Sparctic	11
	Aat-1 with G3p-1	.69	F+M	Brook	15, 27
	Aat-1 with G3p-1	.63	M	Cutbow	11
	G3p-1 with Mdh-1	.74	F+M	Splake	15, 16
	G3p-1 with Mdh-1	.68	M	Sparctic bc	16
	Aat-1 with Aat-2	.86	F+M	Splake	15
	Aat-1 with Aat-2	.72	F+M	Brook	27
	Aat-1 with Aat-2	.92	M	Sparctic	11
	Aat-1 with Aat-2	.82	F+M	Cutthroat	2, 26
	Aat-1 with Aat-2	.85	F+M	Cutbow	27
	Aat-2 with Mdh-1	.71	F+M	Splake	15
	Aat-2 with Hgb-1	.88	F+M	Cutbow	11
	Aat-2 with Pro-1	.77	M	Cutbow	16
	Pro-1 with Pro-3	.81	M	Cutbow	16
	Mdh-1 with PepC-2	.98	F+M	Sparctic	11
	Mdh-1 with PepC-2	.96	M	Splake	16
	Mdh-1 with PepC-2	.97	M	Sparctic bc	16
	Aat-1 with PepC-2	.99	M	Splake	16
	Aat-1 with PepC-2	.67	M	Sparctic bc	16
II.	<u>Linkage Group 3 Pseudolinked with Linkage Group 4</u>				
	PepD-1 with PepD-2	.86	F+M	Splake	10, 16
	PepD-1 with Gpi-2	.83	F+M	Splake	10
	PepD-1 with Sdh-2	.83	F+M	Splake	10
	PepD-1 with Sdh-2	.96	F+M	Sparctic	11
	PepD-1 with Sdh-2	.84	M	Cutbow	11
	Sdh-1 with Sdh-2	.84	M	Cutbow	11
	PepD-2 with Sdh-1	.87	M	Splake	16
III.	<u>Linkage Group 7 Pseudolinked with Linkage Group 8</u>				
	Ldh-3 with Ldh-4	.80	F+M	Splake	4, 19, 15, 16
	Ldh-3 with Ldh-4	.52	F+M	Brook	4
	Ldh-3 with Ldh-4	.59	F+M	Rainbow	25
	Ldh-3 with Ldh-4	.98	F+M	Sparctic bc	16
	Ldh-3 with PepB-2	.96	M	Sparctic bc	16
IV.	<u>Linkage Group 13 Pseudolinked with Linkage Group 14</u>				
	Mdh-3 with Mdh-4	.53	M	Brook	17
	Mdh-3 with Mdh-4	.89	F+M	Splake	17
	Mdh-3 with Mdh-4	.73	M	Rainbow	11, 18
	Mdh-3 with Mdh-4	.90	M	Pink Salmon	3
V.	<u>Linkage Group 5 Pseudolinked with Linkage Group 15</u>				
	Idh-2 with Idh-3	.84	F+M	Cutbow	11
	Idh-2 with Me-2	.83	M	Cutbow	11
VI.	<u>Linkage Group 16 with Sex</u>				
	Sex with Pro-2	.66	M	Splake	14
	Sex with Gpt-2	.65	M	Splake	14

[1]Values for species hybrids were calculated from F_1's only.
[2]All females tested showed random assortment.

Table III. Gene-centromere recombination frequency estimates for female salmonids determined from genotypic distributions of their gynogenetic progeny, where cM= distance in centimorgans, F= number of females tested and N= total number of progeny.

Locus	Species	cM	F	N	Ref.
Aat-(1,2)	Brown	49	2	155	11
Aat-(1,2)	Brown	50	1	35	8
Aat-(1,2)	Rainbow	46	4	138	7
Aat-(1,2)	Sparctic	47	4	162	16
Aat-(1,2)	Splake	45	3	183	16
Aat-(1,2)	Splake	43	2	42	11
Aat-1	Cutbow	44	5	99	11
Aat-2	Cutbow	46	5	99	11
Aat-3	Chum	49	3	168	20
Aat-3	Rainbow	33	2	67	1
Aat-4	Brown	50	2	35	8
Aat-4	Sparctic	46	3	188	16
Aat-4	Splake	47	4	119	16
Aat-5	Splake	36	1	11	16
Aco-2	Rainbow	32	6	269	1
Ada-1	Sparctic	30	1	22	16
Ada-1	Splake	43	2	38	11
Ada-2	Cutbow	48	7	113	11
Adh	Brook	47	1	85	6
Adh	Splake	45	1	85	16
Ak-1	Sparctic	41	1	22	16
Cat	Splake	48	2	77	16
Ck-(1,2)	Brown	42	1	25	8
Ck-1	Rainbow	23	4	126	1
Ck-1	Sparctic	50	1	22	16
Ck-1	Splake	46	2	42	11
Ck-2	Brown	42	1	78	11
CkC-1	Rainbow	43	1	59	1
Dia-1	Splake	41	2	38	11
Dia-1	Sparctic	48	1	22	16
Fdp-1	Brown	40	1	25	8
Est-1	Rainbow	39	9	429	1
Est-1	Rainbow	42	2	177	19
Est-2	Rainbow	19	2	40	1
Fdp	Cutbow	50	5	109	11
Fum-(1,2)	Brown	50	1	35	8
Fum-(1,2)	Sparctic	50	5	214	16
Fum-(1,2)	Splake	49	2	40	16
Gam-1	Sparctic	20	5	205	16
Gam-1	Splake	32	4	119	16
Gold	Rainbow	28	2	316	7
Gpi-2	Brown	50	1	25	8
Gpi-3	Sparctic	49	2	142	16
Gpi-3	Splake	48	1	29	11
Gpt-2	Sparctic	27	1	22	16
G3p	Rainbow	50	1	31	7
G3p-1	Sparctic	46	3	157	16
G3p-1	Splake	50	2	115	16
G3p-1	Splake	48	1	29	11
G3p-1	Rainbow	50	2	20	1
G3p-3	Cutbow	41	6	131	11
G3p-3	Rainbow	43	2	157	11
G3p-3	Sparctic	46	1	22	16
G3p-3	Splake	50	2	115	16
Ha	Rainbow	2	9	207	1
Hgb-1	Cutbow	40	2	87	11
Idh-1	Chum	50	1	47	20
Idh-1	Cutbow	30	8	149	11
Idh-2	Cutbow	40	3	52	11
Idh-2	Rainbow	30	4	156	7
Idh-2	Rainbow	35	2	184	19
Idh-2	Rainbow	38	2	149	11
Idh-2	Rainbow	34	8	346	1
Idh-3	Brown	49	2	50	8
Idh-3	Rainbow	29	1	26	7
Idh-3	Rainbow	29	6	346	19
Idh-3	Rainbow	27	4	395	11
Idh-3	Rainbow	29	21	979	1
Idh-3	Sparctic	45	3	96	16
Idh-3	Splake	44	3	125	16
Idh-3	Splake	38	2	42	11
Idh-4	Chum	44	6	218	20
Idh-4	Cutbow	36	8	153	11
Idh-4	Rainbow	42	1	34	1
Ldh-1	Cutbow	0	1	0	11
Ldh-3	Rainbow	5	4	375	11
Ldh-3	Sparctic	11	1	22	16
Ldh-4	Spleu	7	1	36	6
Ldh-4	Rainbow	1	2	96	19
Ldh-4	Rainbow	1	2	140	11
Ldh-4	Rainbow	1	8	343	1
Ldh-4	Splake	8	2	40	11
Ldh-5	Rainbow	1	3	288	11
Mdh-1	Rainbow	12	1	21	19
Mdh-1	Splake	13	1	87	16
Mdh-1	Splake	11	2	42	11
Mdh-1	Rainbow	12	1	21	1
Mdh-2	Brown	30	1	25	8
Mdh-(3,4)	Brown	50	1	73	11
Mdh-(3,4)	Brown	50	1	25	8
Mdh-(3,4)	Rainbow	49	6[c]	236	7
Mdh-(3,4)	Rainbow	49	5	289	19
Mdh-(3,4)	Rainbow	49	2	239	11
Mdh-(3,4)	Rainbow	49	8	367	1
Mdh-(3,4)	Rainbow	37	12	746	23
Mdh-3,4*	Splake	45	2	42	11
Mdh-M1	Rainbow	31	1	8	1
Me-1	Splake	46	1	11	16
Me-1	Spleu	50	1	38	6
Me-2	Rainbow	48	3	114	1
Me-2	Splake	49	2	42	11
Mpi	Brown	48	2	151	11
Mpi	Brown	38	1	25	8
Mpi	Chum	44	3	85	20
Mpi	Rainbow	4	1	57	1
Mup-2	Sparctic	26	2	103	16
Mup-2	Splake	45	2	41	11
Palb-1	Rainbow	46	1	25	7
Palb-(1,2)	Rainbow	49	2	238	11
Palb-(1,2)	Cutbow	50	1	16	11
PepA-1	Rainbow	25	4	181	19
PepA-1	Rainbow	22	7	307	1
PepB-1	Cutbow	6	8	152	11
PepB-1	Sparctic	13	2	148	16
PepB-2	Rainbow	4	1	72	11
PepB-2	Splake	3	2	31	11
PepC	Splake	29	2	42	11
PepC-2	Sparctic	34	1	22	16
PepD	Rainbow	20	1	22	1
PepD-(1,2)	Cutbow	50	1	12	11
PepD-1,2*	Splake	45	2	42	11
Pgd	Brown	30	1	25	8
Pgk-1	Brown	43	2	157	11
Pgk-2	Cutbow	40	3	43	11
Pgm	Rainbow	6	3	124	7
Pgm	Rainbow	8	14	1061	23
Pgm-1-t	Rainbow	1	5	315	1
Pgm-1	Sparctic	50	1	22	16
Pgm-2	Rainbow	7	2	104	19
Pgm-2	Rainbow	5	2	237	11
Pgm-2	Rainbow	5	3	152	1
Pgm-2	Cutbow	6	6	111	11
Pgm-(3,4)	Sparctic	48	1	59	16
Pgm-(3,4)	Splake	45	2	40	16
Pp-1	Sparctic	50	3	130	16
Pp-1	Splake	50	5	207	16
Sdh-1	Brown	50	1	25	8
Sdh-1	Cutbow	27	5	112	11
Sdh-1	Rainbow	20	1	15	1
Sdh-2	Cutbow	27	2	28	11
Sdh-2	Sparctic	41	1	22	16
Sdh-2	Splake	36	2	36	11
Sod	Cutbow	50	3	103	11
Sod	Rainbow	47	13	1013	23
Sod	Rainbow	49	1	78	11
Sod	Rainbow	50	2	129	19
Sod	Rainbow	50	10	644	1
Sod	Splake	49	1	85	16
Sod	Splake	49	2	42	11
Sod	Spleu	50	1	38	6
Tfn	Rainbow	42	2	151	1
Tpi-1	Sparctic	36	1	22	16

[c] seven tests; one female was doubly heterozygous with cM values of 48 and 49

* both loci heterozygous; assumed to have equal gene-centromere recombination

SYMBOLS

ACO Aconitase, **ADA** Adenosine deaminase, **AK** Adenylate kinase, **ADH** Alcohol dehydrogenase, **AAT** Aspartate aminotransferase, **CAT** Catalase, **CK** Creatine kinase, **DIA** Diaphorase, **EST** Esterase, **FDP** Fructose diphosphatase, **FUM** Fumarase, **GAM** BGalactosaminidase, **GOLD** Golden color, **GPT** Glutamic pyruvic transaminase, **GUS** BGlucuronidase, **GPI** Glucosephosphate isomerase, **G3P** Glycerol3phosphate dehydrogenase, **HA** Hexose aminidase, **IDH** Isocitrate dehydrogenase, **LDH** Lactate dehydrogenase, **MDH** Malate dehydrogenase, **ME** Malic enzyme, **MPI** Mannosephosphate isomerase, **MUP** 4-Methylumbelliferyl phosphatase, **ODH** Octanol dehydrogenase, **PALB** Para-albumin, **PepA** Peptidase with Glycyl leucine, **PepB** Peptidase with Leucyl glycyl glycine, **PepC** Peptidase with Leucyl alanine, **PepD** Peptidase with Phenylalanyl proline, **PGM** Phosphoglucomutase, **PGMT** Regulatory gene, **PGD** Phosphogluconate dehydrogenase, **PGK** Phosphoglycerate kinase, **PP** Pyrophosphatase, **PRO** Unknown protein, **SDH** Sorbitol dehydrogenase, **SEX** Sex determination, **SOD** Superoxide dismutase, **TFN** Transferrin, **TPI** Triosephosphate isomerase.

REFERENCES

1. Allendorf, F.W., J.E. Seeb, K.L. Knudsen, G.H. Thorgaard and R.F. Leary. 1986. J. Hered. 77:307-312.
2. Allendorf, F.W. and F.M. Utter. 1976. Hereditas 82:19-24.
3. Aspinwall, N. 1974. Genetics 76:65-72.
4. Davisson, M.T., J.E. Wright and L.M. Atherton. 1973. Genetics 73:645-658.
5. Forbes, S.H. and F.W. Allendorf. 1989. Isozyme Bulletin. 22.
6. Fujino, K., T. Hosaka, K. Arai, and M. Kawamura. 1989. Nippon Suisan Gakkaishi. 55:1-7.
7. Guyomard, R. 1984. Theor. Appl. Genet. 67:307-316.
8. Guyomard, R. 1986. Genet. Sel. Evol. 18:385-392.
9. Hamilton, K.E. 1987. PhD Thesis, The Queen's University of Belfast, Northern Ireland.
10. Hollister, A., K.R. Johnson and J.E. Wright. 1984. J. Hered. 75:253-259.
11. Johnson, K.R., J.E. Wright and B. May. 1987. Genetics 116:579-591.
12. Lee, G.M. and J.E. Wright. 1981. J. Hered. 72:321-327.
13. May, B. 1980. Ph.D. Thesis, The Pennsylvania State University, 199pp.
14. May, B., K.R. Johnson, and J.E. Wright. 1989. Biochem. Genet. 27:291-301.
15. May, B., M. Stoneking and J.E. Wright. 1980. Genetics 95:707-726.
16. May, B. and J.E. Wright, unpublished data.
17. May, B., J.E. Wright and M. Stoneking. 1979. J. Fish. Res. Bd. Can. 36:1114-1128.
18. May, B., J.E. Wright and K.R. Johnson. 1982. Biochem. Genet. 20:29-40.
19. Morrison, W.J. 1970. Trans. Am. Fish. Soc. 99:193-206.
20. Seeb, J.E. and L.W. Seeb. 1986. J. Hered. 77:399-402.
21. Stoneking, M., B. May and J.E. Wright. 1979. Biochem. Genet. 17:599-619.
22. Taggart, J., and A. Ferguson. 1984. Heredity 53:339-359.
23. Thompson, D. and A.P. Scott. 1984. Heredity 53:441-452.
24. Thorgaard, G.H., F.W. Allendorf and K.L. Knudsen. 1983. Genetics 103:771-783.
25. Wright, J.E., J.R. Heckman and L.M. Atherton. 1975. Isozymes III. Developmental Biology. Academic Press. pp. 375-401.
26. Wright, J.E., K.R. Johnson, A. Hollister and B. May. 1983. Isozymes 10. Genetics and Evolution. Alan R. Liss, Inc. pp. 239-260.
27. Wright, J.E., B. May, M. Stoneking and G. Lee. 1980. J. Hered. 71:223-228.

Multipoint Linkage Groups of Biochemical Loci in Non-Salmonid Fishes

Xiphophorus, Poeciliidae

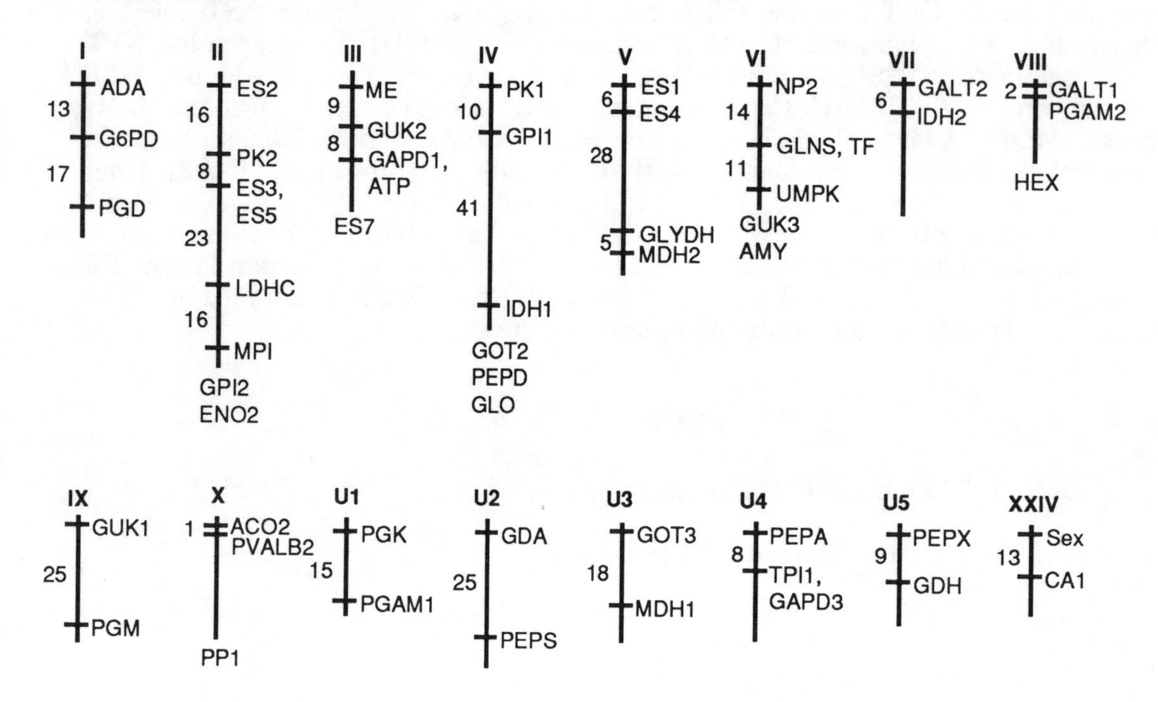

Poeciliopsis, Poeciliidae

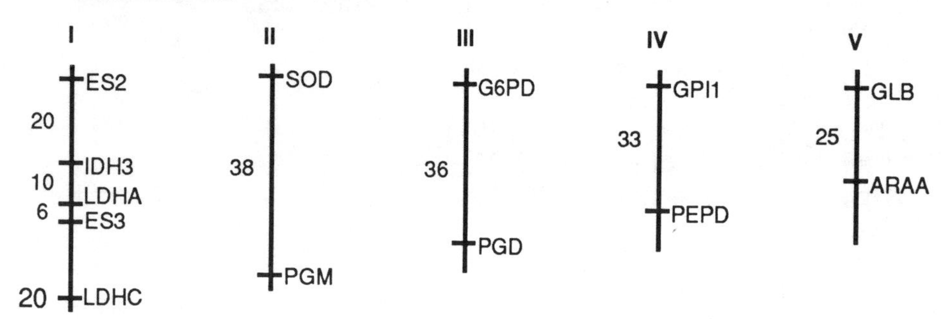

Fundulus, Cyprinodontidae

Lepomis, Centrarchidae

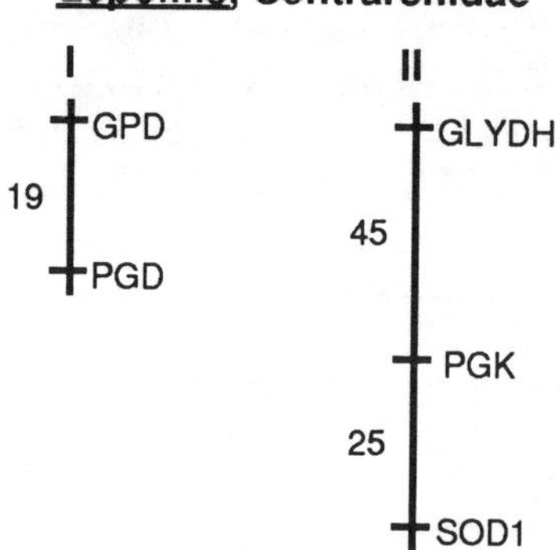

LINKAGE MAPS OF BIOCHEMICAL LOCI IN NON-SALMONID FISHES

2N = 48

August 1989

Donald C. Morizot, University of Texas M. D. Anderson Cancer Center, Science Park, Research Division, P. O. Box 389, Smithville, Texas 78957

Linkage studies in fishes other than salmonids have resulted in descriptions of multipoint linkage groups of protein- and enzyme-coding loci only in poeciliid, cyprinodontid, and centrarchid species. The linkage map of *Xiphophorus* species, constructed primarily from analyses of interspecific backcrosses, is by far the largest with some 54 biochemical loci assigned to 16 linkage groups. Testing of these linkage groups for independent assortment is highly variable; a manuscript summarizing data from more than 2500 backcross hybrid individuals analyzed to date currently is in preparation and will supplant ref. 13 as the primary review. More than 15 additional polymorphic protein-coding loci assort independently from each other and from multipoint group markers tested. In addition to the biochemical loci, ref. 3 reviews sex chromosome linkage (LG XXIV) of a number of pigment pattern loci, and refs. 18, 20 and 22 present evidence for sex chromosome linkage of oncogene-related DNA sequences. Mapping of DNA restriction fragment length polymorphisms is ongoing in several laboratories.

Gene symbols: Symbols generally follow standard human gene nomenclature (6), except in cases where no human gene symbols have been accepted (Table 1). Ref. 19 presents a proposed standard genetic nomenclature for fishes which may become widely used in the future; at present, use of human nomenclature seems preferable to facilitate gene map comparisons among vertebrates. Table 1 provides references to primary literature and comments hopefully useful in assessing homology among taxa. Gene symbols in all fish maps suggest homology with *Xiphophorus* protein-coding loci where known.

Linkage group designations: In *Xiphophorus*, linkage groups designated with Roman numerals assort independently from each other within the limits of the data. Linkage groups U1-U5 (U = unassigned) are so designated due to lack of pairwise recombination tests with at least one other multipoint group. Changes in linkage group designations are intended to supplant earlier versions of the *Xiphophorus* linkage map. Linkage group designations in other fishes are those of the original authors.

Development of the *Xiphophorus* linkage map in my laboratory has been supported by NIH grants CA28909, CA39729, and CA44303, and by NSF grants BSR16569 and BSR19355. Submitted manuscripts and summaries of unpublished data are available from the author upon request.

Table 1. Biochemical loci assigned to multipoint linkage groups

Locus symbol	Locus name	E.C. number	Linkage reference	Comments
ACO2	Aconitase	4.2.1.3	Unpublished	Mitochondrial, muscle
ADA	Adenosine deaminase	3.5.4.4	13,16	Most tissues
AMY	α-Amylase	3.2.1.1	Unpublished	Pancreas
ARAA	α-Arabinosidase	3.2.1.55	Unpublished	Liver; no human locus symbol
ATP	Adenosine triphosphatase	3.6.1.	Unpublished	Mg-dependent
CA1	Carbonic anhydrase-1	4.2.1.1	8	Anodal, brain-eye extracts
ENO2	Enolase-2	4.2.1.11	Unpublished	Muscle isozyme
ES	Carboxylesterase	3.1.1.1	1, 4, 5, 7, 9, 12	ES1-5, 7; see refs. for details
FHA	Fumarate hydratase-A	4.2.1.2	2	Liver, spleen, gut, heart
GALT1	Galactose-1-phosphate uridylyltransferase-1	2.7.7.12	Unpublished	Liver, brain-eye
GALT2	Galactose-1-phosphate uridylyltransferase-2	2.7.7.12	Unpublished	Unusual phenotype; may not be a GALT isozyme
GAPD1	Glyceraldehyde-3-phosphate dehydrogenase-1	1.2.1.12	11,13	Neural tissue-predominant
GAPD3	Glyceraldehyde-3-phosphate dehydrogenase-3	1.2.1.12	Unpublished	Testis- and ovary-specific
GDA	Guanine deaminase	3.5.4.3	Unpublished	Liver; locus symbol from ref. 8
GDH	Glucose dehydrogenase	1.1.1.47	Unpublished	Liver
GLB	β-galactosidase	3.1.1.23	15	Most tissues
GLO	Glyoxalase I	4.4.1.5	Unpublished	Muscle
GLNS	Glutamine synthetase	6.3.1.2	13, 14	Brain
GLYDH	Glycerate dehydrogenase	1.1.1.29	Unpublished	Liver; locus symbol from ref. 16
GOT2	Glutamate-oxaloacetate transaminase-1	2.6.1.1	Unpublished	Mitochondrial, muscle
GOT3	Glutamate-oxaloacetate transaminase-3	2.6.1.1	Unpublished	Cytosolic, liver
GPD	Glycerol-3-phosphate dehydrogenase	1.1.1.8	17,21	Liver
G6PD	Glucose-6-phosphate dehydrogenase	1.1.1.49	13,16	Most tissues
GPI1	Glucose phosphate isomerase-1	5.3.1.9	10,13	Most tissues
GPI2	Glucose phosphate isomerase-2	5.3.1.9	Unpublished	Muscle expression strongest
GUK1	Guanylate kinase-1	2.7.4.8	Unpublished	Eye-specific, highly anodal
GUK2	Guanylate kinase-2	2.7.4.8.	11,13	Brain-eye
GUK3	Guanylate kinase-3	2.7.4.8	Unpublished	Brain-eye, muscle
HEX	Hexosaminidase	3.2.1.52	Unpublished	Liver; single isozyme?
IDH1	Isocitrate dehydrogenase-1	1.1.1.42	10,13	Cytosolic, liver
IDH2	Isocitrate dehydrogenase-2	1.1.1.42	Unpublished	Cytosolic, most tissues
IDH3	Isocitrate dehydrogenase-3	1.1.1.42	4	Mitochondrial, muscle
LDHA	Lactate dehydrogenase-A	1.1.1.27	4	Muscle-predominant
LDHC	Lactate dehydrogenase-C	1.1.1.27	4, 7, 9, 13	Eye-specific
MDH1	Malate dehydrogenase-1	1.1.1.37	Unpublished	Cytosolic, muscle
MDH2	Malate dehydrogenase-2	1.1.1.37	12,13	Cytosolic, muscle
ME	Malic enzyme	1.1.1.40	Unpublished	Subcellular localization unknown
MPI	Mannose phosphate isomerase	5.3.1.8	7, 9, 13	Most tissues
NP2	Nucleoside phosphorylase-2	3.4.2.1	Unpublished	Brain-eye, less anodal at pH 8
PEPA	Peptidase A	3.4.11. or 3.4.13.	Unpublished	Muscle; EDTA-inhibited; glycyl-leucine substrate

Table 1 (cont.)

Locus symbol	Locus name	E.C. number	Linkage reference	Comments
PEPD	Peptidase D	3.4.13.9	15, unpublished	Most tissues; phenylalanyl-proline substrate
PEPS	Peptidase S	3.4.11. or 3.4.13.	Unpublished	Most tissues; not EDTA-inhibited; di- and tripeptides
PEPX	Peptidase X	3.4.11. or 3.4.13	Unpublished	Muscle; EDTA-inhibited; alanyl-methionine substrate
PGAM1	Phosphoglycerate mutase-1	5.4.2.1	Unpublished	Most tissues
PGAM2	Phosphoglycerate mutase-2	5.4.2.1	Unpublished	Muscle; less anodal
PGD	Phosphogluconate dehydrogenase	1.1.1.43	13,16	Most tissues
PGK	Phosphoglycerate kinase	2.7.2.3	Unpublished	Most tissues
PGM	Phosphoglucomutase	5.4.2.2	Unpublished	Most tissues
PGMA	Phosphoglucomutase-A	5.4.2.2	2	Liver; less anodal; high activity
PK1	Pyruvate kinase-1	2.7.1.40	10,13	Most tissues
PK2	Pyruvate kinase-2	2.7.1.40	7,13	Muscle-predominant
PP1	Pyrophosphatase-1 (inorganic)	3.6.1.1	Unpublished	Brain-eye extracts
PVALB2	Parvalbumin-2		Unpublished	Muscle, nonspecific protein
SOD	Superoxide dismutase	1.15.1.1	4	Subcellular localization unknown
SOD1	Superoxide dismutase-1	1.15.1.1	17	Cytosolic
TF	Transferrin		13,14	Blood plasma, brain-eye
TPI1	Triosephosphate isomerase-1	5.3.1.1	Unpublished	Eye-specific in advanced teleosts
UMPK	Uridine monophosphate kinase	2.7.4.	13,14	Brain-eye extracts

REFERENCES

1. Ahuja, M. R., M. Schwab and F. Anders. 1980. J. Hered. 71:403-407.
2. Brown, D. C., I. J. Ropson and D. A. Powers. 1988. J. Hered. 79:359-365.
3. Kallman, K. D. 1984. In B. J. Turner (ed.), Evolutionary Genetics of Fishes, pp. 95-171.Plenum, New York.
4. Leslie, J. F. 1982. J. Hered. 73:19-23.
5. Leslie, J. F. and P. J. Pontier. 1980. Biochem. Genet. 18:103-115.
6. McAlpine, P. J., C. Boucheix, A. J. Pakstis, L. C. Stranc, T. G. Berent and T. B. Shows. 1988.Cytogenet. Cell Genet. 49:4-38.
7. Morizot, D. C. 1983. J. Hered. 74:413-416.
8. Morizot, D. C. and K. D. McEntire. 1988. Isozyme Bull. 21:201.
9. Morizot, D. C. and M. J. Siciliano. 1979. Genetics 93:947-960.
10. Morizot, D. C. and M. J. Siciliano. 1982. Biochem. Genet. 20:505-518.
11. Morizot, D. C. and M. J. Siciliano. 1982. J. Hered. 73:163-167.
12. Morizot, D. C. and M. J. Siciliano. 1983. J. Nat. Cancer Inst. 71:809-813.
13. Morizot, D. C. and M. J. Siciliano. 1984. In B. J. Turner (ed.), Evolutionary Genetics of Fishes,pp. 173-234. Plenum, New York.
14. Morizot, D. C., J. A. Greenspan and M. J. Siciliano. 1983. Biochem. Genet. 21:1041-1049.
15. Morizot, D. C., R. S. Wells and R. J. Schultz. 1989. Submitted.
16. Morizot, D. C., D. A. Wright and M. J. Siciliano. 1977. Genetics 86:645-656.
17. Pasdar, M., D. P. Philipp and G. S. Whitt. 1984. Genetics 107:435-446.
18. Schartl, M. 1988. Genetics 119:679-685.
19. Shaklee, J. B., F. W. Allendorf, D. C. Morizot and G. S. Whitt. 1989. Trans. Am. Fish. Soc. 118:218-227.
20. Vielkind, J. R. and E. Dippel. 1984. Can. J. Genet. Cytol. 26:607-614.
21. Wheat, T. E., G. S. Whitt and W. F. Childers. 1973. Genetics 71:343-350.
22. Zechel, C., U. Schleenbecker, A. Anders and F. Anders. 1988. Oncogene 3:605-617.

FIGURE 1 - LINKAGE GROUPS IN RANA PIPIENS

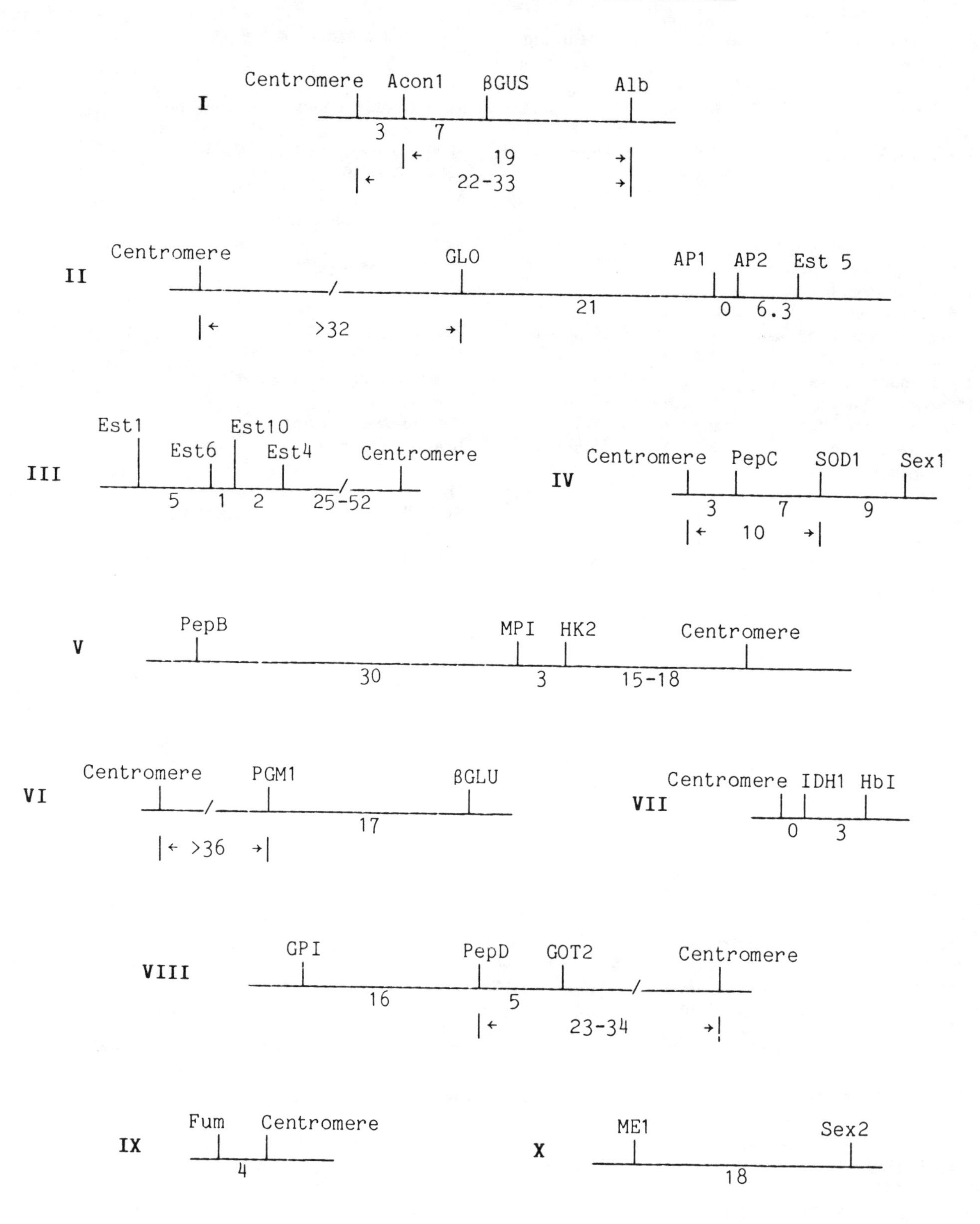

LINKAGE GROUPS IN THE LEOPARD FROG, RANA PIPIENS

2N = 26 October, 1986

David A. Wright
Department of Genetics
M. D. Anderson Hospital and
Tumor Institute
Houston, Texas 77030

and

Christina M. Richards
Department of Biology
Wayne State University
Detroit, Michigan 48202

Linkage maps in amphibians are still rather sparse. This is partially due to the long generation time of most amphibians, few inbred strains, and only few loci characterized in those inbred strains. The methods for study of linkage in amphibians have been discussed in detail elsewhere (28,31). In general the inheritance of loci is studied in progeny of wild caught frogs that are heterozygous for multiple loci (28) or backcross progeny of F_1 hybrids between species (24,31). Genotypes are determined by starch gel electrophoresis followed by histochemical staining for enzymes (9,22). Biparental test crosses have been used to look for linkage between pairs of loci (28,29) and sex-linkage of loci (3-6,30-31). Gynogenetic progeny of heterozygous females allows examination of linkage to the centromere (1,14,19,25,31).

Species specific cytological markers on lampbrush chromosomes have been used to map inheritance of enzyme loci in backcross progeny of species hybrids (16,17,18) and some hyperdiploid species hybrids allow for testing linkage to individual somatic cell chromosomes (10). Other technical advances such as *in situ* hybridization of cloned genes (8) and chromsome mediated gene transfer (20) are being applied to amphibian systems and offer great promise.

The linkage map derived from pure *Rana pipiens* crosses is summarized in Figure 1. Table 1 contains a listing of linkage information from other *Rana* species and information from a few other amphibians as well. It appears that in *Rana pipiens* multiple genes for sex determination exist. Where known, homologous linkages or lack of linkage of homologous loci in mammals are indicated for linked loci in frogs.

ABBREVIATIONS USED:

Acid Phosphatase (AP) E.C. 3.1.3.3.; Aconitase (Acon) E.C. 4.2.1.3.; Alcohol dehydrogenase (ADH) E.C. 1.1.1.1.; Fructose 1,6 diphosphatase (F16DP) E.C. 3.1.3.11.;Fumerase (Fum) E.C. 4.2.1.2.; Glucose phosphate isomerase (GPI) E.C. 5.3.1.9.; βGlucosidase (βGLU) E.C. 2.3.1.21.; βGlucuronidase (βGUS) E.C. 3.2.1.31.; Glutamate-oxaloacetate Transaminase (GOT) E.C. 2.6.1.1.; αGlycerophosphate dehydrogenase (αGPDH) E.C. 1.1.1.8.; Gyloxalase I (GLO) E.C. 4.4.1.5.; Hexokinase (HK) E.C. 2.7.1.1.; Isocitrate dehydrogenase (IDH) E.C. 1.1.1.42.; Lactate dehydrogenase (LDH) E.C. 1.1.1.27.; Malate dehydrogenase (MDH) E.C. 1.1.1.37.; Malic enzyme (ME) E.C. 1.1.1.40.; Mannose phosphate isomerase (MPI) E.C. 5.3.1.8.; αMannosidase (αMan) E.C. 3.2.1.24.; Peptidase (Pep A, B, C, D, S) E.C. 3.4.3.1.; Phosphoglucomutase (PGM) E.C. 2.7.5.1.; Superoxide dismutase (SOD) E.C. 1.15.1.1.; Hemoglobin (Hb); Albumin (Alb); chromosome (chr.)

TABLE 1

LINKED LOCI	SPECIES	RANA PIPIENS POSSIBLE LINKAGE GROUP	POSSIBLE OTHER VERTEBRATE HOMOLOGY
βGUS-Acon 1-Alb	R. pipiens (28,31)	I	Acon1 and Alb not linked in mammals (27)
Alb-F16DP-PGM1	R. palustris (31)	I or VI	PGM1 and Alb linked in mouse (27)
Sex-Alb-F16DP-ADH-PGM1-βGLU	R. berlandieri (31)	IV,I or VI	ADH and PGM1 not linked in mouse (27)
Sex-Acon 1	R. sphenocephala (31)	IV or I	
Sex-Acon 1	R. clamitans (4)	IV or I	
Alb-ADH-albino b-chr.1	R. brevipoda, R. nigromaculata (16,17)	I	
GLO-AP1-AP2-Est 5	R. pipiens (28)	II	GLO and AP not linked in human (15)
Est1-Est6-Est10-Est4	R. pipiens (28)	III	Group of Est loci linked in rat (2,26), mouse (27) and fishes (12,13)
PepC-SOD1-Sex1	R. pipiens (29,30)	IV	These loci unlinked in mammals (15,27)
MDH2-PEPC-SOD1	R. berlandieri (31)	IV	"
SOD1-Sex	R. blairi (31)	IV	"
αGPDH-ME2(A)-Albino a-PepC--albino c- SOD1?(B)-chr.2	R. nigromaculata (16,17)	IV	
PepB-αMan-MPI-HK2	R. pipiens (31)	V	αMan (MANA) & MPI to human chr.15
MPI-LDHB	R. palustris (31)	V	LDH1 & MPI linked in Xiphophorus (12,13);
MPI-LDHB	R. clamitans (4)	V	LDHB-PepB linked to human chr.12 (15)
Sex-LDHB	R. catesbeiana (3)	IV or V	LDHB not sex linked in mammals
LDHB-chr.4	R. brevipoda, R. nigromaculata (16)		
PGM1-βGLU	R. pipiens (31)	VI	
IDH1-HB1	R. pipiens, R. palustris (31)	VII	IDH1 and Hb not linked in mammals (15)
IDH1-Hb-chr.6	R. brevipoda, R. nigromaculata (16)	VII	
GOT2-PepD-GPI	R. pipiens (Fig. 1)	VIII	PepD & GPI linked in mouse (27) man (15) and Chinese hamster (23)
PepD-GPI	R. berlandieri (31)	VIII	GPI and GOT2 linked in Xiphorporus (12)
GPI-GOT1-TPI	R. palustris (31)	VIII	
ME1-Sex2	R.pipiens (Fig.1)	X	
MDH1(B)-olive-albino e--ME1(B)-chr.3	R. brevipoda, R. nigromaculata (16,17,18)	X	
PGM2-chr.13	R. pipiens (31)	?	
PepA (Pep 1)-Sex	Pleurodeles waltlii (6,7)	?	
5s RNA (somatic type)-chr.9 (oocyte type)-many chr.	Xenopus laevis and X. borealis (8)	?	

REFERENCES

1 Armstrong, J. B. 1984. Can. J. Genet. Cytol. 26:1-6.
2. Cramer, D. V., 1982. Isozyme Bull. 15:65-73.
3. Elinson, R. P., 1981. Develop. Biol. 81:167-176.
4. Elinson, R. P., 1983. Biochem. Genet. 21:435-442.
5. Ferrier, V., A. Jaylet, C. Cayrol. F. Gasser and J. J. Buisan, 1980. Comptes Rendus 290:571-574.
6. Ferrier, V., F. Gasser, A. Jaylet and C. Cayrol, 1983. Biochem. Genet. 21:535-550.
7. Gasser, F., V. Ferrier, A. Jaylet and C. Cayrol, 1983. Biochem. Genet. 21:527-534.
8. Harper, M. E., J. Price and L. J. Korn, 1983. Nucleic Acids Res. 11:2313-2323.
9. Harris, H., and D. A. Hopkinson, 1976. North Holland Publishing Co., Amsterdam.
10. Kobel, H. R. and L. DuPasquier, 1979. Nature 279:157-158.
11. Lalley, P. A., J. D. Minna and U. Francke, 1978. Nature 274:160-163.
12. Morizot, D., 1983. Isozyme Bull. 16:7-15.
13. Morizot, D. C. and M. J. Siciliano, 1983. Current Topics in Biological and Medical Research, Vol. 10: Genetics and Evolution (M. C. Rattazzi, J. G. Scandalios and G. S. Whitt, eds.) Alan R. Liss, Inc., New York, pp. 261-285.
14. Nace, G. W., C. M. Richards and J. H. Asher Jr., 1970. Genetics 66:349-368.
15. Naylor, S. L., 1983. Isozyme Bull. 16:16-41.
16. Nishioka, M., H. Ohtani and M. Sumida, 1980. Sci. Rep. Lab. Amphib. Biol. Hiroshima Univ. 4:127-184.
17. Nishioka, M., H. Ohtani and M. Sumida, 1987. Sci. Rep. Lab. Amphib. Biol. Hiroshima Univ. 9: (in press)
18. Nishioka, M. and H. Ohtani, 1986. Sci. Rep. Lab. Amphib. Biol. Hiroshima Univ. 8:1-27.
19. Reinschmidt, D., J. Friedman, J. Hauth, E. Ratner, M. Cohen, M. Miller, D. Krotoski, and R. Tompkins, 1985. J. Heredity 76:345-347.
20. Rosenstein, B. S., 1982. Transition into Biotechnology Miami Winter Symposia, Vol. 19, Academic Press, New york, p. 554.
21. Shows, T. B., 1983. Current Topics in Biological and medical Research, Vol. 10: Genetics and Evolution (M. C. Rattazzi, J. G. Scandalios and G. S. Whitt, eds), Alan R. Liss, Inc., New York, pp. 323-339.
22. Siciliano, M. J. and C. R. Shaw, 1976. Chromatographic and Electrophoretic Techniques", (Ivor Smith, ed), Vol. 2, 4th edition, Wm Heinemann Medical Books, London, pp. 184-209.
23. Stallings, R. l. and M. j. Siciliano, 1983. Current Topics in Biological and medical Research, Vol. 10: Genetics and Evolution (M. C. Rattazzi, J. G. Scandalios and G. S. Whitt, eds), Alan R. Liss, Inc. New York, pp. 313-321.
24. Szymura, J. M. and I. Farana, 1978. Biochem. Genet. 16:307-320.
25. Volpe, E. P., 1970 Genetics 64:11-21.
26. Womack, J. E. and . Sharp, 1976. Genetics 82:665-675.
27. Womack, J. e., 198 . Isozyme Bull. 15:56-64.
28. Wright, D. A., C. M. Richards and G. W. Nace, 1980. Biochem. Genet. 8:591-616.
29. Wright, D. A. and C. M. Richards, 1982. J. Exp. Zool. 221:283-293.
30. Wright, D. A. and C. M. Richards, 1983. Genetics 103:249-261.
31. Wright, D. A., C. M. Richards, J. S. Frost, a. M. Camozzi and B. J. Kunz, 1983. Current Topics in Biological and Medical Research, Vol. 10: Genetics and Evolution (M. C. Rattazzi, J. G. Scandalios and S. Whitt, eds), Alan R. Liss, Inc., New York, pp. 287-311.

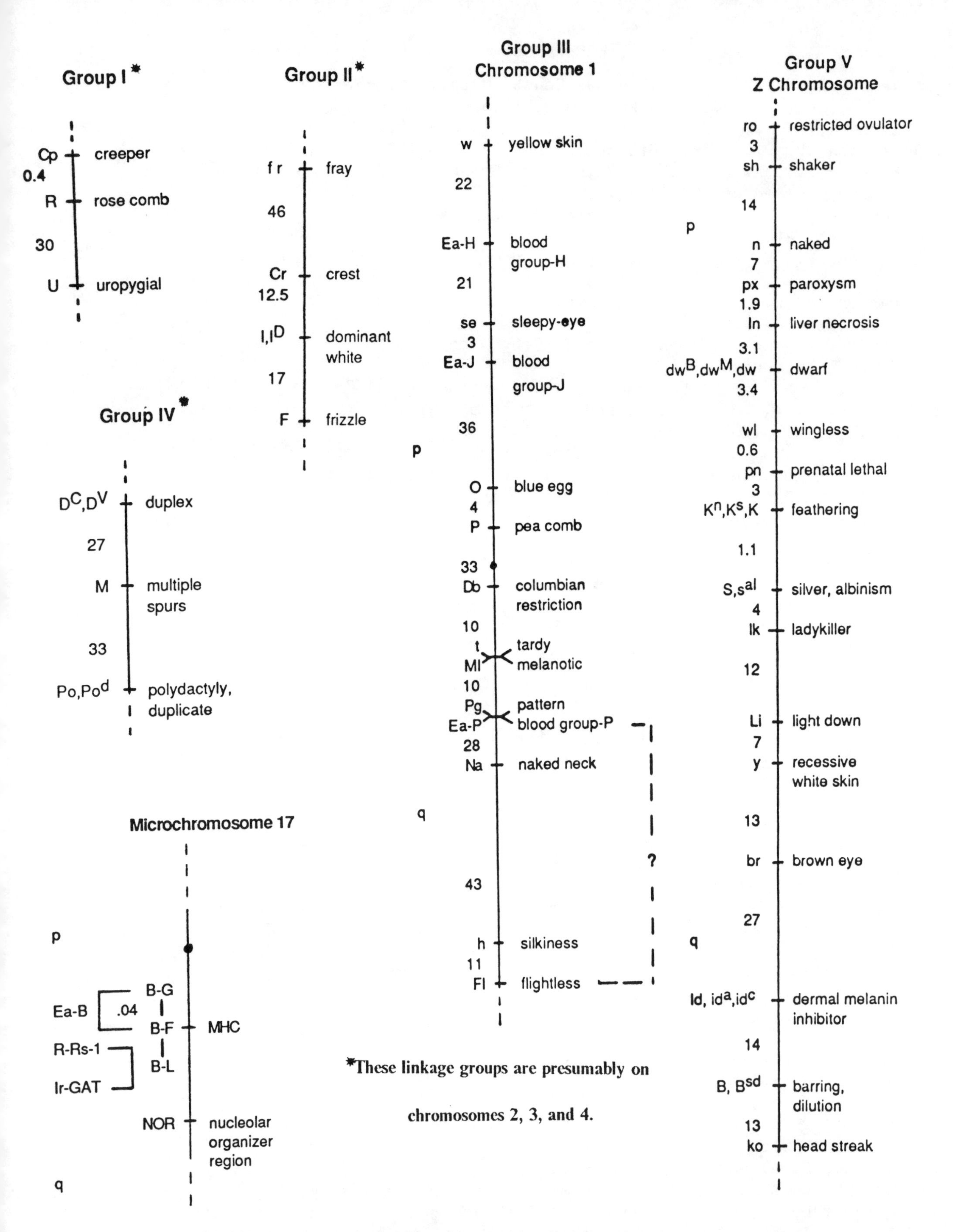

*These linkage groups are presumably on chromosomes 2, 3, and 4.

Fig. 1. Linkage Map of Domestic Chicken

GENE MAP OF THE CHICKEN (GALLUS GALLUS or G. DOMESTICUS) 2N = 78

July 1989

Ralph G. Somes, Jr.
Nutritional Sciences Department
University of Connecticut
Storrs, Connecticut 06268

J. James Bitgood
Department of Poultry Science
University of Wisconsin
Madison, Wisconsin 53706

Although the chicken has 39 pairs of chromosomes {ref. 20}, only the first five pairs (macrochromosomes) are large and distinctive enough to permit easy identification. Chromosome pairs 6-10 can be identified in photomicrographs. The remaining 29 pairs of microchromosomes have not been assigned numerical designations but are shown in karyotypes in decreasing size order. The female is the heterogametic sex, ZW, while the male is ZZ. The Z chromosome is the fifth largest in the karyotype, while the W chromosome is similar in size to chromosome 9 {ref. 20}. Linkage group numbering has been changed in this map to comply with that recommended by Etches and Hawes {ref. 49}.

The first linkage map was published in 1936 by F.B. Hutt {ref. 67} and consisted of 18 loci in five linkage groups. As new data have become available, revised linkage maps have been published {refs. 49, 59, 68, 70, 72, 73, 130, 131, 133, 134, 135, 137}. A summary of all the linkage relationships in the chicken, including linkage distances, has been published {ref. 132}. Several summaries of linkage data showing independent segregation have also been published and are invaluable to those engaged in this area of research {refs. 3, 49, 50, 156}.

Table 1 presents the traits that are depicted on the present chromosome map (Fig. 1). Traits that have been identified as being on microchromosomes are listed in Table 2. Individual traits that have been identified with a particular chromosome but have not yet been mapped are listed in Table 3. Several traits have been shown to be linked to each other, and in some cases have also been identified with a particular chromosome but are as yet unmapped. These traits are listed in Table 4. A number of chromosomal rearrangement break points have been particularly useful in chicken linkage work; these data are presented in Table 5.

TABLE 1 -- Traits that are part of the chromosomal map.

Gene symbol	Trait name	Chromosome size	References
B, B^{Sd}	Sex-linked barring, dilution	Z	61, 109, 115, 151
br	Brown eye	Z	86, 109
dw, dw^M, dw^B	Sex-linked dwarfism	Z	41, 64, 71, 72
Id, id^a, id^c	Dermal melanin inhibitor	Z	63, 86, 109, 115
K, K^s, K^n	Rate of feathering	Z	63, 72, 90, 127
ko	Head streak	Z	61, 109
Li	Light down	Z	63, 109
lk	Ladykiller	Z	118, 119
ln	Sex-linked lethal liver necrosis	Z	39
n	Sex-linked naked	Z	36, 39, 72
pn	Prenatal lethal	Z	126, 127, 139
px	Paroxysm	Z	36
ro	Restricted ovulator	Z	91
S, s^{al}	Silver, imperfect albinism	Z	38, 72, 109, 127
sh	Shaker	Z	72, 98
wl	Sex-linked wingless	Z	81
y	Recessive white skin	Z	15, 92
Db	Columbian-like restriction	1	33, 35, 62, 123
Ea-H	Erythrocyte alloantigen H	1	12, 26, 49
Ea-J	Erythrocyte alloantigen J	1	24, 26, 49
Ea-P	Erythrocyte alloantigen P	1	12, 13, 26, 49
Fl	Flightless	1	152, 154, 156
h	Silkiness	1	152, 154, 156
Ml[1)]	Melanotic	1	32, 33, 34, 40, 96
Na	Naked neck	1	12, 13, 35, 62
O	Blue egg	1	14, 27
P	Pea comb	1	14, 34, 62, 158
Pg	Pattern gene	1	31, 34, 96, 97

TABLE 1 -- Traits that are part of the chromosomal map. (cont.)

Gene symbol	Trait name	Chromosome size	References
se	Sleepy-eye	1	124, 125
t	Tardy feather growth	1	17, 49
w	Yellow skin	1	12, 26, 49, 124
-----[2)]	Centromere, chromosome 1	1	14
B-F	MHC-cell antigen B-F	17	79, 101, 106
B-G	MHC-erythrocyte specific antigen B-G	17	79, 101, 106
B-L	MHC-leucocyte specific antigen B-L	17	101, 106
Ea-B	Erythrocyte alloantigen B	17	21, 22, 101, 107
Ir-GAT	Immune response-GAT	17	11, 52, 101, 104
R-Rs-1	Rous sarcoma regression	17	51, 52, 114
-----[2)]	Major histocompatibility complex (MHC)	17	21, 23, 58
-----[2)]	Nucleolar organizer region (NOR)	17	7, 21, 22, 23, 65
Cp	Creeper	(I)[3)]	82, 83, 144
R	Rose comb	(I)[3)]	67, 82, 83, 144
U	Uropygial	(I)[3)]	67
Cr	Crest	(II)[3)]	141, 153, 157
F	Frizzling	(II)[3)]	141, 157
fr	Fray	(II)[3)]	153
I, I^D	Dominant white, dun	(II)[3)]	66, 157, 161
D^v, D^c	Duplex comb, v-shaped, cup-shaped	(IV)[3)]	69, 74, 136, 155
M	Multiple spurs	(IV)[3)]	69, 74, 155
Po, Po^d	Polydactyly, duplicate	(IV)[3)]	74, 155

1) There are two conflicting reports on the *P*___*Ml* linkage, see Table 4.
2) Not single gene loci.
3) The chromosomes containing each of these linkage groups are unknown, but they are presumably chromosomes 2, 3, and 4.

TABLE 2 -- Loci linked to microchromosomes.

Gene symbol	Trait name	Chromosome size	References
c-erb-A	Cellular oncogene of avian erythroblastosis virus A	10-14	142, 159
c-ets	Cellular oncogene of avian E-26 leukemia virus	9-16	143
c-fps	Cellular oncogene of avian Fujinami sarcoma virus	9-16	142
c-mil/mht	Cellular oncogene of avian Mill Hill 2 leukemia virus	9-16	143
c-myb[1)]	Cellular oncogene of avian myeloblastosis virus	2, 3, or 13-16	143, 150
c-myc[1)]	Cellular oncogene of avian myelocytomatosis virus	2 or 13-16	117, 143, 150
c-src	Cellular oncogene of Rous sarcoma virus	10-12	65, 141, 147, 148
ev3	Defective ALV provirus, gs, chf^+, V^-	micro.	146
Hprt	Hypoxanthine phosphoribosyltransferase	micro.	110
Ovm	Ovomucoid	10-15	65
Tf	Transferrin	9-12	65
Tk-F	Cytosol thymidine kinase F	micro.	85
-----[2)]	α^A-globin	10-15	43, 47, 65
-----[2)]	α^D-globin	10-15	43, 47, 65
-----[2)]	Embryonic α-like globin	10-15	47, 48
-----[2,3)]	β-actin	2 or 9-12	116

1) Two conflicting reports: one places *c-myc* on chromosome 2 and *c-myb* on either chromosome 2 or 3 while the other report places both of these loci on microchromosomes of the size 13-16.
2) No gene symbol assigned.
3) β-actin gene is located either on the long arm of chromosome 2 or on a chromosome in the size range of 9-12. The other site is apparently occupied by another gene in the multigenic actin family.

TABLE 3 -- Traits that have been linked to a chromosome.

Gene symbol	Trait name	Chromosome size	References
cd	Cerebellar degeneration	Z	88
chz	Sex-linked chondrodystrophy	Z	87
cm	Sex-linked coloboma	Z	2
dp-4	Diplopodia-4	Z	1
ev7	Defective ALV provirus codes for 15-ILV	Z	122, 148
ga	Gasper	Z	108
H-Z	Z-linked histoantigen	Z	8
j	Jittery	Z	56
pr	Protoporphyrin inhibitor	Z	120
Pw_1	"Pw*1*" agglutinogen	Z	95
Pw_2	"Pw*2*" agglutinogen	Z	95
rg	Recessive sex-linked dwarf	Z	55
sex	Sex-linked lethal-Bernier	Z	129
sln	Sex-linked nervous disorder	Z	77
St_1	"St*1*" agglutinogen	Z	94
St_2	"St2" agglutinogen	Z	93
xl	Sex-linked lethal	Z	57
Z	Dominant sex-linked dwarf	Z	89
H-W[1]	W-linked histoantigen	W	9, 53
c-erb-B	Cellular oncogene of avian erythroblastosis virus B	2	142
ev2	ALV provirus, codes for RAV-O	2	148
shl	Shankless	2	84
----[2,3]	β-actin	2 or 9-12	116
c-myc[4]	Cellular oncogene of avian myelocytomatosis virus	2 or 13-16	117, 143, 150
c-myb[4]	Cellular oncogene of avian myeloblastosis virus	2, 3 or 13-16	143, 150
ev14	Endogenous retrovirus	3	148
Ade-A	Adenine synthesis A	6	76
Alb	Serum albumin	6	75, 102
Gc	Vitamin D binding protein	6	75, 102
Pgm-2	Phosphoglucomutase	6	102
Ade-B	Adenine synthesis B	7	76

1) The presence of this locus being on the W chromosome is now in question {ref. 99}.
2) No gene symbol assigned.
3) β-actin gene is located either on the long arm of chromosome 2 or on a chromosome in the size range of 9-12. The other site is apparently occupied by another gene in the multigenic actin family.
4) Two conflicting reports: one places *c-myc* on chromosome 2 and *c-myb* on either chromosome 2 or 3 while the other report places both of these loci on microchromosomes of the size 13-16.

TABLE 4 -- Unmapped linked loci.

Linked genes and distances	Linked traits	Chromosome size	References
ev21 close *K*	Endogenous retrovirus ___ feathering	Z	10
ba 7.6 s^{al}	Congenital baldness ___ imperfect albinism	Z	128, 138
pop 35 *K*	Popeye ___ feathering	Z	16
pe 27 *se*	Perosis ___ sleepyeye	1	125
P 0.3 *Ml*[1]	Pea comb ___ melanotic	1	32, 33, 34, 40, 95
Mb 44.3 *Ea-H*	Muffs & beard ___ erythrocyte alloantigen H	1	12, 26, 49
Ea-C 45 *Ea-J*	Erythrocyte alloantigen C __ erythrocyte alloantigen J	1	24, 26, 49, 107
Ea-E 0.5 *Ea-A*	Erythrocyte alloantigen E __ erythrocyte alloantigen A	1	12, 24, 25, 49, 111
Ea-A 45 *Ea-J*	Erythrocyte alloantigen A __ erythrocyte alloantigen J	1	12, 24, 25, 26, 49
Ea-C __ev locus[2]	Erythrocyte alloantigen C __ ALV provirus locus	1	107

TABLE 4 -- Unmapped linked loci. (cont.)

Linked genes and distances	Linked traits	Chromosome size	References
Ea-D 38 *t*	Erythrocyte alloantigen D ___ tardy feather growth	1	19
Ea-I 36.7 *w*	Erythrocyte alloantigen I ___ skin color	1	19
Ea-I 32.9 *P*	Erythrocyte alloantigen I ___ pea comb	1	19
Pti-?[3] 40 *w*	Shank feathering gene ___ skin color	1	12, 80
ev4 20 *ev5* __ *ev1* __ *ev6* __ *ev13* __ *ev8*[2]	Six linked ALV provirus loci	1	4, 5, 6, 146, 147, 149
--------------[2,5]	β-globin ___ embryonic β-like globin	1 or 2	44, 65
G_3 7 *Ov*	Ovoglobulin G_3 ___ ovalbumin	2 or 3	29, 30, 46, 54, 140
--------------[2,5]	α^A-globin __ α^D-globin __ embryonic α-globin	10-15	47, 48
Ea-B __ *ev* locus[5]	Erythrocyte alloantigen B __ ALV provirus locus	17	107
lav 32.5 *R*	Lavender ___ rose comb	(I)[4]	28
Mp 16 *R*	Ametapodia ___ rose comb	(I)[4]	37
Ea-M 5.4 *Ea-Q*	Erythrocyte alloantigen M __ erythrocyte alloantigen Q	UNK	113
Ea-O 2.6 *Ea-S*	Erythrocyte alloantigen O __ erythrocyte alloantigen S	UNK	113
Es-1 .04 *Es-2*	Esterase-1 ___ esterase-2	UNK	78
IgG-1 2.8 *IgM-1*	7S-1 immunoglobulin heavy chain __ immunoglobulin M	UNK	42, 105
Tv-A .17 *Tv-C*	ALV subgroup A ___ ALV subgroup C	UNK	45, 103
Tv-B close *Tv-E*	ALV subgroup B ___ ALV subgroup E	UNK	112
--------------[2,5]	δ-subunit ___ γ-subunit of nicotine acetylcholine receptor	UNK	100
--------------[2,5]	δ1-crystallin ___ δ2-crystallin	UNK	60

1) There are two conflicting reports on the *P* __ *Ml* linkage, see Table 1.
2) Distance not known.
3) There are several shank feathering genes, the one involved here has not been identified.
4) The chromosome containing this linkage group is unknown; it is presumably either chromosome 2, 3, or 4.
5) No gene symbol assigned.

TABLE 5 -- Genes showing linkage with chromosomal rearrangement break points (R_B).

R_B	Distance	Gene	Trait name	Chromosome	Arm	Reference
OH t(Z;micro)	23	*K*	slow feathering	Z	short	145
MN t(Z;1)	22	*B*	barring	Z	long	12, 18
NM 7092 t(Z;1)	6.3	*S*	silver	Z	short	12
MN t(Z;3)	0	*S*	silver	Z	short	15
NM 7092 t(Z;1)	22.3	*pop*	popeye	Z	short	16
NM 7092 t(Z;1)	0	*pr*	protoporphyrin	Z	short	121
MN t(Z;1)	17	*y*	rec. white skin	Z	long	15, 18
MN inv(1)	4.4	*P*	pea comb	1	short	12
MN t(1;micro)	5.9	*P*	pea comb	1	short	12
MN t(Z;1)	6.7	*P*	pea comb	1	short	12
NM 7092 t(Z;1)	15.9[1]	*P*	pea comb	1	long	160
NM 7659 t(Z;1)	10.4	*P*	pea comb	1	short	14, 160
MN t(Z;1)	4.5	*O*	blue egg	1	short	12
NM 7659 t(Z;1)	13.5	*O*	blue egg	1	short	14
NM 7092 t(Z;1)	30.6	*Ea-D*	blood group D	1	long	19

1) This report conflicts with others as to the arm location of the pea comb locus.

REFERENCES

1. Abbott, U.K., and M. Kieny. 1961. C.R. Acad. Sci. 242:1863-1865.
2. Abbott, U.K., et al. 1970. J. Hered. 61:95-102.
3. Abbott, U.K., and G.W. Yee. 1975. In: Handbook of Genetics. R.C. King, ed. Plenum, New York, v. 4, pp. 151-200.
4. Astrin, S.M. 1978. Proc. Natl. Acad. Sci. 75:5941-5945.
5. Astrin, S.M., and H.L. Robinson. 1979. J. Virol. 32:420-425.
6. Astrin, S.M., et al. 1979. Virology 99:1-9.
7. Auer, H., et al. 1987. Cytogenet. Cell Genet. 45:218-221.
8. Bacon, L.D. 1970. Transplantation 10:126-129.
9. Bacon, L.D., and J.V. Craig. 1969. Transplantation 7:387-393.
10. Bacon, L.D., et al. 1985. Poultry Sci. 64 (Suppl. 1):60.
11. Benedict, A.A., et al. 1975. Immunogenetics 2:313-324.
12. Bitgood, J.J., et al. 1980. Poultry Sci. 59:1686-1693.
13. Bitgood, J.J., et al. 1984. Poultry Sci. 63:592-594.
14. Bitgood, J.J. 1985. Poultry Sci. 64:1411-1414.
15. Bitgood, J.J. 1985. Poultry Sci. 64:2234-2238.
16. Bitgood, J.J., and R.D. Whitley. 1986. J. Hered. 77:123-125.
17. Bitgood, J.J., et al. 1987. J. Hered. 78:329-330.
18. Bitgood, J.J. 1988. Poultry Sci. 67:530-533.
19. Bitgood, J.J., et al. 1989. Poultry Sci. 68 (Suppl. 1):(in press).
20. Bloom, S.E. 1969. J. Hered. 60:217-220.
21. Bloom, S.E., and R.K. Cole. 1978. Poultry Sci. 57:1119.
22. Bloom, S.E., et al. 1978. Genetics 88 (suppl.):s13.
23. Bloom, S.E., and L.D. Bacon. 1985. J. Hered. 76:146-154.
24. Briles, W.E. 1964. Z. Tierz. Zuechtungsbiol. 79:371-391.
25. Briles, W.E. 1968. Genetics 60:164.
26. Briles, W.E., et al. 1967. Poultry Sci. 46:1238.
27. Bruckner, J.H., and F.B. Hutt. 1939. Science 90.88-89.
28. Brumbaugh, J.A., et al. 1972. J. Hered. 63:19-25.
29. Buvanendran, V. 1964. Genet. Res. 5:330-332.
30. Buvanendran, V. 1967. Br. Poultry Sci. 8:9-12.
31. Carefoot, W.C. 1985. Br. Poultry Sci. 26:409-412.
32. Carefoot, W.C. 1986. Br. Poultry Sci. 27:71-74.
33. Carefoot, W.C. 1987. Br. Poultry Sci. 28:69-73.
34. Carefoot, W.C. 1987. Br. Poultry Sci. 28:347-350.
35. Carefoot, W.C. 1988. Br. Poultry Sci. 29:785-790.
36. Cole, R.K. 1958. Poultry Sci. 37:1194-1195.
37. Cole, R.K. 1967. J. Hered. 58:141-146.
38. Cole, R.K., and T.K. Jeffers. 1963. Nature (London) 200:1238-1239.
39. Cole, R.K., and D.G. Jones. 1972. Poultry Sci. 51:1795.
40. Crawford, R.D. 1985. Poultry Sci. 64 (Suppl. 1):83.
41. Custodio, R.W.S., and R.G. Jaap. 1973. Poultry Sci. 52:204-210.
42. Derka, J., and K. Hala. 1979. Folia Biol., Praha 25:266-270.
43. Dodgson, J.B. 1981. Proc. Natl. Acad. Sci. 78:5998-6002.
44. Dodgson, J.B. et al. 1979. Cell. 17:879-887.
45. Dren, C.N., and P.K. Pani. 1977. J. Gen. Virol. 35:13-23.
46. Durand, L., and P. Merat. 1982. Annls. Genet. Sel. Anim. 14:49-66.
47. Engel, J.D., and J.B. Dodgson. 1980. Proc. Natl. Acad. Sci. 77:2596-2600.
48. Engel, J.D., et al. 1983. Proc. Natl. Acad. Sci. 80:1392-1396.
49. Etches, R.J., and R.O. Hawes. 1973. Can. J. Genet. Cytol. 15:553-570.
50. Etches, R.J., and R.O. Hawes. 1979. In: Inbred and Genetically Defined Strains of Laboratory Animals. P.L. Altman and D.D. Katz, eds. FASEB, Bethesda, v. 2, pp. 628-637.
51. Gebriel, G.M., et al. 1979. Immunogenetics 9:327-334.
52. Gebriel, G.M., and A.W. Nordskog. 1983. J. Immunogenet. 10:231-235.
53. Gilmour, D.G. 1967. Transplantation 5:699-706.
54. Gintovt, V.E., et al. 1976. Genetika, USSR 12:61-71.
55. Godfrey, E.F. 1953. Poultry Sci. 32:248-259.
56. Godfrey, E.F., et al. 1953. J. Hered. 44:108-112.
57. Goodwin, K., et al. 1950. Science 112:460-461.

58. Hala, K. 1977. In: The Major Histocompatibility System in Man and Animals. D. Gotze, ed. Springer-Verlag. New York. pp. 291-312.
59. Hann, C.M. 1966. Ministry of Agr. Fisheries and Food, Bull. #38, Her Majesty's Stationary Office, London.
60. Hawkins, J.W., et al. 1984. J. Biol. Chem. 259:9821-9825.
61. Hertwig, P. 1930. Biol. Zentralbl. 50:333-341.
62. Hertwig, P. 1933. Verhandl. Dtsch. Zool. Ges.:112-118.
63. Hertwig, P., and T. Rittershaus. 1929. Z. Indukt. Abstamm. Verebungsl. 51:354-372.
64. Hsu, P.L., et al. 1975. Poultry Sci. 54:1315-1319.
65. Hughes, S., et al. 1979. Proc. Natl. Acad. Sci. 76:1348-1352.
66. Hutt, F.B. 1933. Genetics 18:82-94.
67. Hutt, F.B. 1936. Neue Forsch. Tierz. Abstammungsl. (Fetschr.):105-112.
68. Hutt, F.B., and W.F. Lamoreux. 1940. J. Hered. 31:231-235.
69. Hutt, F.B. 1941. J. Hered. 32:356-364.
70. Hutt, F.B. 1949. Genetics of the Fowl. McGraw-Hill Book Co., Inc., New York.
71. Hutt, F.B. 1959. J. Hered. 50:209-221.
72. Hutt, F.B. 1960. Heredity 15:97-110.
73. Hutt, F.B. 1964. Animal Genetics. The Ronald Press, Co., New York.
74. Hutt, F.B., and C.D. Mueller. 1943. Am. Nat. 77:70-78.
75. Juneja, R.K., et al. 1982. Genetical Research 40:95-98.
76. Kao, F. 1973. Proc. Natl. Acad. Sci. 70:2893-2898.
77. Kawahara, T. 1955. Ann. Rep. Natl. Inst. Genet. Japan 5:26-27.
78. Kimura, M., and N. Kameyama. 1970. Japanese Poultry Sci. 7:39-41.
79. Kock, C., et al. 1983. Tissue Antigens 21:129-137.
80. Lambert, W.V., and C.W. Knox. 1929. Poultry Sci. 9:51-64.
81. Lancaster, F.M. 1968. Heredity 23:257-262.
82. Landauer, W. 1932. J. Genet. 26:285-290.
83. Landauer, W. 1933. Nature (London) 132:606.
84. Langhorst, L.J., and N.S. Fechheimer. 1985. J. Hered. 76:182-186.
85. Leung, W.C., et al. 1975. Exp. Cell Res. 95:320-326.
86. MacArthur, J.W. 1933. Genetics 18:210-220.
87. Mann, G.E. 1963. NAAS Poultry Sect. A.J. 69:1-2.
88. Markson, L.M. 1959. J. Comp. Pathol. 69:223-229.
89. Maw, A.J.G. 1935. Sci. Agric. 16:85-112.
90. McGibbon, W.H. 1977. Poultry Sci. 56:872-875.
91. McGibbon, W.H. 1977. Genetics 86 (Suppl.):s43-s44.
92. McGibbon, W.H. 1981. J. Hered. 72:139-140.
93. Mizutani, M., et al. 1979. An. Blood Grp. Res. Inform. 7:24-27.
94. Mizutani, M., et al. 1980. An. Blood Grp. Res. Inform. 8:15-17.
95. Mizutani, M., et al. 1984. An. Blood Grp. Res. Inform. 12:16-19.
96. Moore, J.W., and J.R. Smyth, Jr. 1972. J. Hered. 63:179-184.
97. Moore, J.W., and J.R. Smyth, Jr. 1972. Poultry Sci. 51:1149-1156.
98. Mueller, C.D. 1952. Poultry Sci. 31:1105-1106.
99. Mueller, U., et al. 1979. Nature (London) 280:142-144.
100. Nef, P., et al. 1984. Proc. Natl. Acad. Sci. 81:7975-7979.
101. Palmer, D.K., and A.W. Nordskog. 1982. Proc. XVIII Inter. Conf., Anim. Blood Grps., Univ. of Ottawa, Canada.
102. Palmer, D.K., and C. Jones. 1986. J. Hered. 77:106-108.
103. Pani, P.K. 1974. Nature (London) 248:592-594.
104. Pevzner, I.Y., et al. 1978. Immunogenetics 7:25-33.
105. Pink, J.R.L., and J. Ivanyi. 1975. Eur. J. Immunol. 5:506-507.
106. Pink, J.R.L., et al. 1977. Immunogenetics 5:203-216.
107. Plachy, J., et al. 1985. Folia Biol., Czecho. 31:353-356.
108. Price, D.J., et al. 1966. Poultry Sci. 45:423-424.
109. Punnett, R.C. 1940. J. Genet. 39:335-342.
110. Rasko, I., et al. 1979. Cytogenet. Cell Genet. 24:129-137.
111. Reevey, C.M., et al. 1969. Poultry Sci. 48:1862-1863.
112. Robinson, H.L., and W.F. Lamoreux. 1976. Virology 69:50-62.
113. Scheinberg, S.L. 1956. Genetics 41:834-844.
114. Schierman, L.W. 1977. Immunogentics 5:325-332.
115. Serebrovsky, A.S., and S.G. Petrov. 1930. J. Exp. Biol. 6:157-179.

116. Shaw, E.M., et al. 1988. Poultry Sci. 67(Suppl. 1):154.
117. Sheiness, D.K., et al. 1980. Virology 105:415-424.
118. Sheridan, A.K. 1964. Proc. Aust. Poultry Sci. Conv. 1:87-90.
119. Sheridan, A.K. 1979. Br. Poultry Sci. 20:571-573.
120. Shoffner, R.N. 1978. Poultry Sci. 57:1163.
121. Shoffner, R.N., et al. 1982. Poultry Sci. 61:817-820.
122. Smith, E., and L. Crittenden. 1981. Virology 112:370-373.
123. Smyth, J.R., Jr., and G.W. Malone. 1979. Poultry Sci. 58:1108-1109.
124. Somes, R.G., Jr. 1968. J. Hered. 59:375-378.
125. Somes, R.G., Jr. 1969. J. Hered. 60:163-166.
126. Somes, R.G., Jr. 1969. J. Hered. 60:185-186.
127. Somes, R.G., Jr. 1969. J. Hered. 60:281-286.
128. Somes, R.G., Jr. 1970. Poultry Sci. 49:1440.
129. Somes, R.G., Jr. 1971. Conn. Agric. Exp. Stn. Storrs Bull. 420.
130. Somes, R.G., Jr. 1973. J. Hered. 64:217-221.
131. Somes, R.G., Jr. 1978. J. Hered. 69:401-403.
132. Somes, R.G., Jr. 1979. In: Inbred and Genetically Defined Strains of Laboratory Animals. P.L. Altman and D.D. Katz, eds. FASEB, Bethesda. v. 2. pp. 622-626.
133. Somes, R.G., Jr. 1980. Genetic Maps 1:210-217.
134. Somes, R.G., Jr. 1982. Genetic Maps 2:310-318.
135. Somes, R.G., Jr. 1984. Genetic Maps 3:465-473.
136. Somes, R.G., Jr. 1986. Poultry Sci. 65 (Suppl. 1):128.
137. Somes, R.G., Jr. 1987. Genetic Maps 4:422-429.
138. Somes, R.G., Jr. 1989. J. Hered. (in press).
139. Somes, R.G., Jr., and J.R. Smyth, Jr. 1976. J. Hered. 58:25-29.
140. Stratil, A. 1970. Anim. Blood Groups Biochem. Genet. 1:83-88.
141. Suttle, A.D., and G.R. Sipe. 1932. J. Hered. 23:135-142.
142. Symonds, G., et al. 1984. Mol. Cell. Biol. 4:1627-1630.
143. Symonds, G., et al. 1986. J. Virol. 59:172-175.
144. Taylor, L.W. 1934. J. Hered. 25:205-206.
145. Telloni, R.V., et al. 1976. Poultry Sci. 55:1886-1896.
146. Tereba, A. 1981. J. Virol. 40:920-926.
147. Tereba, A., and S.M. Astrin. 1980. J. Virol. 35:888-894.
148. Tereba, A., et al. 1981. J. Virol. 39:282-289.
149. Tereba, A., and S.M. Astrin. 1982. J. Virol. 43:737-740.
150. Tereba, A., and M.M.C. Lai. 1982. Virology 116:654-657.
151. Van Albada, M., and A.R. Kuit. 1960. Genen Phaenen 5:1-9.
152. Warren, D.C. 1935. Am. Nat. 69:82.
153. Warren, D.C. 1938. Genetics 23:174.
154. Warren, D.C. 1940. Am. Nat. 74:93-95.
155. Warren, D.C. 1944. Genetics 29:217-231.
156. Warren, D.C. 1949. Genetics 34:333-350.
157. Warren, D.C., and F.B. Hutt. 1936. Am. Nat. 70:379-394.
158. Washburn, K.W., and J.R. Smyth, Jr. 1967. J. Hered. 58:131-134.
159. Wong, T., et al. 1981. Virology 111:418-426.
160. Zartman, D.L. 1973. Poultry Sci. 52:1455-1462.
161. Ziehl, M.A., and W.F. Hollander. 1987. Iowa State J. of Res. 62:337-342.

INDEX